AF868864

Birkhäuser

Progress in Probability

Volume 76

Progress in Probability is designed for the publication of workshops, seminars and conference proceedings on all aspects of probability theory and stochastic processes, as well as their connections with and applications to other areas such as mathematical statistics and statistical physics.

More information about this series at http://www.springer.com/series/4839

Uta Freiberg • Ben Hambly • Michael Hinz •
Steffen Winter
Editors

Fractal Geometry and Stochastics VI

Birkhäuser

Editors
Uta Freiberg
Fakultät für Mathematik
Technische Universität Chemnitz
Chemnitz, Germany

Ben Hambly
Mathematical Institute
University of Oxford
Oxford, UK

Michael Hinz
Fakultät für Mathematik
Universität Bielefeld
Bielefeld, Germany

Steffen Winter
Institut für Stochastik
Karlsruher Institut für Technologie
Karlsruhe, Germany

ISSN 1050-6977 ISSN 2297-0428 (electronic)
Progress in Probability
ISBN 978-3-030-59648-4 ISBN 978-3-030-59649-1 (eBook)
https://doi.org/10.1007/978-3-030-59649-1

This book is published under the imprint Birkhäuser, www.birkhauser-science.com, by the registered company Springer Nature Switzerland AG.
The registered company address is: Gewerbestrasse 11, 6330 Cham, Switzerland

We dedicate this book to Christoph Bandt and Martina Zähle. They have been the main forces behind the series 'Fractal Geometry and Stochastics' since its beginning in 1994, and major drivers of the work on fractal geometry in Germany and beyond for over 30 years. Both their mathematics and their engagement with the community continue to be a great inspiration for all of us.

Preface

The conference 'Fractal Geometry and Stochastics VI' with 122 participants from 20 different countries took place in Bad Herrenalb, Baden-Württemberg, Germany, from September 30 to October 6, 2018. It was the sixth in a series of conferences, initiated by Christoph Bandt, Siegfried Graf and Martina Zähle with the first conference in 1994. Since then, the mathematics of fractal structures has experienced a rapid expansion and a growing diversification. Aiming to cover most recent developments while representing a broad spectrum of topics, 'Fractal Geometry and Stochastics' has become widely recognized as one of the world's leading conference series in the field, and it continues to provide a vibrant platform for the exchange of new ideas. The main contributions of each single conference have been published by Birkhäuser in their series 'Progress in Probability'.

Continuing the tradition of the earlier conferences, we invited the representatives of particularly active areas of research to give keynote and invited talks, including promising young colleagues. The chapters collected in this volume address a wide range of different topics. Some are expositional, while others contain original results, but in style they all follow the philosophy of these conference proceedings to present material highly interesting for specialists while remaining as accessible as possible to newcomers in the field and to experts from related disciplines.

We express our gratitude to the Deutsche Forschungsgemeinschaft for their essential financial support for the conference and gratefully acknowledge additional support from the cluster of excellence SimTech, University of Stuttgart, and the Karlsruhe Institute of Technology, Karlsruhe. We thank our Scientific Committee (Christoph Bandt, Kenneth Falconer, Jun Kigami, Marc Pollicott and Martina Zähle) for their advice, constant support and encouragement, and we thank a number of referees for their generous help in preparing this volume.

Chemnitz, Germany — Uta Freiberg
Oxford, UK — Ben Hambly
Bielefeld, Germany — Michael Hinz
Karlsruhe, Germany — Steffen Winter
July 2020

Contents

Part I Fractal Dimensions and Measures

Interpolating Between Dimensions 3
Jonathan M. Fraser

Assouad Type Dimensions in Geometric Analysis 25
Juha Lehrbäck

A Survey on Prescription of Multifractal Behavior 47
Stéphane Seuret

Renewal Theorems and Their Application in Fractal Geometry 71
Sabrina Kombrink

Part II Random Graphs and Complexes

Fractal Dimension of Discrete Sets and Percolation 101
Markus Heydenreich

Asymptotics of Integrals of Betti Numbers for Random Simplicial Complex Processes 125
Masanori Hino

Part III Trees and Hyperbolicity

The Continuum Self-Similar Tree 143
Mario Bonk and Huy Tran

***p*-Hyperbolicity of Ends and Families of Paths in Metric Spaces** 191
Nageswari Shanmugalingam

Part IV Physical Models and Fractals

Breaking of Continuous Scale Invariance to Discrete Scale Invariance: A Universal Quantum Phase Transition 209
Omrie Ovdat and Eric Akkermans

The Random Conductance Model with Heavy Tails on Nested Fractal Graphs 239
David A. Croydon

Space-Time Duality for Semi-Fractional Diffusions 255
Peter Kern and Svenja Lage

From Fractals in External DLA to Internal DLA on Fractals 273
Ecaterina Sava-Huss

Contents of the Previous Volumes of the 'Fractal Geometry and Stochastics' Series 299

Introduction

This book presents some of the recent developments in various areas of modern mathematics naturally connected to 'Fractal Geometry and Stochastics', although the variety of ideas and results collected here goes well beyond the scope of this modest label. The book consists of four parts, which are detailed in what follows.

Part I of the book contains four chapters on topics at the heart of fractal geometry. The chapter by J. Fraser discusses new ideas in dimension theory, namely the Assouad spectrum, which interpolates between the upper box and the Assouad dimension, and the intermediate dimensions, which interpolate between the Hausdorff and box dimensions. It is followed by a chapter by J. Lehrbäck on upper and lower Assouad dimensions and their connections to the integrability of distance functions, to Muckenhoupt weights and to thickness and thinness conditions for the validity of Hardy–Sobolev inequalities on Euclidean open sets. The chapter by S. Seuret explains some of the latest results related to the idea of finding objects with prescribed multifractal properties, such as local dimensions for measures, or singularity or multifractal spectra for functions and measures. Some connections to function spaces are highlighted. A panorama of classical renewal theorems in probability and the discussion of a contemporary renewal theorem in symbolic dynamics are the subjects of the chapter by S. Kombrink, along with applications to counting problems and Minkowski measurability in fractal geometry.

Part II of the book consists of two chapters relating to random discrete structures. The first one, by M. Heydenreich, reviews different dimension concepts for integer lattices and more general graphs, such as fractal dimension (in the sense of volume growth), spectral dimension and mass dimension. It also characterizes the various dimensions for the incipient infinite cluster of (bond) percolation on integer lattices. In the second one, M. Hino surveys recent results and new ideas in the homology theory of random simplicial complexes. A particular result is the asymptotic behaviour of time integrals of Betti numbers for Linial–Meshulam complex processes, which may be seen as higher-dimensional analogues of Erdös–Renyi graph processes.

Part I
Fractal Dimensions and Measures

Interpolating Between Dimensions

Jonathan M. Fraser

Abstract Dimension theory lies at the heart of fractal geometry and concerns the rigorous quantification of how large a subset of a metric space is. There are many notions of dimension to consider, and part of the richness of the subject is in understanding how these different notions fit together, as well as how their subtle differences give rise to different behaviour. Here we survey a new approach in dimension theory, which seeks to unify the study of individual dimensions by viewing them as different facets of the same object. For example, given two notions of dimension, one may be able to define a continuously parameterised family of dimensions which interpolates between them. An understanding of this 'interpolation function' therefore contains more information about a given object than the two dimensions considered in isolation. We pay particular attention to two concrete examples of this, namely the *Assouad spectrum*, which interpolates between the box and (quasi-)Assouad dimension, and the *intermediate dimensions*, which interpolate between the Hausdorff and box dimensions.

Keywords Dimension theory · Hausdorff dimension · Box dimension · Assouad dimension · Assouad spectrum · Intermediate dimensions

Mathematics Subject Classifications (2010) Primary: 28A80, Secondary: 37C45

1 Dimension Theory and a New Perspective

Roughly speaking, a *fractal* is an object which exhibits complexity on arbitrarily small scales. Such objects are hard to analyse, and cannot be easily measured. Dimension theory is the study of how to measure fractals, specifically aimed at quantifying how they fill up space on small scales. This is done by developing

J. M. Fraser (✉)
School of Mathematics and Statistics, The University of St Andrews, St Andrews, Scotland
e-mail: jmf32@st-andrews.ac.uk

U. Freiberg et al. (eds.), *Fractal Geometry and Stochastics VI*,
Progress in Probability 76, https://doi.org/10.1007/978-3-030-59649-1_1

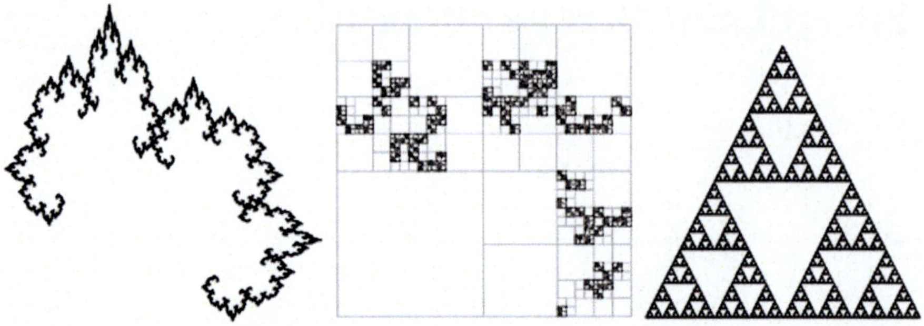

Fig. 1 Three fractals: a self-affine set (left), a random set generated by *Mandelbrot percolation* (centre), and the self-similar *Sierpiński triangle* (right)

precise mathematical formulations of dimension and then developing techniques which can be used to compute these dimensions in specific settings, such as, for sets invariant under a dynamical system or generated by a random process, see Fig. 1. There are many ways to define dimension which naturally extend our intuitive idea that lines have dimension 1 and squares have dimension 2, etc. The box dimension is a particularly natural and easily digested notion of dimension, which comes from understanding how a coarse measure of size behaves as the resolution increases. More precisely, given a bounded set $F \subseteq \mathbb{R}^d$ and a scale (resolution) $r > 0$, let $N_r(F)$ denote the minimum number of sets of diameter r required to cover F, see Fig. 2. This should increase as $r \to 0$ and it is natural to expect $N_r(F) \approx r^{-\delta}$ for some $\delta > 0$, which can be readily interpreted as the 'dimension' of F. As such, the *upper box dimension* of F is defined by

$$\overline{\dim}_{\mathrm{B}} F = \limsup_{r \to 0} \frac{\log N_r(F)}{-\log r}.$$

If the lim sup is replaced by lim inf, one gets the *lower box dimension* $\underline{\dim}_{\mathrm{B}} F$. However, often the lim sup and lim inf agree, in which case we refer to the common value as the *box dimension*, denoted by $\dim_{\mathrm{B}} F$. Despite how convenient and natural this definition is, it has some theoretical disadvantages, such as not being countably stable, see [8, page 40]. A more sophisticated notion, which is similar in spirit, is the *Hausdorff dimension*. This can be defined, for any set $F \subseteq \mathbb{R}^d$, by

$$\dim_{\mathrm{H}} F = \inf \Big\{ \alpha > 0 : \text{for all } \varepsilon > 0 \text{ there exists a cover } \{U_i\} \text{ of } F \text{ such that } \sum_i |U_i|^\alpha < \varepsilon \Big\}.$$

The key difference here is that sets with vastly different diameters are permitted in the covers and their contribution to the 'dimension' is weighted according to their

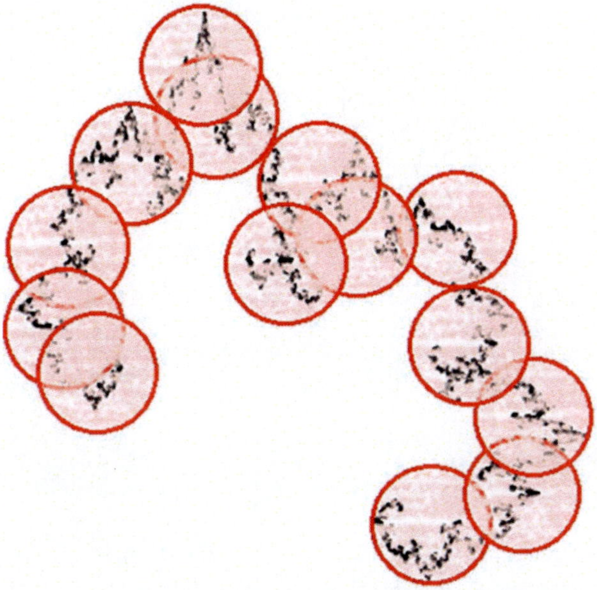

Fig. 2 An efficient covering of the self-affine set from Figure 1 by balls of the same radius. Counting the number of balls required for such a cover as the radius tends to 0 gives rise to the *box dimension*

diameter, denoted by $|U_i|$, see Fig. 3. In particular, it is easily seen that the Hausdorff dimension is countably stable. Both the Hausdorff and box dimension measure the size of the whole set, giving rise to an "average dimension". It is often the case that more extremal information is required, for example in embedding theory, see [29]. The *Assouad dimension* is designed to capture this information and is defined, for any set $F \subseteq \mathbb{R}^d$, by

$$\dim_A F = \inf\Big\{ \alpha > 0 : \text{ there exists a constant } C > 0 \text{ such that, for all } 0 < r < R \text{ and } x \in F \text{ we have } N_r\big(B(x, R) \cap F\big) \leq C\left(\frac{R}{r}\right)^{\alpha} \Big\}.$$

The key point here is that one does not seek covers of the whole space, but only a small ball, and the expected covering number is appropriately normalised, see Fig. 4. The Assouad dimension has many useful applications outside the realm of fractal geometry. For example, see the survey [23] (also published in these proceedings) which considers applications of the Assouad dimension to problems in geometric analysis. One of the joys of dimension theory is in understanding how these different

Fig. 3 An efficient covering of the self-affine set from Fig. 1 by balls of arbitrarily varying radii. Understanding the weighted sum of diameters of the sets in such a cover gives rise to the *Hausdorff dimension*

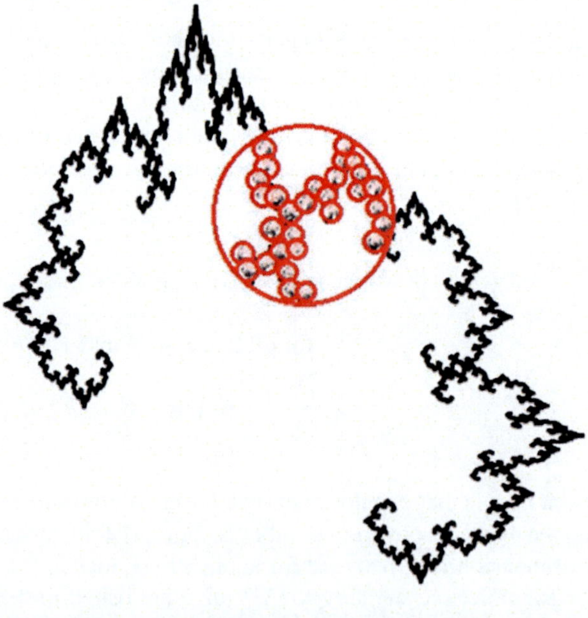

Fig. 4 An efficient covering of a particular ball in the self-affine set from Fig. 1 by smaller balls of the same radius. Counting the number of balls required for such a cover, optimised over all larger balls and all pairs of scales, gives rise to the *Assouad dimension*

notions of dimension relate to each other and how they behave in different settings. It is a simple exercise to demonstrate that

$$\dim_{\mathrm{H}} F \leq \underline{\dim}_{\mathrm{B}} F \leq \overline{\dim}_{\mathrm{B}} F \leq \dim_{\mathrm{A}} F$$

for any bounded $F \subseteq \mathbb{R}^d$, and that these inequalities can be strict inequalities or equalities in any combination. Equality throughout can be interpreted as a manifestation of 'strong homogeneity'. For example, if F is Ahlfors-David regular then $\dim_{\mathrm{H}} F = \dim_{\mathrm{B}} F = \dim_{\mathrm{A}} F$.

There are of course many other notions of dimension, each important in its own right and motivated by particular questions or applications. We omit discussion of these, but other examples include the packing, lower, quasi-Assouad, modified box, topological, Fourier, among many others. We refer the reader to [2, 7, 8, 27, 29] for more background on dimension theory, including a thorough investigation of the basic properties of the various notions of dimension.

The main purpose of this article is to motivate a new perspective in dimension theory. Rather than view these notions of dimension in isolation, we should try to view them as different facets of the same object. This approach will give rise to a continuum of dimensions, which fully describes the scaling structure of the space, both locally and globally. Moreover, this will yield a more nuanced understanding of the individual notions of dimensions as well as insight into the somewhat philosophical question of how to define dimension itself. This sounds rather grand and ambitious, but by focusing our attention slightly and applying this philosophy in particular settings, an interesting and workable theory has started to emerge.

More concretely, given dimensions dim and Dim which generally satisfy $\dim F \leq \mathrm{Dim}\, F$, we wish to introduce a parameterised family of dimensions d_θ, with parameter $\theta \in [0, 1]$, which (ideally) satisfies:

- $d_0 = \dim$
- $d_1 = \mathrm{Dim}$
- $\dim F \leq d_\theta(F) \leq \mathrm{Dim}\, F$, for all $\theta \in (0, 1)$ and all reasonable sets F
- for a given F, $d_\theta(F)$ varies continuously in θ.

Moreover,

- the definition of d_θ should be natural, sharing the philosophies of both dim and Dim
- d_θ should give rise to a rich and workable theory.

The most important of these points are the final two. One can achieve the first four in any number of trivial and meaningless ways, but the key idea is that the function $\theta \mapsto d_\theta(F)$ should be ripe with easily interpreted, meaningful, and nuanced information regarding the set F. If this can be achieved then the rewards are likely to include:

- a better understanding of dim and Dim
- an explanation of one type of behaviour changing into another

- more information, leading to better applications
- a (large) new set of questions
- fun.

In the following subsections we describe two concrete examples of this philosophy in action.

1.1 The Assouad Spectrum

The *Assouad spectrum*, introduced by Fraser and Yu in 2016 [14], aims to interpolate between the upper box dimension and the Assouad dimension. The parameter $\theta \in (0, 1)$ serves to fix the relationship between the two scales $r < R$ used to define the Assouad dimension, by setting $R = r^{\theta}$. As such, the Assouad spectrum of $F \subseteq \mathbb{R}^d$ is defined by

$$\dim_{\mathrm{A}}^{\theta} F = \inf\Bigg\{ \alpha > 0 : \text{ there exists a constant } C > 0 \text{ such that, for all } 0 < r < 1 \text{ and } x \in F \text{ we have } N_r\big(B\big(x, r^{\theta}\big) \cap F\big) \leq C\left(\frac{r^{\theta}}{r}\right)^{\alpha} \Bigg\}.$$

At this point it might seem equally natural to bound the two scales away from each other by considering all $0 < r \leq R^{1/\theta}$ rather than fixing $r = R^{1/\theta}$. Rather than go into details here, we simply observe that fixing the relationship between the scales is both easier to work with and provides strictly more information than the alternative, see [18]. We also note that in [14] the scales were denoted by $R^{1/\theta}$ and R, rather than r and r^{θ}. These two formulations are clearly equivalent but the notation we use here seems a little less cumbersome, however, in certain situations it is more natural to use $R^{1/\theta}$ and R. It was established in [14] that $\dim_{\mathrm{A}}^{\theta} F$ is:

- continuous in $\theta \in (0, 1)$, see [14, Corollary 3.5]
- Lipschitz on any closed subinterval of $(0, 1)$, see [14, Corollary 3.5]
- not necessarily monotonic (but often is), see [14, Proposition 3.7 and Section 8].

Moreover, we have the following general bounds, adapted from [14, Proposition 3.1].

Lemma 1.1 *For any bounded set $F \subseteq \mathbb{R}^d$,*

$$\overline{\dim}_{\mathrm{B}} F \leq \dim_{\mathrm{A}}^{\theta} F \leq \min\left\{ \frac{\overline{\dim}_{\mathrm{B}} F}{1-\theta},\ \dim_{\mathrm{A}} F \right\}.$$

Proof Let $s > \overline{\dim}_B F$, $x \in F$ and $r \in (0, 1)$. By definition there exists $C > 0$ depending only on s such that

$$N_r\big(B(x, r^\theta) \cap F\big) \leq N_r(F) \leq Cr^{-s} = C\left(\frac{r^\theta}{r}\right)^{s/(1-\theta)}$$

which implies $\dim_A^\theta F \leq s/(1-\theta)$ and since $s > \overline{\dim}_B F$ was arbitrary, the upper bound follows, noting that $\dim_A^\theta F \leq \dim_A F$ is trivial.

For the lower bound, we may assume $\overline{\dim}_B F > 0$ and let $0 < t < \overline{\dim}_B F < s$. Covering F with r^θ-balls and then covering each of these r^θ-balls with r-balls, we obtain

$$N_r(F) \leq N_{r^\theta}(F) \left(\sup_{x \in F} N_r\big(B(x, r^\theta) \cap F\big)\right).$$

Again, by definition, there exist arbitrarily small $r > 0$ such that

$$\sup_{x \in F} N_r\big(B(x, r^\theta) \cap F\big) \geq \frac{N_r(F)}{N_{r^\theta}(F)} \geq \frac{r^{-t}}{r^{-s\theta}} = \left(\frac{r^\theta}{r}\right)^{\frac{s\theta - t}{\theta - 1}}$$

which establishes $\dim_A^\theta F \geq \frac{t-s\theta}{1-\theta}$ and, since s and t can be made arbitrarily close to $\overline{\dim}_B F$, the lower bound follows. □

A useful consequence of Lemma 1.1 is that $\dim_A^\theta F \to \overline{\dim}_B F$ as $\theta \to 0$ for any bounded F. However, $\dim_A^\theta F$ may *not* approach $\dim_A F$ as $\theta \to 1$. In fact, it was proved in [18] that $\dim_A^\theta F \to \dim_{qA} F$ as $\theta \to 1$, where $\dim_{qA} F$ is the *quasi-*Assouad dimension. In many cases the quasi-Assouad dimension and Assouad dimension coincide and so the intended interpolation is achieved. Moreover, the appearance of Assouad dimension in Lemma 1.1 may be replaced by the quasi-Assouad dimension.

Generally, one has $\dim_{qA} F \leq \dim_A F$ and if this inequality is strict, then the intended interpolation is not achieved. However, an approach for "recovering" the interpolation was outlined in [14]. Let $\phi : [0, 1] \to [0, 1]$ be an increasing continuous function such that $\phi(R) \leq R$ for all $R \in [0, 1]$. The *ϕ-Assouad dimension*, introduced in [14], is defined by

$$\begin{aligned} \dim_A^\phi F = \inf\Big\{ \alpha > 0 :\ & \text{there exists a constant } C > 0 \text{ such that,} \\ & \text{for all } 0 < r \leq \phi(R) \leq R \leq 1 \text{ and } x \in F \text{ we have} \\ & N_r\big(B(x, R) \cap F\big) \leq C\left(\frac{R}{r}\right)^\alpha \Big\}. \end{aligned}$$

The goal is now to identify precise conditions on ϕ which guarantee $\dim_A^\phi F = \dim_A F$. Resolution of this problem for a particular F gives precise information on how the Assouad dimension of F can be *witnessed* and, moreover, completes the interpolation between the upper box and Assouad dimension in a precise sense. Often $\dim_A^\theta F = \dim_A F$ for some $\theta \in (0, 1)$, in which case the threshold for witnessing the Assoaud dimension is provided by the function $\phi(R) = R^{1/\theta}$. The *ϕ-Assouad dimension* has been considered in detail by García, Hare, and Mendivil [19, 20] and Troscheit [32].

Various other dimension spectra are introduced in [14], including the *lower spectrum*, which is the natural dual to the Assouad spectrum and lives in between the lower dimension and the lower box dimension. This has been investigated, in conjunction with the Assouad spectrum, by Chen [5], Chen et al. [6], Hare and Troscheit [21] and Fraser and Yu [15].

1.2 Intermediate Dimensions

The *intermediate dimensions*, introduced by Falconer, Fraser and Kempton in 2018 [9], aim to interpolate between the Hausdorff and box dimensions. The parameter $\theta \in (0, 1)$ serves to restrict the discrepancy between the size of covering sets in the definition of the Hausdorff dimension by insisting that $|U_i| \leq |U_j|^\theta$ for all i, j. As such, the θ-intermediate dimensions of a bounded set $F \subseteq \mathbb{R}^d$ are defined by

$$\dim_\theta F = \inf\Big\{ \alpha > 0 : \text{for all } \varepsilon > 0 \text{ there exists a cover } \{U_i\} \text{ of } F \\ \text{with } |U_i| \leq |U_j|^\theta \text{ for all } i, j \text{ such that } \sum_i |U_i|^\alpha < \varepsilon \Big\}.$$

In fact, [9] considers upper and lower intermediate dimensions, but we restrict our attention here to the lower version. It was proved in [9] that $\dim_\theta F$ is:

- continuous in $\theta \in (0, 1)$, see [9, Proposition 2.1]
- monotonically increasing
- bounded between the Hausdorff and lower box dimension, that is, for bounded F

$$\dim_H F \leq \dim_\theta F \leq \underline{\dim}_B F$$

- and satisfies appropriate versions of the *mass distribution principle* and *Frostman's lemma*, see [9, Propositions 2.2–2.3].

Next we establish general lower bounds for the intermediate dimensions which involve the Assouad dimension, see [9, Proposition 2.4]. In the proof we rely on the following mass distribution principle, first proved in [9, Proposition 2.2]. The main difference between Lemma 1.2 and the usual mass distribution principle, see [8, 4.2], is that a family of measures $\{\mu_r\}$ is used instead of a single measure.

Lemma 1.2 *Let F be a Borel subset of $\mathbb{R}^d$, $0 \le \theta \le 1$ and $s \ge 0$. Suppose that there are numbers $a, c, r_0 > 0$ such that for all $0 < r \le r_0$ we can find a Borel measure μ_r supported by F with $\mu_r(F) \ge a$, such that*

$$\mu_r(U) \le c|U|^s \tag{1.1}$$

for all Borel sets $U \subseteq \mathbb{R}^d$ with $r \le |U| \le r^\theta$. Then $\dim_\theta F \ge s$.

Proof Let $\{U_i\}$ be a cover of F such that $r \le |U_i| \le r^\theta$ for all i and some $r \le r_0$. We may clearly assume the U_i are Borel (even closed). Then

$$a \le \mu_r(F) \le \mu_r\Big(\bigcup_i U_i\Big) \le \sum_i \mu_r(U_i) \le c \sum_i |U_i|^s,$$

so that $\sum_i |U_i|^s \ge a/c > 0$ for every admissible cover (by sets with sufficiently small diameters) and therefore $\dim_\theta F \ge s$. □

Lemma 1.3 *For bounded $F \subseteq \mathbb{R}^d$ and $\theta \in (0, 1)$, we have*

$$\dim_\theta F \ge \dim_{\mathrm{A}} F - \frac{\dim_{\mathrm{A}} F - \underline{\dim}_{\mathrm{B}} F}{\theta}.$$

Proof Fix $\theta \in (0, 1)$ and assume that $\underline{\dim}_{\mathrm{B}} F > 0$, since otherwise there is nothing to prove. Let

$$0 < s < \underline{\dim}_{\mathrm{B}} F \le \dim_{\mathrm{A}} F < t < \infty$$

and $r \in (0, 1)$ be given. Since $s < \underline{\dim}_{\mathrm{B}} F$, there exists a constant C_0 such that there is an r-separated set of points in F of cardinality at least $C_0 r^{-s}$. Let μ_r be a uniformly distributed probability measure supported on this set of points.

Let $U \subseteq \mathbb{R}^d$ be a Borel set with $|U| = r^\gamma$ for some $\gamma \in [\theta, 1]$. Since $\dim_{\mathrm{A}} F < t$, there exists a constant C_1 such that U intersects at most $C_1(r^\gamma/r)^t$ points in the support of μ_r. Therefore

$$\mu_r(U) \le C_1 r^{(\gamma-1)t} C_0^{-1} r^s = C_1 C_0^{-1}\, |U|^{(\gamma t - t + s)/\gamma} \le C_1 C_0^{-1}\, |U|^{(\theta t - t + s)/\theta},$$

which, using Lemma 1.2, implies that

$$\dim_\theta F \ge (\theta t - t + s)/\theta = t - \frac{t-s}{\theta}.$$

Letting $t \to \dim_{\mathrm{A}} F$ and $s \to \underline{\dim}_{\mathrm{B}} F$ yields the desired result. □

It follows from this lemma that $\dim_\theta F \to \underline{\dim}_{\mathrm{B}} F$ as $\theta \to 1$. In contrast, it was shown in [9] that $\dim_\theta F$ does not *necessarily* approach $\dim_{\mathrm{H}} F$ as $\theta \to 0$. Moreover, a mechanism for constructing such examples is provided by the above lemma since if $\underline{\dim}_{\mathrm{B}} F = \dim_{\mathrm{A}} F$, then $\dim_\theta F = \underline{\dim}_{\mathrm{B}} F = \dim_{\mathrm{A}} F$ for all $\theta \in (0, 1)$.

Lemma 1.3 should be compared with Lemma 1.1. For example, combining the two results, one sees that if $F \subseteq \mathbb{R}^d$ and either $\dim_{\mathrm{B}} F = 0$ or $\dim_{\mathrm{B}} F = d$, then both the intermediate dimensions and Assouad spectrum are constant (and equal to $\dim_{\mathrm{B}} F$).

2 Examples

2.1 *Countable Sets*

Fix $p > 0$, and let $F_p = \{n^{-p} : n \in \mathbb{N}\}$. It is straightforward to show that

$$\dim_{\mathrm{H}} F_p = 0 < \dim_{\mathrm{B}} F_p = \frac{1}{1+p} < \dim_{\mathrm{A}} F_p = 1.$$

Moreover, it was shown in [14, Corollary 6.4] that

$$\dim_{\mathrm{A}}^{\theta} F_p = \min\left\{\frac{1}{(1+p)(1-\theta)}, 1\right\}$$

and in [9, Proposition 3.1] that

$$\dim_\theta F_p = \frac{\theta}{\theta + p},$$

see Fig. 5. Therefore these simple examples provide a clear exposition of dimension interpolation in action, noting that genuine continuous interpolation between the dimensions considered is achieved in each case.

2.2 *Self-Affine Sets*

One of the most natural and important families of set which exhibit distinct Hausdorff, box and Assouad dimensions are the self-affine carpets introduced by Bedford [1] and McMullen [28]. These sets are constructed as follows. Divide the unit square $[0, 1]^2$ into an $m \times n$ grid, for integers $n > m \geq 2$, and select a collection of $N \geq 2$ rectangles formed by the grid. Label the rectangles $1, \dots, N$ and, for each rectangle i, let S_i denote the affine map which maps $[0, 1]^2$ onto i by first applying

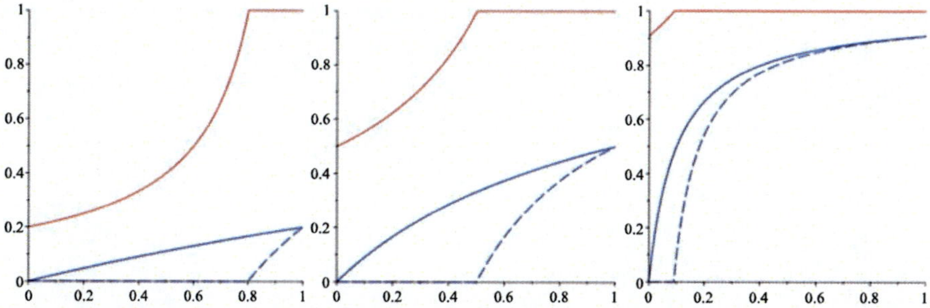

Fig. 5 Plots of $\dim_{\mathrm{A}}^{\theta} F_p$ (red) and $\dim_\theta F_p$ (solid blue) as functions of θ for different values of p. On the left, $p = 4$, in the centre $p = 1$, and on the right $p = 1/10$. For reference, the general lower bounds from Lemma 1.3 for the intermediate dimensions are shown as a dashed blue line. The general upper bounds from Lemma 1.1 for the Assouad spectrum are achieved

the map $(x, y) \mapsto (x/m, y/n)$ and then translating. The *Bedford-McMullen carpet* is defined to be the unique non-empty compact set F satisfying

$$F = \bigcup_{i=1}^{N} S_i(F).$$

The fact that this formula defines such a set uniquely is a well-known result in fractal geometry concerning *iterated function systems*, see [8, Chapter 9] for the details.

In order to state known dimension formulae for F, let $M \in [1, m]$ denote the number of distinct columns in the grid containing chosen rectangles i, $C_j \in [1, n]$ denote the number of chosen rectangles in the jth nonempty column for $j \in \{1, \ldots, M\}$, and $C_{\max} = \max_j C_j$. Bedford and McMullen independently computed the box and Hausdorff dimensions of F in 1984 [1, 28] and the Assouad dimension was computed by Mackay in 2011 [26]. The respective formulae are

$$\dim_{\mathrm{H}} F = \frac{\log \sum_j C_j^{\log m/\log n}}{\log m},$$

$$\dim_{\mathrm{B}} F = \frac{\log M}{\log m} + \frac{\log(N/M)}{\log n},$$

and

$$\dim_{\mathrm{A}} F = \frac{\log M}{\log m} + \frac{\log C_{\max}}{\log n}.$$

Note that if $C_j < C_{\max}$ for some j, then the Hausdorff, box and Assouad dimensions are all distinct. This is called the *non-uniform fibres* case and is the case of interest. In fact, in the *uniform* fibres case, the three dimensions

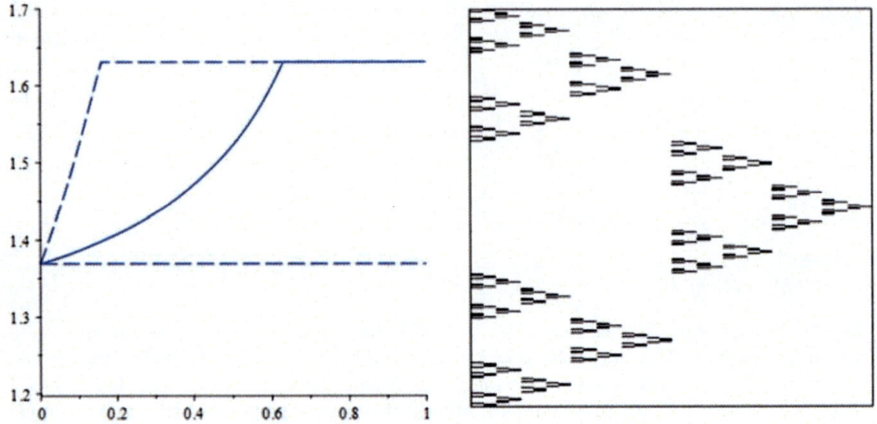

Fig. 6 Left: a plot of $\dim_{\mathrm{A}}^{\theta} F$ (solid blue) as a function of θ where $n = 3$, $m = 2$, $N = 3$, $M = 2$, $C_1 = 2$, $C_2 = 1$ and $C_{\max} = 2$. For reference, the general upper and lower bounds for the Assouad spectrum from Lemma 1.1 are shown as dashed blue lines. Right: an example of a self-affine carpet constructed with the same data

coincide. Therefore, from now on we restrict our attention to the non-uniform fibres setting, where computation of the Assouad spectrum and intermediate dimensions is relevant. It was recently proved in [15, Corollary 3.5] that, for $\theta \in (0, \log m/\log n]$,

$$\dim_{\mathrm{A}}^{\theta} F = \frac{\log M - \theta \log(N/C_{\max})}{(1-\theta)\log m} + \frac{\log(N/M) - \theta \log C_{\max}}{(1-\theta)\log n}$$

and for $\theta \in [\log m/\log n, 1)$

$$\dim_{\mathrm{A}}^{\theta} F = \dim_{\mathrm{A}} F,$$

see Fig. 6. In particular, a single phase transition occurs at $\theta = \log m/\log n$, and a short calculation reveals that this is strictly greater than

$$1 - \frac{\overline{\dim}_{\mathrm{B}} F}{\dim_{\mathrm{A}} F}$$

which is where the single phase transition occurs in the general upper bound from Lemma 1.1. Therefore, the general upper bound for $\dim_{\mathrm{A}}^{\theta} F$ is never achieved by a Bedford-McMullen carpet in the non-uniform fibres setting.

The intermediate dimensions of F were considered in [9], where it was established that $\dim_{\theta} F \to \dim_{\mathrm{H}} F$ as $\theta \to 0$. Recall that this 'genuine interpolation' is not satisfied for all sets. A precise formula for $\dim_{\theta} F$ currently seems out of

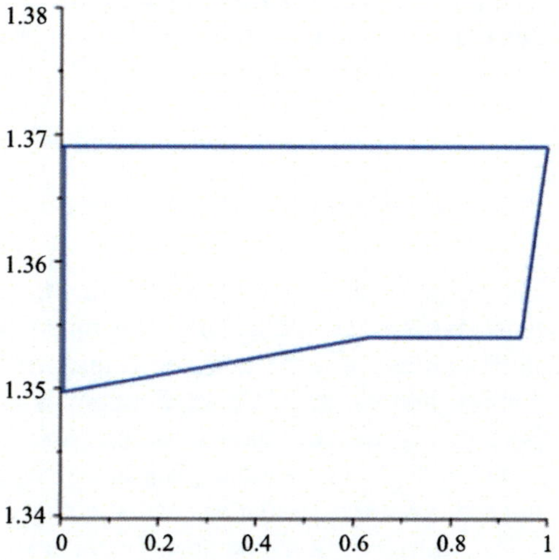

Fig. 7 Plots of the upper and lower bounds for $\dim_\theta F$ for the carpet shown in Fig. 6. The upper bound combines the bound from [9] and the trivial upper bound $\dim_B F$. The lower bound is a piecewise function, with the first increasing part coming from the bounds established in [9], the constant part coming from monotonicity, and the final increasing part coming from the general bounds from Lemma 1.3

reach, but the following bounds were established in [9, Propositions 4.1 and 4.3], see Fig. 7. For $0 < \theta < \left(\frac{\log m}{2\log n}\right)^2$ we have the upper bound

$$\dim_\theta F \leq \dim_H F + \frac{2(\log C_{\max})\log\left(\frac{\log n}{\log m}\right)}{-(\log n)\log\theta},$$

which importantly establishes $\dim_\theta F \to \dim_H F$ as $\theta \to 0$, but only improves on the trivial bound of $\dim_\theta F \leq \dim_B F$ for very small values of θ. For example, for the carpet considered in Fig. 6 this improvement is only achieved for θ smaller than around 10^{-13}. Also, for all $\theta \in (0, \log m/\log n)$ we have the lower bound

$$\dim_\theta F \geq \dim_H F + \theta\frac{\log N - h}{\log n},$$

where

$$h = -m^{-\dim_H F}\sum_j C_j^{\log m/\log n}\left(\left(\frac{\log m}{\log n} - 1\right)\log C_j - \dim_H F \log m\right)$$

is the entropy of the McMullen measure. A short calculation shows that $0 < h \leq \log N$ with $h = \log N$ if and only if F has uniform fibres. Therefore, in the non-uniform fibres case we have $\dim_H F < \dim_\theta F$ for all $\theta \in (0, 1)$. This lower bound improves on the general lower bound from Lemma 1.3 for the carpet considered in Fig. 6 for $\theta \leq 0.95$. In the absence of a precise formula, we ask the following questions.

Question 2.1 For F a Bedford-McMullen carpet with non-uniform fibres, is it true that $\dim_\theta F < \dim_B F$ for all $\theta \in (0, 1)$? Moreover, is it true that $\dim_\theta F$ is strictly increasing, differentiable, or analytic?

2.3 *Self-Similar Sets and Random Sets*

The examples discussed so far (the countable sets, and self-affine carpets with non-uniform fibres) are particularly well-suited to the models of interpolation we discuss in this article. In particular, the Hausdorff, box, and Assouad dimensions are all distinct, and the intermediate dimensions and Assouad spectrum achieve genuine interpolation between these three dimensions. Recall that this is not always the case. Here we discuss two natural families of sets, for which the desired interpolation is not achieved: self-similar sets with overlaps, and Mandelbrot percolation.

We restrict our attention to self-similar sets in $\mathbb{R}$, but interesting questions remain open in higher dimensions. Let $\{S_i\}_i$ be a finite collection of contracting orientation preserving similarities mapping $[0, 1]$ into itself. That is, for each i, there are constants $c_i \in (0, 1)$ and $t_i \in [0, 1 - c_i]$ such that S_i is given by $S_i(x) = c_i x + t_i$. Similar to the setting of self-affine carpets, there exists a unique non-empty compact set $F \subseteq [0, 1]$ satisfying

$$F = \bigcup_i S_i(F).$$

Such sets F are known as *self-similar*, see [8, Chapter 9]. It is well-known that if there exists a non-empty open set $U \subseteq [0, 1]$ such that $\cup_i S_i(U) \subset U$ and the sets $S_i(U)$ are pairwise disjoint, then

$$\dim_H F = \dim_B F = \dim_A F = s$$

where $s \in (0, 1]$ is the unique solution to *Hutchinson's formula* $\sum_i c_i^s = 1$. In particular, this 'separation condition', known as the *open set condition* (OSC), guarantees that the pieces $S_i(F)$ do not overlap too much and thus the images of F under iterates of the defining maps directly give rise to efficient covers of F, facilitating calculation of dimension. It also guarantees sufficient homogeneity to ensure equality of the three dimensions we discuss. In particular, self-similar sets satisfying the OSC are not interesting from our dimension interpolation perspective. However, if the OSC fails, then the Assouad dimension can strictly exceed the box dimension, see [11, 16]. On the other hand, the Hausdorff and box dimension always coincide for self-similar sets, see [7, Corollary 3.3]. Thus, the natural object to consider here is the Assouad spectrum. The following result was proved in [15, Corollary 4.2].

Theorem 2.2 *Let $F \subseteq \mathbb{R}$ be a self-similar set which does not have 'super-exponential concentration of cylinders'. Then for all $\theta \in (0, 1)$*

$$\dim_{\mathrm{A}}^{\theta} F = \dim_{\mathrm{B}} F.$$

In particular, this result implies that genuine interpolation between the box dimension and the Assouad dimension is *not* achieved for these self-similar sets whenever the Assouad dimension strictly exceeds its box dimension. It remains open whether the conclusion of the above result is true for *all* self-similar sets. This theorem was proved using a recent result of Shmerkin [30] and we refer the reader to this paper for more details on the 'super-exponential concentration' assumption. We note, however, that this assumption is satisfied if the semigroup generated by the defining maps is free (that is, there are no 'exact overlaps') and the parameters t_i and c_i defining the maps are algebraic.

Mandelbrot percolation is a natural random process giving rise to fractals which are *statistically* self-similar, see [8, Section 15.2]. We begin with the unit cube $M_0 = [0, 1]^d$, a fixed integer $m \geq 2$, and a probability $p \in (0, 1)$. At the first step of the construction we divide M_0 into m^d (closed) cubes of side length m^{-1} and for each cube we independently choose to 'keep it' with probability p, or 'throw it away' with probability $(1 - p)$. We let M_1 be the collection of kept cubes and we then repeat this process inside each kept cube independently, denoting the collection of kept cubes at stage n by M_n. The limit set is then defined by $M = \cap_n M_n$, see Fig. 1 for an example with $d = m = 2$. It is well-known that if $p > m^{-d}$, then M is non-empty with positive probability. Moreover, if we condition on M being non-empty, then

$$\dim_{\mathrm{H}} M = \dim_{\mathrm{B}} M = d + \frac{\log p}{\log m} \in (0, d)$$

almost surely. It was shown in [17, Theorem 5.1] that, conditioned on M being non-empty,

$$\dim_{\mathrm{A}} M = d \tag{2.1}$$

almost surely, and therefore it is natural to consider the Assouad spectrum of M. However, it was proved in [15, 31, 33] that, conditioned on M being non-empty, almost surely

$$\dim_{\mathrm{A}}^{\theta} M = \dim_{\mathrm{B}} M \tag{2.2}$$

for all $\theta \in (0, 1)$. Therefore, again we see that genuine interpolation between the box dimension and Assouad dimension is *not* achieved by the Assouad spectrum for Mandelbrot percolation. However, using the finer analysis introduced in [14] and discussed in Sect. 1.1, it is possible to observe the interpolation by considering $\dim_{A}^{\phi} M$ for different functions ϕ. Troscheit proved the following dichotomy in [32].

Theorem 2.3 *If*

$$\frac{\log(R/\phi(R))}{\log|\log R|} \to 0$$

as $R \to 0$*, then, conditioned on* M *being non-empty, almost surely*

$$\dim_{\mathrm{A}}^{\phi} M = d = \dim_{\mathrm{A}} M.$$

Moreover, if

$$\frac{\log(R/\phi(R))}{\log|\log R|} \to \infty$$

as $R \to 0$*, then, conditioned on* M *being non-empty, almost surely*

$$\dim_{\mathrm{A}}^{\phi} M = \dim_{\mathrm{B}} M = d + \frac{\log p}{\log m}.$$

Note that this result implies (2.2) by considering $\phi(R) = R^{1/\theta}$ and (2.1) by considering $\phi(R) = R$. A similar dichotomy, with the same threshold on ϕ, was obtained in a different random setting in [20]. The Assouad spectrum of random self-affine carpets was considered in [13].

3 Applications: Bi-Lipschitz and Bi-Hölder Distortion

A key aspect of this new perspective in dimension theory is in its applications. The idea is that if we can interpolate between two given dimensions in a meaningful way, then we will get strictly better information than when the dimensions are considered in isolation. This better information should, in turn, yield stronger applications. For example, see the recent papers [3, 4] which use the intermediate dimensions to obtain new results concerning the *box* dimensions of orthogonal projections and images under fractional Brownian motion, respectively.

A common application of dimension theory is derived from the fact that dimensions are often invariant, or approximately invariant in a quantifiable sense, under a family of transformations. For example, the Hausdorff, box and Assouad dimensions are all invariant under bi-Lipschitz maps and therefore provide useful invariants in the problem of classification up to bi-Lipschitz image. The Assouad spectrum and intermediate dimensions are also invariant under bi-Lipschitz maps and therefore provide a continuum of invariants in the same context. Recall that an injective map $f : X \to \mathbb{R}^d$ is *bi-Lipschitz* if there exists a constant $C \geq 1$ such that for all distinct $x, y \in X$

$$C^{-1}|x - y| \leq |f(x) - f(y)| \leq C|x - y|. \tag{3.1}$$

Here we assume that X is a bounded subset of $\mathbb{R}^d$. In particular, for such f we have

$$\dim_A^\theta X = \dim_A^\theta f(X) \qquad \text{and} \qquad \dim_\theta X = \dim_\theta f(X)$$

for all $\theta \in (0, 1)$. This was proved for the Assouad spectrum in [14] and we prove it for the intermediate dimensions here.

Lemma 3.1 *For any bounded set $X \subseteq \mathbb{R}^d$ and bi-Lipschitz map $f : X \to \mathbb{R}^d$, we have $\dim_\theta X = \dim_\theta f(X)$ for all $\theta \in (0, 1)$.*

Proof Let $s > \dim_\theta X$ and $\varepsilon > 0$. It follows that there exists a cover $\{U_i\}$ of X with $|U_i| \leq |U_j|^\theta$ for all i, j such that $\sum_i |U_i|^s < \varepsilon$. It follows that $\{f(U_i)\}$ is a cover of $f(X)$ and that $|f(U_i)| \leq C|U_i| \leq C|U_j|^\theta \leq C^{1+\theta}|f(U_j)|^\theta$ for all i, j, where C is the constant from (3.1). Let $\delta = \inf_j |f(U_j)|$. For all i such that $\delta^\theta < |f(U_i)| \leq C^{1+\theta}\delta^\theta$, cover the set $f(U_i)$ with balls of diameter δ^θ and replace the covering set $f(U_i)$ by these balls. Note that we can always do this with fewer than $c_d C^{d(1+\theta)}$ balls where $c_d \geq 1$ is a constant depending only on d. This yields an allowable cover $\{V_l\}$ of $f(X)$ and we have

$$\sum_l |V_l|^s \leq c_d C^{d(1+\theta)} \sum_i C^s |U_i|^s \leq c_d C^{d(1+\theta)+s} \varepsilon$$

which proves $\dim_\theta f(X) \leq \dim_\theta X$ by letting $s \to \dim_\theta X$. The reverse inequality follows by replacing f by f^{-1} in the above. □

An immediate consequence of the bi-Lipschitz invariance of the Assouad spectrum is that if F_1 and F_2 are Bedford-McMullen carpets associated with $m_1 \times n_1$ and $m_2 \times n_2$ grids, respectively, and there exists a bi-Lipschitz map between F_1 and F_2, then

$$\frac{\log m_1}{\log n_1} = \frac{\log m_2}{\log n_2}.$$

This is because this ratio corresponds to the phase transition in the spectrum, and is therefore a bi-Lipschitz invariant. This is not at all surprising, but serves as a simple example of the spectrum yielding applications which are not immediate when considering the dimensions in isolation. Classification of self-affine sets up to bi-Lipschitz equivalence is an interesting problem in general, see [24].

Bi-Hölder maps are a natural generalisation of bi-Lipschitz maps where more distortion is allowed. We say an injective map $f : X \to \mathbb{R}^d$ is (α, β)-Hölder, or bi-Hölder, for $0 < \alpha \leq 1 \leq \beta < \infty$ if there exists a constant $C \geq 1$ such that for all distinct $x, y \in X$

$$C^{-1}|x - y|^\beta \leq |f(x) - f(y)| \leq C|x - y|^\alpha.$$

We note that being (1, 1)-Hölder is the same as being bi-Lipschitz. Dimensions are typically not preserved under bi-Hölder maps, but one can often control the distortion. For example, if dim is the Hausdorff, or upper or lower box dimension, and f is (α, β)-Hölder, then

$$\frac{\dim X}{\beta} \leq \dim f(X) \leq \frac{\dim X}{\alpha}, \tag{3.2}$$

see [8, Proposition 3.3]. Notably, the Assouad dimension does not satisfy such bounds, see [25, Proposition 1.2]. The Assouad spectrum, which is inherently more regular than the Assouad dimension, *can* be controlled in this context but the control is more complicated than (3.2). The following lemma is adapted from [14, Proposition 4.7].

Lemma 3.2 *Suppose $f : X \to \mathbb{R}^d$ is (α, β)-Hölder. Then, for all $\theta \in (0, 1)$,*

$$\frac{1 - \beta\theta/\alpha}{\beta(1 - \theta)} \dim_{\mathrm{A}}^{\beta\theta/\alpha} X \leq \dim_{\mathrm{A}}^{\theta} f(X) \leq \frac{1 - \alpha\theta/\beta}{\alpha(1 - \theta)} \dim_{\mathrm{A}}^{\alpha\theta/\beta} X$$

where $\dim_{\mathrm{A}}^{\beta\theta/\alpha} X$ is taken to equal 0 if $\beta\theta/\alpha \geq 1$.

In order to motivate this result, we consider the *winding problem*. Given $p \geq 1$, let

$$\mathcal{S}_p = \{x^{-p} \exp(ix) : 1 < x < \infty\}$$

which is a polynomially winding spiral with focal point at the origin. The winding problem concerns quantifying how little distortion is required to map (0, 1) onto $\mathcal{S}_p$. For example, if x^{-p} is replaced by e^{-cx} for some $c > 0$, then it is possible to map (0, 1) onto the corresponding spiral via a bi-Lipschitz map, see [22]. However, this is not possible for the spirals $\mathcal{S}_p$, see [10]. Therefore, it is natural to consider bi-Hölder winding functions, and attempt to optimise the Hölder exponents.

Here there is a possible application of dimension theory: if the dimensions of $\mathcal{S}_p$ can be computed, and strictly exceed 1, then (3.2) (or similar) will directly lead to bounds on the possible Hölder exponents for winding functions $f : (0, 1) \to \mathcal{S}_p$. However, since $\mathcal{S}_p$ can be broken up into a countable collection of bi-Lipschitz curves, it follows that $\dim_{\mathrm{H}} \mathcal{S}_p = 1$. Moreover, it was proved in [12] that $\dim_{\mathrm{B}} \mathcal{S}_p = 1$. This does not follow from the countable decomposition since box dimension is not countably stable. Therefore, neither the Hausdorff nor box dimensions give any information on the Hölder exponents. It was proved in [12] that $\dim_{\mathrm{A}} \mathcal{S}_p = 2$, but despite this being strictly greater than $\dim_{\mathrm{A}}(0, 1) = 1$, we also get no information from the Assouad dimension since the change in dimension cannot be controlled by the Hölder exponents. It was proved in [12] that

$$\dim_{\mathrm{A}}^{\theta} \mathcal{S}_p = 1 + \frac{\theta}{p(1 - \theta)}$$

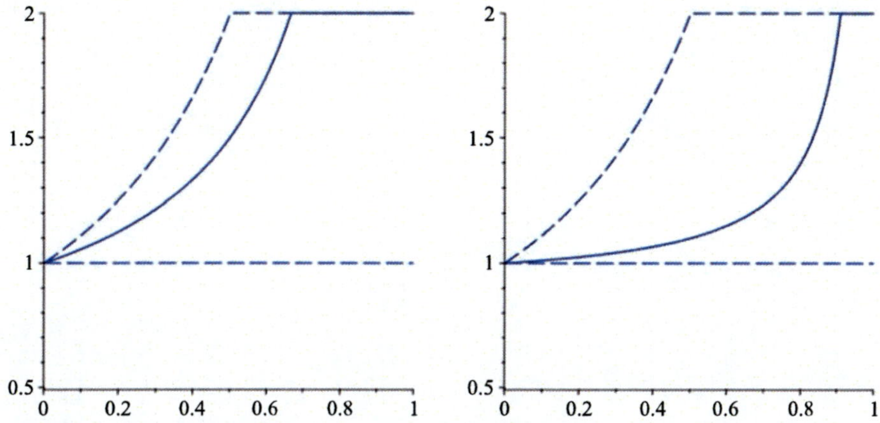

Fig. 8 Plots of $\dim_{\mathrm{A}}^{\theta} \mathcal{S}_p$ (solid blue) as a function of θ. On the left $p = 2$ and on the right $p = 10$. For reference, the general upper and lower bounds for the Assouad spectrum from Lemma 1.1 are shown as dashed blue lines

for $0 < \theta < \frac{p}{1+p}$, and

$$\dim_{\mathrm{A}}^{\theta} \mathcal{S}_p = 2$$

for $\frac{p}{1+p} \le \theta < 1$, see Fig. 8. Therefore, since we do have some control on how the Assouad spectrum distorts under bi-Hölder maps, this dimension formula *does* yield non-trivial information. Specifically, we get that if $f : (0, 1) \to \mathcal{S}_p$ is an (α, β)-Hölder map, then

$$\alpha \le \frac{p\beta + \beta}{p + 2\beta}. \tag{3.3}$$

This follows by applying the first inequality in Lemma 3.2 to f^{-1} for $\theta = \alpha p/(\beta p + \beta)$. In particular, if $\beta = 1$, then $\alpha \le \frac{p+1}{p+2} < 1$, which is a stronger, quantitative, analogue of the fact that $(0, 1)$ cannot be mapped to $\mathcal{S}_p$ via a bi-Lipschitz map.

It turns out that the bounds (3.3) are *not* sharp. The sharp relationship between α and β is given by

$$\alpha \le \frac{p\beta}{p + \beta},$$

see [12] and Fig. 9. We note the amusing resemblance of this relationship to that of *Sobolev conjugates*. Recall the *Sobolev embedding theorem* which says that, for $1 \le p < d$, one has

$$W^{1,p}(\mathbb{R}^d) \subset L^q(\mathbb{R}^d)$$

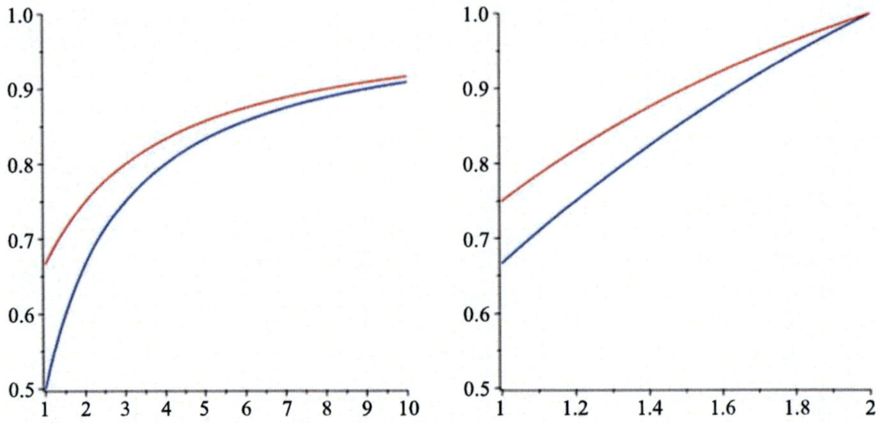

Fig. 9 Left: a plot of the upper bounds for α as a function of p where $\beta = 1$ is fixed. The sharp upper bound is shown in blue and the upper bound given by the Assouad spectrum is shown in red. Right: a plot of the upper bounds for α as a function of β where $p = 2$ is fixed. The sharp upper bound is shown in blue and the upper bound given by the Assouad spectrum is shown in red

where q is defined by

$$p = \frac{dq}{d+q},$$

that is, q is the Sobolev conjugate of p.

A further application of the Assouad spectrum in this context is that $\dim_{\mathrm{A}}^{\theta} \mathcal{S}_p$, distinguishes spirals with different winding rates p. Note that this is *not* achieved by the Hausdorff, box, or Assouad dimensions, since these (somewhat surprisingly) do not depend on p. In particular, the Assouad spectrum shows that $\mathcal{S}_p$ and $\mathcal{S}_q$ are not bi-Lipschitz equivalent for $p \neq q$.

4 Further Remarks

We note that the Assouad spectrum of the spirals considered in the previous section exhibits a single phase transition at $\frac{p}{p+1}$. Similar to the self-affine carpets, it is easy to see that this phase transition occurs strictly to the right of the phase transition in the general upper bound, provided $p > 1$, and therefore the general upper bound is not realised by these spirals. This gives rise to a similar form for the spectrum of the carpets and the spectrum of the spirals. We observe that this similarity goes a little deeper. In fact, in both cases we have the formula

$$\dim_{\mathrm{A}}^{\theta} E = \min\left\{\dim_{\mathrm{B}} E + \frac{(1-\rho)\theta}{(1-\theta)\rho}\left(\dim_{\mathrm{A}} E - \dim_{\mathrm{B}} E\right),\ \dim_{\mathrm{A}} E\right\}, \qquad (4.1)$$

where ρ is a constant which holds particular geometric significance for the object E. Specifically, for carpets $\rho = \frac{\log m}{\log n}$, and for spirals $\rho = \frac{p}{p+1}$. Also note that ρ is the value of θ at which the unique phase transition occurs. In both cases ρ captures some fundamental scaling property of the set. For carpets, the kth level rectangles in the standard construction of F are of size $m^{-k} \times n^{-k}$ and therefore ρ is the "logarithmic eccentricity". For spirals, the *kth revolution*, given by

$$\{x^{-p} \exp(ix) : 1 + 2\pi(k-1) < x \leq 1 + 2\pi k\},$$

has diameter comparable to k^{-p}, while the distance between the end points (or, outer radius minus inner radius) is comparable to $k^{-(p+1)}$. These are fundamental measurements considered in the winding problem, see [12], and measure how big the kth revolution is and how tightly it is wound, respectively. Again the "logarithmic eccentricity" is

$$\frac{\log\left(k^{-p}\right)}{\log\left(k^{-(p+1)}\right)} = \frac{p}{p+1} = \rho.$$

We wonder if this is a coincidence, or whether it is reflective of a more general phenomenon. It would be interesting to identify other natural classes of set for which this formula holds for a particular choice of "fundamental ratio" ρ. Finally, we note that the Assouad spectrum does *not* generally satisfy an equation of the form (4.1), see [14, 15, 18].

Acknowledgments The author thanks Stuart Burrell, Kenneth Falconer, Kathryn Hare, Kevin Hare, Antti Käenmäki, Tom Kempton, Sascha Troscheit, and Han Yu for many interesting discussions relating to dimension interpolation. He was financially supported in part by the EPSRC Standard Grant EP/R015104/1. He is also grateful to the Leverhulme Trust for funding his project *New Perspectives in the dimension theory of fractals* (2019–2023), which is largely focused on the concept of dimension interpolation. Finally, he thanks Stuart Burrell and Kenneth Falconer for making several helpful comments on an earlier version of this article.

References

1. Bedford, T.: Crinkly curves, Markov partitions and box dimensions in self-similar sets. Ph.D. Thesis, University of Warwick (1984)
2. Bishop, C.J., Peres, Y.: Fractals in Probability and Analysis. Cambridge Studies in Advanced Mathematics, vol. 162. Cambridge University Press, Cambridge (2017)
3. Burrell, S.A.: Dimensions of fractional Brownian images (2020). https://arxiv.org/abs/2002.03659
4. Burrell, S.A., Falconer, K.J., Fraser, J.M.: Projection theorems for intermediate dimensions. J. Fractal Geom. (2019). https://arxiv.org/abs/1907.07632
5. Chen, H.: Assouad dimensions and spectra of Moran cut-out sets. Chaos Sol. Fract. **119**, 310–317 (2019)

6. Chen, H., Wu, M., Chang, Y.: Lower Assouad type dimensions of uniformly perfect sets in doubling metric spaces. Fract. **28**(2), (2020).
7. Falconer, K.J.: Techniques in Fractal Geometry. Wiley, New York (1997)
8. Falconer, K.J.: Fractal Geometry: Mathematical Foundations and Applications, 3rd. edn. Wiley, New York (2014)
9. Falconer, K.J., Fraser, J.M., Kempton, T.: Intermediate dimensions. Math. Z. **296**, 813–830 (2020)
10. Fish, A., Paunescu, L.: Unwinding spirals. Methods and Applications of Analysis **25**(3), 225–232 (2018)
11. Fraser, J.M.: Assouad type dimensions and homogeneity of fractals. Trans. Am. Math. Soc. **366**, 6687–6733 (2014)
12. Fraser, J.M.: On Hölder solutions to the spiral winding problem (2019). https://arxiv.org/abs/1905.07563
13. Fraser, J.M., Troscheit, S.: The Assouad spectrum of random self-affine carpets. Ergodic Theory Dyn. Syst. (to appear). https://arxiv.org/abs/1805.04643
14. Fraser, J.M., Yu, H.: New dimension spectra: finer information on scaling and homogeneity. Adv. Math. **329**, 273–328 (2018)
15. Fraser, J.M., Yu, H.: Assouad type spectra for some fractal families. Ind. Univ. Math. J. **67**, 2005–2043 (2018)
16. Fraser, J.M., Henderson, A.M., Olson, E.J., Robinson, J.C.: On the Assouad dimension of self-similar sets with overlaps. Adv. Math. **273**, 188–214 (2015)
17. Fraser, J.M., Miao, J.J.,Troscheit, S.: The Assouad dimension of randomly generated fractals. Ergodic Theory Dyn. Syst. **38**, 982–1011 (2018)
18. Fraser, J.M., Hare, K.E., Hare, K.G., Troscheit, S., Yu, H.: The Assouad spectrum and the quasi-Assouad dimension: a tale of two spectra. Ann. Acad. Sci. Fenn. Math. **44**, 379–387 (2019)
19. García, I., Hare, K.E., Mendivil, F.: Intermediate Assouad-like dimensions (2019). https://arxiv.org/abs/1903.07155
20. García, I., Hare, K.E., Mendivil, F.: Almost sure Assouad-like dimensions of complementary sets (2019). https://arxiv.org/abs/1903.07800
21. Hare, K.E., Troscheit, S.: Lower Assouad dimension of measures and regularity. Math. Proc. Cambridge Phil. Soc. 1–37 (2019)
22. Katznelson, Y., Nag, S., Sullivan, D.: On conformal welding homeomorphisms associated to Jordan curves. Ann. Acad. Sci. Fenn. Math. **15**, 293–306 (1990)
23. Lehrbäck, J.: Assouad type dimensions in geometric analysis. In: Fractal Geometry and Stochastics VI. Progress in Probability. Birkhäuser, Basel (2020)
24. Li, B., Li, W., Miao, J.J.: Lipschitz equivalence of McMullen sets. Fractals **21**, 1350022 (2013)
25. Lü, F., Xi, L.: Quasi-Assouad dimension of fractals. J. Fractal Geom. **3**, 187–215 (2016)
26. Mackay, J.M.: Assouad dimension of self-affine carpets. Conform. Geom. Dyn. **15**, 177–187 (2011)
27. Mattila, P.: Geometry of sets and measures in Euclidean spaces. Cambridge Studies in Advanced Mathematics, vol. 44. Cambridge University Press, Cambridge (1995)
28. McMullen, C.T.: The Hausdorff dimension of general Sierpiński carpets. Nagoya Math. J. **96**, 1–9 (1984)
29. Robinson, J.C.: Dimensions, Embeddings, and Attractors. Cambridge University Press, Cambridge (2011)
30. Shmerkin, P.: On Furstenberg's intersection conjecture, self-similar measures, and the L^q norms of convolutions. Ann. Math. **189**, 319–391 (2019)
31. Troscheit, S.: The quasi-Assouad dimension of stochastically self-similar sets. Proc. Roy. Soc. Edinburgh. **150**(1), 261–275 (2020)
32. Troscheit, S.: Assouad spectrum thresholds for some random constructions. Can. Math. Bull. **63**(2), 434–453 (2020)
33. Yu, H.: Assouad type dimensions and dimension spectra for some fractal families. Ph.D. Thesis, The University of St Andrews (2019)

Assouad Type Dimensions in Geometric Analysis

Juha Lehrbäck

Abstract We consider applications of the dual pair of the (upper) Assouad dimension and the lower (Assouad) dimension in analysis. We relate these notions to other dimensional conditions such as a Hausdorff content density condition and an integrability condition for the distance function. The latter condition leads to a characterization of the Muckenhoupt A_p properties of distance functions in terms of the (upper) Assouad dimension. It is also possible to give natural formulations for the validity of Hardy–Sobolev inequalities using these dual Assouad dimensions, and this helps to understand the previously observed dual nature of certain cases of these inequalities.

Keywords Assouad dimension · Lower dimension · Aikawa condition · Muckenhoupt weight · Hardy–Sobolev inequality

Mathematics Subject Classifications (2010) Primary: 28A75; Secondary: 28A80, 35A23

1 Introduction

Mathematicians working in fractal geometry and related fields are well aware of the fact that there can not be a unique definition for the concept of dimension of a set, since different problems require different ways to deal with dimensional information. In fact, what sometimes may seem like a negligible nuance in the definition might actually lead to interesting discoveries concerning the fine structure of sets. On the flip side, the multitude of the notions of dimension may easily create confusion, and thus it is important to be able to justify the existence of all these concepts via natural applications.

J. Lehrbäck (✉)
Department of Mathematics and Statistics, University of Jyvaskyla, Jyvaskyla, Finland
e-mail: juha.lehrback@jyu.fi

U. Freiberg et al. (eds.), *Fractal Geometry and Stochastics VI*,
Progress in Probability 76, https://doi.org/10.1007/978-3-030-59649-1_2

The purpose of this article is to describe some recent observations concerning the applications of the dual pair of the upper and lower Assouad dimension, often simply called the Assouad dimension and the lower dimension, respectively. These notions provide geometric information which is relevant not only in fractal geometry, but also for instance in harmonic analysis, potential theory, and partial differential equations. One manifestation of these connections can be seen via the validity of the so-called Hardy–Sobolev inequalities. Our aim is not so much in presenting any novelties on the level of the details or techniques, but rather in trying to illustrate how a new point of view in terms of dimensional conditions may offer clarity and reveal connections between known results. On the other hand, we do give proofs for some basic results, hoping that these will help the reader to gain familiarity with the relevant concepts.

We begin in Sect. 2 by recalling the definitions of the upper and lower Assouad dimension and relating them to the more familiar Hausdorff dimension. In particular, we explain the connection between the lower Assouad dimension and a Hausdorff content density condition. In Sect. 3 we study integrability conditions for distance functions $w(x) = \operatorname{dist}(x, E)^{-\alpha}$, where $E \subset \mathbb{R}^n$ and (usually) $0 < \alpha < n$. Such conditions, originally introduced by Aikawa, can be used to characterize the upper Assouad dimension, see Theorem 3.5. Next, in Sect. 4, we ask when a distance function w as above belongs to the important class of Muckenhoupt A_p weights. As it turns out, the answer can be given in terms of the upper Assouad dimension, using the integrability conditions from Sect. 3 as a helpful stepping stone. Finally, Sect. 5 completes the circle by showing how both upper and lower Assouad dimension play an important role when examining the validity of the Hardy–Sobolev inequalities in an open set $\Omega \subset \mathbb{R}^n$. In particular, a previously observed duality between certain cases of such inequalities becomes more transparent and natural when the conditions are formulated in terms of suitable dimensions.

Much of the theory presented in this survey can be extended to more general metric spaces satisfying standard structural assumptions. We give some comments and remarks related to such extensions, but for simplicity we focus on the case of the n-dimensional Euclidean space $\mathbb{R}^n$.

Notation

The open ball with center $x \in \mathbb{R}^n$ and radius $r > 0$ is

$$B(x, r) = \{y \in \mathbb{R}^n : |y - x| < r\},$$

and $\overline{B}(x, r)$ is the corresponding closed ball. When $A \subset \mathbb{R}^n$, we write $\operatorname{diam}(A)$ for the diameter of A, and $\operatorname{dist}(x, A)$ denotes the distance from a point $x \in \mathbb{R}^n$ to the set A. The complement of A is $A^c = \mathbb{R}^n \setminus A$. If A is (Lebesgue) measurable, then the Lebesgue measure of A is denoted by $|A|$. If $0 < |A| < \infty$ and $f \in L^1(A)$, then the mean value integral of f over A is

$$\fint_A f(x)\,dx = \frac{1}{|A|}\int_A f(x)\,dx.$$

As usual, C denotes a constant whose exact value may change at each occurrence.

For simplicity, we use the following versions of Hausdorff contents and measures. It is easy to see that these are comparable to the more standard definitions in e.g. [9, 30].

Definition 1.1 Let $E \subset \mathbb{R}^n$ and $\lambda \geq 0$. For $0 < \delta \leq \infty$, the λ-dimensional Hausdorff δ-content of E is

$$\mathcal{H}^\lambda_\delta(E) = \inf\left\{\sum_{i=1}^{\infty} r_i^\lambda : E \subset \bigcup_{i=1}^{\infty} B(x_i, r_i),\ 0 < r_i \leq \delta\right\}.$$

(In the case $\lambda = 0$ we allow also finite summations.) Then the (spherical) λ-dimensional Hausdorff measure of E is

$$\mathcal{H}^\lambda(E) = \lim_{\delta \to 0+} \mathcal{H}^\lambda_\delta(E) = \sup_{\delta > 0} \mathcal{H}^\lambda_\delta(E),$$

and the Hausdorff dimension of E is defined as

$$\dim_{\mathrm{H}}(E) = \inf\{\lambda \geq 0 : \mathcal{H}^\lambda(E) = 0\} = \inf\{\lambda \geq 0 : \mathcal{H}^\lambda_\infty(E) = 0\}.$$

2 Assouad Type Dimensions

The definitions of the Assouad type dimensions of a set $E \subset \mathbb{R}^n$ are based on simple and natural local covering properties of E: we consider pieces $E \cap \overline{B}(x, R)$, with $x \in E$ and $0 < R < \operatorname{diam}(E)$, and ask how many balls of radius $0 < r < R$ are needed at most (upper Assouad), or respectively at least (lower Assouad), to cover such pieces. Thus these concepts reveal the most "extreme" local behavior of sets, whereas other notions of dimension usually tell more about the "average" properties of sets.

When $A \subset \mathbb{R}^n$ is a bounded set and $r > 0$, we let $N(A, r)$ denote the minimal number of open balls of radius r that are needed to cover the set A.

Definition 2.1 Let $E \subset \mathbb{R}^n$. The upper Assouad dimension $\overline{\dim}_{\mathrm{A}}(E)$ is the infimum of $\lambda \geq 0$ for which there exists a constant C such that

$$N\big(E \cap \overline{B}(x, R), r\big) \leq C\Big(\frac{r}{R}\Big)^{-\lambda} = C\Big(\frac{R}{r}\Big)^{\lambda} \tag{2.1}$$

for every $x \in E$ and $0 < r < R < \operatorname{diam}(E)$.

In particular, the estimate in (2.1) holds whenever $\lambda > \overline{\dim}_A(E)$, and possibly also when $\lambda = \overline{\dim}_A(E)$. If $E \subset E'$, then clearly $\overline{\dim}_A(E) \leq \overline{\dim}_A(E')$. It is also easy to see that $0 \leq \overline{\dim}_A(E) \leq n$ for every $E \subset \mathbb{R}^n$.

In the literature, the upper Assouad dimension is often called the Assouad dimension and denoted by $\dim_A(E)$. This concept was used by Assouad in connection with the bi-Lipschitz embedding problem between metric and Euclidean spaces, see e.g. [4]. A nice account on the basic properties and history of the Assouad dimension is given in [29]. See also the survey by Fraser [11] in this same volume (and the references therein) for recent fractal geometric applications of the (upper) Assouad dimension and its generalizations.

We illustrate the definition by proving the fact that the Hausdorff dimension always gives a lower bound for the upper Assouad dimension.

Lemma 2.2 *Let $E \subset \mathbb{R}^n$. Then* $\dim_H(E) \leq \overline{\dim}_A(E)$.

Proof By the countable stability of the Hausdorff dimension it suffices to show that

$$\dim_H\big(E \cap \overline{B}(x, R)\big) \leq \overline{\dim}_A(E)$$

for every $x \in E$ and $R > 0$. Let $s > \overline{\dim}_A(E)$, choose λ satisfying $\overline{\dim}_A(E) < \lambda < s$, and fix $x \in E$ and $R > 0$. Then $E \cap \overline{B}(x, R)$ can be covered by

$$N \leq C\Big(\frac{R}{r}\Big)^{\lambda}$$

balls of radius r, for every $0 < r < R$. Thus, by the definition of Hausdorff content,

$$\mathcal{H}^s_r\big(E \cap \overline{B}(x, R)\big) \leq Nr^s \leq C_1 R^{\lambda} r^{s-\lambda}.$$

Letting $r \to 0$ gives $\mathcal{H}^s\big(E \cap \overline{B}(x, R)\big) = 0$, and we conclude that

$$\dim_H\big(E \cap \overline{B}(x, R)\big) \leq \overline{\dim}_A(E).$$

□

Definition 2.3 Let $E \subset \mathbb{R}^n$. The lower Assouad dimension $\underline{\dim}_A(E)$ is the supremum of $\lambda \geq 0$ for which there exists a constant C such that

$$N\big(E \cap \overline{B}(x, R), r\big) \geq C\Big(\frac{r}{R}\Big)^{-\lambda} = C\Big(\frac{R}{r}\Big)^{\lambda} \tag{2.2}$$

for every $x \in E$ and $0 < r < R < \operatorname{diam}(E)$.

In particular, the estimate in (2.2) holds whenever $0 \leq \lambda < \underline{\dim}_A(E)$, and possibly also when $\lambda = \underline{\dim}_A(E)$. In the case $E = \{x_0\}$, $x_0 \in \mathbb{R}^n$, we remove the requirement $R < \operatorname{diam}(E)$ from the definition and hence $\underline{\dim}_A(\{x_0\}) = 0$. It is easy to verify that $0 \leq \underline{\dim}_A(E) \leq \overline{\dim}_A(E) \leq n$ for every $E \subset \mathbb{R}^n$. However, it should be noted that, unlike (most) other natural concepts of dimension, the lower Assouad dimension is not monotone. For instance, $\underline{\dim}_A(\{0\} \cup [1, 2]) = 0$, due to the isolated point 0, but for the subset $[1, 2]$ we have $\underline{\dim}_A([1, 2]) = 1$.

The lower Assouad dimension is often called the lower dimension and denoted by $\dim_L(E)$. Thus the pair of Assouad-type dimensions can be referred to as the (upper) Assouad dimension $\overline{\dim}_A(E) = \dim_A(E)$ and the lower (Assouad) dimension $\underline{\dim}_A(E) = \dim_L(E)$. Also other names, such as (uniform) metric dimension and minimal dimensional number, respectively, have been used. An early reference concerning the lower (Assouad) dimension is [21], and more recently some basic properties of this dimension have been discussed e.g. in [10] and [18].

Remark 2.4 It should be noted that in the literature there are some slight differences in the definitions of the upper and lower Assouad dimensions. In particular, sometimes the covering inequalities in (2.1) and (2.2) are required to hold only for $0 < r < R \leq R_0$, for some fixed $R_0 < \infty$. This change may affect the dimensions of unbounded sets. Notice also that in (2.1) we may omit the upper bound $R < \operatorname{diam}(E)$ without altering the value of the upper Assouad dimension. On the other hand, if we omit this upper bound in (2.2), then all bounded sets would have lower Assouad dimension equal to zero, which is perhaps not so desirable.

Recall that a closed set $E \subset \mathbb{R}^n$ is called (Ahlfors–David) λ-regular, or a λ-set, for $0 \leq \lambda \leq n$, if there is a constant $C \geq 1$ such that

$$C^{-1}r^{\lambda} \leq \mathcal{H}^{\lambda}\big(E \cap \overline{B}(x,r)\big) \leq Cr^{\lambda} \tag{2.3}$$

for every $x \in E$ and $0 < r < \operatorname{diam}(E)$; for $\lambda = 0$ the upper bound $r < \operatorname{diam}(E)$ is omitted.

Examples of λ-regular sets include subspaces of $\mathbb{R}^n$ and self-similar fractals satisfying the open set condition. It is not hard to see that for a λ-regular set $E \subset \mathbb{R}^n$ the upper and lower Assouad dimensions agree. More precisely, if $E \subset \mathbb{R}^n$ is λ-regular then

$$\overline{\dim}_A(E) = \underline{\dim}_A(E) = \dim_H(E) = \lambda.$$

In order to examine the relation between the lower Assouad dimension and the Hausdorff dimension for more general sets, we consider the following density condition for Hausdorff contents.

Definition 2.5 Let $0 \leq \lambda \leq n$. We say that a set $E \subset \mathbb{R}^n$ satisfies the λ-Hausdorff content density condition if there exists a constant C such that

$$\mathcal{H}^{\lambda}_{\infty}\big(E \cap \overline{B}(x,R)\big) \geq CR^{\lambda} \tag{2.4}$$

for every $x \in E$ and $0 < R < \operatorname{diam}(E)$.

Sometimes the upper bound $R < \operatorname{diam}(E)$ is omitted in Definition 2.5, but then a bounded set can not satisfy this condition for any $\lambda > 0$.

The λ-Hausdorff content density condition holds for a set $E \subset \mathbb{R}^n$ if and only if there is a constant C such that if $\{B(x_i, r_i) : i \in \mathbb{N}\}$ is a cover of $E \cap \overline{B}(x, R)$, for

$x \in E$ and $0 < R < \operatorname{diam}(E)$, then

$$\sum_{i=1}^{\infty} r_i^{\lambda} \geq C R^{\lambda}. \tag{2.5}$$

If we only use balls $B(x_i, r)$ having a fixed radius $0 < r < R$, then (2.5) reads as

$$\sum_{i=1}^{N} r^{\lambda} \geq C R^{\lambda}, \quad \text{or equivalently, } N \geq C\left(\frac{R}{r}\right)^{\lambda}, \tag{2.6}$$

which is exactly (2.2) for $E \cap \overline{B}(x, R)$.

Condition (2.6) might seem *a priori* much weaker than (2.5). However, when required to hold uniformly for every $x \in E$ and $0 < R < \operatorname{diam}(E)$, these conditions are almost equivalent for closed sets. That is, the estimate in (2.7), for covers using balls of fixed radii r, yields a corresponding estimate (2.8) for covers where balls of all radii are allowed. The price to pay is a small drop in the dimensional parameter λ.

Lemma 2.6 *Let $E \subset \mathbb{R}^n$ be a closed set. Assume that there exist $0 < \lambda_0 \leq n$ and a constant C_1 such that*

$$N\big(E \cap \overline{B}(x, R), r\big) \geq C_1\left(\frac{R}{r}\right)^{\lambda_0} \tag{2.7}$$

for every $x \in E$ and $0 < r < R < \operatorname{diam}(E)$. Then, for every $0 < \lambda < \lambda_0$, there exists a constant C such that

$$\mathcal{H}^{\lambda}_{\infty}\big(E \cap \overline{B}(x, R)\big) \geq C R^{\lambda} \tag{2.8}$$

for every $x \in E$ and $0 < R < \operatorname{diam}(E)$.

The proof of Lemma 2.6 requires a bit work. Roughly speaking, the idea is to construct a Cantor-type set $F \subset E \cap \overline{B}(x, R)$ by using (2.7) iteratively, and then deduce (2.8) with the help of the equally distributed probability measure μ on F. We omit the details, which are similar to those in [17, Theorem 3.1] and [23, Lemma 4.1].

Lemma 2.6 has several important consequences. The following theorem shows that the lower Assouad dimension of closed sets can be characterized using the Hausdorff content density condition.

Theorem 2.7 *Let $E \subset \mathbb{R}^n$ be a closed set and assume that $0 \leq \lambda < \underline{\dim}_{\mathrm{A}}(E)$. Then E satisfies the λ-Hausdorff content density condition. Moreover, $\underline{\dim}_{\mathrm{A}}(E)$ is the supremum of the exponents $\lambda \geq 0$ for which E satisfies the λ-Hausdorff content density condition.*

Proof Choose λ_0 satisfying $0 \le \lambda < \lambda_0 < \underline{\dim}_A(E)$. The definition of the lower Assouad dimension implies that (2.7) holds with a constant C_1 for every $x \in E$ and $0 < r < R < \operatorname{diam}(E)$. Thus we obtain from Lemma 2.6 that

$$\mathcal{H}^\lambda_\infty\big(E \cap \overline{B}(x, R)\big) \ge CR^\lambda$$

for every $x \in E$ and $0 < R < \operatorname{diam}(E)$; that is, E satisfies the λ-Hausdorff content density condition.

Assume then that E satisfies the λ-Hausdorff content density condition. Fix $x \in E$ and $0 < r < R < \operatorname{diam}(E)$, and let $\{B(x_i, r) : i = 1, \dots, N\}$ be a cover of $E \cap \overline{B}(x, R)$. Then

$$R^\lambda \le C\mathcal{H}^\lambda_\infty\big(E \cap \overline{B}(x, R)\big) \le C\sum_{i=1}^{N} r^\lambda = CNr^\lambda,$$

and so $N \ge C\big(\frac{R}{r}\big)^\lambda$. Since this holds for all such covers, we have

$$N\big(E \cap \overline{B}(x, R), r\big) \ge C\Big(\frac{R}{r}\Big)^\lambda.$$

Thus $\underline{\dim}_A(E) \ge \lambda$, and the proof is complete. □

Theorem 2.7 yields a comparison between the Hausdorff dimension and the lower Assouad dimension of a closed set. Such a comparison was first obtained in [21].

Corollary 2.8 *Let $E \subset \mathbb{R}^n$ be a closed set. Then*

$$\underline{\dim}_A(E) \le \dim_H\big(E \cap \overline{B}(x, r)\big) \le \dim_H(E)$$

for every $x \in E$ and $r > 0$.

Proof The second inequality follows from the monotonicity of the Hausdorff dimension. For the first inequality we may clearly assume that $\underline{\dim}_A(E) > 0$ and $0 < r < \operatorname{diam}(E)$. Fix $0 \le \lambda < \underline{\dim}_A(E)$. By Theorem 2.7, we then have $\mathcal{H}^\lambda_\infty\big(E \cap \overline{B}(x, r)\big) > 0$. Hence $\lambda \le \dim_H\big(E \cap \overline{B}(x, r)\big)$, and the claim follows. □

The assumption that E is closed is necessary in Corollary 2.8. Indeed, it is easy to see that $\underline{\dim}_A(\overline{E}) = \underline{\dim}_A(E)$ for all $E \subset \mathbb{R}^n$, and hence for instance

$$\underline{\dim}_A(\mathbb{Q}^n) = \underline{\dim}_A(\mathbb{R}^n) = n \not\le 0 = \dim_H\big(\mathbb{Q}^n \cap B(x, r)\big)$$

for every $x \in \mathbb{Q}^n$ and $r > 0$.

For comparison, we recall also the definitions of the Minkowski (or box-counting) dimensions of bounded sets. As before, we let $N(E, r)$ be the minimal number of open balls of radius r that are needed to cover the bounded set $E \subset \mathbb{R}^n$.

Then the upper Minkowski dimension of E, $\overline{\dim}_{\mathrm{M}}(E)$, can be defined as the infimum of all $\lambda \geq 0$ for which there exists a constant C such that $N(E,r) \leq Cr^{-\lambda}$ for every $0 < r < \operatorname{diam}(E)$. Correspondingly, the lower Minkowski dimension of E, $\underline{\dim}_{\mathrm{M}}(E)$, is the supremum of all $\lambda \geq 0$ for which there exists a constant C such that $N(E,r) \geq Cr^{-\lambda}$ for every $0 < r < \operatorname{diam}(E)$.

It follows easily from these definitions that

$$\underline{\dim}_{\mathrm{A}}(E) \leq \underline{\dim}_{\mathrm{M}}(E) \leq \overline{\dim}_{\mathrm{M}}(E) \leq \overline{\dim}_{\mathrm{A}}(E)$$

for all bounded sets $E \subset \mathbb{R}^n$. Moreover, if $E \subset \mathbb{R}^n$ is compact, then

$$\underline{\dim}_{\mathrm{A}}(E) \leq \dim_{\mathrm{H}}(E) \leq \underline{\dim}_{\mathrm{M}}(E) \leq \overline{\dim}_{\mathrm{M}}(E) \leq \overline{\dim}_{\mathrm{A}}(E).$$

A typical example with strict inequalities is the set $E = \left\{\frac{1}{k} : k \in \mathbb{N}\right\} \cup \{0\} \subset \mathbb{R}$, for which $\underline{\dim}_{\mathrm{A}}(E) = \dim_{\mathrm{H}}(E) = 0$, $\underline{\dim}_{\mathrm{M}}(E) = \overline{\dim}_{\mathrm{M}}(E) = \frac{1}{2}$, and $\overline{\dim}_{\mathrm{A}}(E) = 1$.

3 The Aikawa Condition

The following integrability condition for the distance function creates a natural link between the (upper) Assouad dimension and the Muckenhoupt A_p properties of distance weights, see Sect. 4. This condition was introduced and used by Aikawa in connection with the so-called quasiadditivity property of Riesz capacities in [1], see also [2, Part II, Section 7]. In [20] and [22] this condition was applied in the context of Hardy inequalities.

Definition 3.1 Let $E \subset \mathbb{R}^n$ be a non-empty set. We say that E satisfies the Aikawa condition for $\alpha \in \mathbb{R}$, if there exists a constant C (depending on α) such that

$$\int_{B(x,r)} \operatorname{dist}(y,E)^{-\alpha}\,dy \leq Cr^{n-\alpha} \tag{3.1}$$

or, equivalently,

$$\fint_{B(x,r)} \operatorname{dist}(y,E)^{-\alpha}\,dy \leq Cr^{-\alpha} \tag{3.2}$$

for every $x \in E$ and $r > 0$. Here we use the convention that $0^0 = 1$, and if $\alpha > 0$ then we also require that $|\overline{E}| = 0$.

We let $\mathcal{A}(E)$ denote the set of all $\alpha \in \mathbb{R}$ for which E satisfies the Aikawa condition.

It is easy to see that a non-empty set $E \subset \mathbb{R}^n$ satisfies the Aikawa condition for all $\alpha \le 0$. On the other hand, if $\alpha \ge n$, then

$$\int_{B(x,r)} \operatorname{dist}(y, E)^{-\alpha}\, dy \ge \int_{B(x,r)} |y - x|^{-\alpha}\, dy = \infty$$

for every $x \in E$ and $r > 0$, and thus E does not satisfy the Aikawa condition for any $\alpha \ge n$. Hence we may restrict our attention to the range $0 < \alpha < n$ in the Aikawa condition.

We now begin to examine the close connections between the upper Assouad dimension and the Aikawa condition.

Lemma 3.2 *Let $E \subset \mathbb{R}^n$. If $\alpha \in \mathcal{A}(E)$, then $\overline{\dim}_{\mathrm{A}}(E) \le n - \alpha$.*

Proof If $\alpha \le 0$, then the claim is clear since $\overline{\dim}_{\mathrm{A}}(E) \le n$. Hence we may assume that $0 < \alpha < n$. Fix $x \in E$ and $0 < r < R$, and write $F = E \cap \overline{B}(x, R)$. By the existence of maximal packings there are pairwise disjoint open balls $B(x_i, \frac{r}{2})$, $i = 1, \ldots, N$, with $x_i \in F$, such that $F \subset \bigcup_{i=1}^N B(x_i, r)$.

Let F_r be the r-neighborhood of F, that is,

$$F_r = \{y \in \mathbb{R}^n : \operatorname{dist}(y, F) < r\} \subset B(x, 2R).$$

Using the pairwise disjointness of the balls $B(x_i, \frac{r}{2}) \subset F_r$, the fact that $\operatorname{dist}(y, E) \le \operatorname{dist}(y, F) < r$ for all $y \in F_r$, and the assumed Aikawa condition (3.1), we obtain

$$\begin{aligned} NCr^n &\le \sum_{i=1}^N \left|B(x_i, \tfrac{r}{2})\right| \le |F_r| \le r^\alpha \int_{F_r} d(y, E)^{-\alpha}\, dy \\ &\le r^\alpha \int_{B(x,2R)} d(y, E)^{-\alpha}\, dy \le r^\alpha C R^{n-\alpha} = Cr^n \left(\frac{R}{r}\right)^{n-\alpha}. \end{aligned}$$

Thus

$$N\big(E \cap \overline{B}(x, R), r\big) = N(F, r) \le N \le C\left(\frac{R}{r}\right)^{n-\alpha},$$

and the claim $\overline{\dim}_{\mathrm{A}}(E) \le n - \alpha$ follows since $n - \alpha > 0$. □

For the converse direction we need to assume a strict upper bound for the dimension. See, however, also Theorem 3.5 below concerning the strict inequality in the previous Lemma 3.2.

Lemma 3.3 *Let $E \subset \mathbb{R}^n$ be a non-empty set. If $\alpha \in \mathbb{R}$ and $\overline{\dim}_{\mathrm{A}}(E) < n - \alpha$, then $\alpha \in \mathcal{A}(E)$.*

Proof Again, the claim is clear if $\alpha \le 0$, and so we may assume that $\alpha > 0$. Choose $\overline{\dim}_{\mathrm{A}}(E) < \lambda < n - \alpha$, and let $x \in E$ and $r > 0$. Define

$$F_j = \{y \in B(x,r) : d(y,E) < 2^{-j+1}r\} \quad \text{and} \quad A_j = F_j \setminus F_{j+1},$$

for $j \in \mathbb{N}$. Since $\lambda > \overline{\dim}_{\mathrm{A}}(E)$, there is a constant C_1 such that the set $E \cap \overline{B}(x,2r)$ can be covered by $N_j \le C_1 2^{j\lambda}$ balls of radius $2^{1-j}r$, for every $j \in \mathbb{N}$. It follows that each F_j can be covered by at most N_j balls of radius $2^{2-j}r$. If B_i^j, $i = 1, \dots, N_j$, are such balls, then

$$|F_j| \le \sum_{i=1}^{N_j} |B_i^j| \le N_j C (2^{2-j}r)^n \le C(2^{-j})^{n-\lambda} r^n. \tag{3.3}$$

Since $\overline{E} \cap B(x,r) \subset F_j$ for all $j \in \mathbb{N}$ and $\lambda < n - \alpha < n$, by letting $j \to \infty$ we see in particular that $|\overline{E} \cap B(x,r)| = 0$. Here $r > 0$ is arbitrary, and thus $|\overline{E}| = 0$.

If $y \in A_j$, then $2^{-j}r \le d(y,E) < 2^{-j+1}r$. In addition, $A_j \subset F_j$ for all $j \in \mathbb{N}$ and the sets A_j cover $B(x,r)$ up to the set $\overline{E} \cap B(x,r)$, which has measure zero. By using estimate (3.3) we obtain

$$\begin{aligned}\int_{B(x,r)} d(y,E)^{-\alpha}\,dy &\le C \sum_{j=1}^{\infty} \int_{A_j} d(y,E)^{-\alpha}\,dy \le C \sum_{j=1}^{\infty} |F_j| (2^{-j}r)^{-\alpha} \\ &\le C r^{n-\alpha} \sum_{j=1}^{\infty} (2^{-j})^{n-\lambda-\alpha} \le C r^{n-\alpha},\end{aligned}$$

where the geometric series converges since $\lambda < n - \alpha$. This together with the fact $|\overline{E}| = 0$ shows that $\alpha \in \mathcal{A}(E)$. □

In order to combine the two lemmas above into a characterization, we need the following improvement property for the Aikawa condition, observed in [20]. It is easy to see that the Aikawa condition, for $0 < \alpha < n$, implies a reverse Hölder inequality, see (3.4) below. After that we can apply a suitable version of the so-called Gehring lemma, see [13, Lemma 3], which is a deep result concerning the improvement of reverse Hölder inequalities. This leads to the Aikawa condition for an exponent larger than α. (Notice that conversely it is easy to see that the Aikawa condition, for $0 < \alpha < n$, implies Aikawa conditions for all exponents smaller than α.)

Theorem 3.4 *Let $E \subset \mathbb{R}^n$ and $0 < \alpha < n$. If $\alpha \in \mathcal{A}(E)$, then there exists $\alpha < \alpha' < n$ such that $\alpha' \in \mathcal{A}(E)$.*

Proof Fix a ball $B(x,r) \subset \mathbb{R}^n$ and assume first that $B(x,2r) \cap E \neq \emptyset$. Then $\operatorname{dist}(y,E) \le 3r$ for every $y \in B(x,r)$, and thus the assumed Aikawa condition (3.2) implies

$$\fint_{B(x,r)} \operatorname{dist}(y,E)^{-\alpha}\,dy \le Cr^{-\alpha} = C\big(r^{-\frac{\alpha}{2}}\big)^2 \le C\bigg(\fint_{B(x,r)} \operatorname{dist}(y,E)^{-\frac{\alpha}{2}}\,dy\bigg)^2.$$

It is easy to see that the same conclusion holds also in the case $B(x,2r) \cap E = \emptyset$. Writing $f(y) = d(y,E)^{-\frac{\alpha}{2}}$, we obtain the reverse Hölder inequality

$$\bigg(\fint_{B(x,r)} f(y)^2\,dy\bigg)^{\frac{1}{2}} \le C\fint_{B(x,r)} f(y)\,dy, \tag{3.4}$$

for every ball $B(x,r) \subset \mathbb{R}^n$.

By the Gehring lemma, there exists $p > 2$ such that

$$\bigg(\fint_{B(x,r)} f(y)^p\,dy\bigg)^{\frac{1}{p}} \le C\fint_{B(x,r)} f(y)\,dy \le C\bigg(\fint_{B(x,r)} f(y)^2\,dy\bigg)^{\frac{1}{2}},$$

for every ball $B(x,r) \subset \mathbb{R}^n$, where the second inequality is just the usual Hölder's inequality. Choose $\alpha' = \frac{p}{2}\alpha > \alpha$. Then the estimate above and the assumed Aikawa condition give

$$\bigg(\fint_{B(x,r)} \operatorname{dist}(y,E)^{-\alpha'}\,dy\bigg)^{\frac{\alpha}{2\alpha'}} \le C\bigg(\fint_{B(x,r)} \operatorname{dist}(y,E)^{-\alpha}\,dy\bigg)^{\frac{1}{2}} \le Cr^{-\frac{\alpha}{2}},$$

for every $x \in E$ and $r > 0$, and this implies the Aikawa condition for $\alpha' > \alpha$. □

We are now prepared to characterize the upper Assouad dimension in terms of the Aikawa condition. This result is essentially from [26], where corresponding characterizations were obtained also in more general metric spaces.

Theorem 3.5 *Let $E \subset \mathbb{R}^n$ be a non-empty set and let $\alpha > 0$. Then $\alpha \in \mathcal{A}(E)$ if and only if $\overline{\dim}_{\mathrm{A}}(E) < n - \alpha$.*

Proof If $\overline{\dim}_{\mathrm{A}}(E) < n - \alpha$, then $\alpha \in \mathcal{A}(E)$ by Lemma 3.3.

Assume then that $0 < \alpha \in \mathcal{A}(E)$. Since $\alpha < n$, by Theorem 3.4 there is $\alpha' > \alpha$ such that also $\alpha' \in \mathcal{A}(E)$. Thus Lemma 3.2 yields $\overline{\dim}_{\mathrm{A}}(E) \le n - \alpha' < n - \alpha$, as desired. □

Notice that the assumption $\alpha > 0$ in Theorem 3.5 is essential: if $E \subset \mathbb{R}^n$ and $\overline{\dim}_{\mathrm{A}}(E) = n$, then $0 \in \mathcal{A}(E)$, but $\overline{\dim}_{\mathrm{A}}(E) \not< n - 0$.

4 Muckenhoupt Weights

A measurable function $w\colon \mathbb{R}^n \to \mathbb{R}$ is called a weight in $\mathbb{R}^n$ if $w(x) > 0$ for almost every $x \in \mathbb{R}^n$ and $\int_B w(x)\,dx < \infty$ for all balls $B \subset \mathbb{R}^n$. When w is a weight in $\mathbb{R}^n$ and $E \subset \mathbb{R}^n$ is a measurable set, we write

$$w(E) = \int_E w(x)\,dx.$$

The following classes of Muckenhoupt weights are important tools for instance in harmonic analysis; we refer to [12, Chapter IV] for a thorough discussion. Muckenhoupt weighted $\mathbb{R}^n$ is also an example of a metric space with a doubling measure and supporting a p-Poincaré inequality, which are the standard assumptions in analysis on metric spaces; see for instance [6, 14] and the references therein for more information.

Definition 4.1 Let w be a weight in $\mathbb{R}^n$. We say that w belongs to the Muckenhoupt class

(a) A_p, for $1 < p < \infty$, if there is a constant C such that

$$\left(\fint_B w(x)\,dx\right)\left(\fint_B w(x)^{-\frac{1}{p-1}}\,dx\right)^{p-1} \le C \tag{4.1}$$

for every ball $B \subset \mathbb{R}^n$.

(b) A_1, if there is a constant C such that

$$\left(\fint_B w(x)\,dx\right) \operatorname*{ess\,sup}_{x\in B} \frac{1}{w(x)} \le C, \tag{4.2}$$

for every ball $B \subset \mathbb{R}^n$.

(c) A_∞, if there are constants $C, \delta > 0$ such that

$$\frac{w(E)}{w(B)} \le C\left(\frac{|E|}{|B|}\right)^{\delta}$$

whenever $B \subset \mathbb{R}^n$ is a ball and $E \subset B$ is a measurable set.

It is easy to verify directly from the A_p condition (4.1) that if $1 < p < \infty$ and w is a weight in $\mathbb{R}^n$, then

$$w \in A_p \quad \text{if and only if} \quad w^{-\frac{1}{p-1}} \in A_{\frac{p}{p-1}}. \tag{4.3}$$

Moreover, an application of Hölder's inequality shows that if $1 \le p < q < \infty$, then $A_p \subset A_q$.

The class A_∞ can be characterized as the union of all A_p, for $1 \le p < \infty$, that is,

$$A_\infty = \bigcup_{1 \le p < \infty} A_p. \tag{4.4}$$

Neither of the inclusions in (4.4) is trivial. The main tool for establishing both of them is a reverse Hölder inequality, but we omit the details; see e.g. [12, Chapter IV, Section 2]. We do not really need the class A_∞ below, since all statements "$w \in A_\infty$" could be replaced by the statement "$w \in A_p$ for some $1 \le p < \infty$".

Example 4.2 Consider the weight $w(y) = |y|^{-\alpha}$ for every $y \in \mathbb{R}^n \setminus \{0\}$. It is straightforward to verify by direct computations that $w \in A_1$ if and only if $0 \le \alpha < n$, and $w \in A_p$, for $1 < p < \infty$, if and only if $(1-p)n < \alpha < n$.

Our main interest in this section is in the generalizations of Example 4.2 to more general distance functions, that is, for weights of the type $w(y) = \operatorname{dist}(y, E)^{-\alpha}$, with $E \subset \mathbb{R}^n$ satisfying $|\overline{E}| = 0$. The Aikawa condition is tailor-made for the study of this problem; see [1, 2], in particular [2, p. 151].

Theorem 4.3 *Let $E \subset \mathbb{R}^n$ and $\alpha \in \mathbb{R}$, and define $w(y) = \operatorname{dist}(y, E)^{-\alpha}$ for every $y \in \mathbb{R}^n$. Then the following assertions hold.*

1. *If $0 \le \alpha \in \mathcal{A}(E)$, then $w \in A_p$ for every $1 \le p \le \infty$.*
2. *If $\alpha < 0$ and $1 < p < \infty$ are such that $\frac{-\alpha}{p-1} \in \mathcal{A}(E)$, then $w \in A_p$.*

Proof Consider first part 1. If $\alpha = 0$, then $w(y) = 1$ for every $y \in \mathbb{R}^n$, and it follows that $w \in A_p$ for every $1 \le p \le \infty$. Assume then that $0 < \alpha < n$ and that (3.1) holds with a constant C_1, that is,

$$\int_{B(x,r)} w(y)\, dy \le C_1 r^{n-\alpha} < \infty$$

for every $x \in E$ and $r > 0$. This implies that w is locally integrable. Since $\alpha \in \mathcal{A}(E)$ and $\alpha > 0$, we have $|\overline{E}| = 0$. Therefore $w(x) > 0$ for almost every $x \in \mathbb{R}^n$, and thus w is a weight.

Since $A_1 \subset A_p$ for every $p \ge 1$, it suffices to show that $w \in A_1$. Fix a ball $B(x,r) \subset \mathbb{R}^n$ and assume first that $B(x, 2r) \cap E \ne \emptyset$. Then $B(x,r) \subset B(z, 3r)$, for some $z \in E$, and so the assumed Aikawa condition (3.2) implies

$$\fint_{B(x,r)} w(y)\, dy \le C \fint_{B(z,3r)} \operatorname{dist}(y, E)^{-\alpha}\, dy \le C(3r)^{-\alpha} = Cr^{-\alpha}.$$

On the other hand, if $y \in B(x,r) \setminus \overline{E}$, then

$$\frac{1}{w(y)} = \operatorname{dist}(y, E)^{\alpha} \le d(y, z)^{\alpha} \le (3r)^{\alpha} = Cr^{\alpha},$$

since $\alpha > 0$. By combining the estimates above and recalling that $|\overline{E}| = 0$, we obtain

$$\left(\fint_{B(x,r)} w(y)\,dy\right) \operatorname*{ess\,sup}_{y\in B(x,r)} \frac{1}{w(y)} \le C.$$

This shows that the A_1 condition (4.2) holds for the ball $B(x,r)$ if $B(x,2r)\cap E \neq \emptyset$.

Assume then that $B(x,2r)\cap E = \emptyset$. In this case

$$\tfrac{1}{2}\operatorname{dist}(y,E) \le \operatorname{dist}(x,E) \le 2\operatorname{dist}(y,E)$$

for every $y \in B(x,r)$, and thus

$$\left(\fint_{B(x,r)} w(y)\,dy\right) \operatorname*{ess\,sup}_{y\in B(x,r)} \frac{1}{w(y)} \le C\operatorname{dist}(x,E)^{-\alpha}\operatorname{dist}(x,E)^{\alpha} \le C.$$

Hence (4.2) holds also in the case $B(x,2r)\cap E = \emptyset$, and the proof of part 1 is complete.

In part 2, we let

$$\sigma(y) = w(y)^{-\frac{1}{p-1}} = \operatorname{dist}(y,E)^{\frac{\alpha}{p-1}}$$

for every $y \in \mathbb{R}^n$. By part 1 we have $\sigma \in A_1 \subset A_{\frac{p}{p-1}}$, and the claim $w \in A_p$ follows from the duality property (4.3) of A_p weights. □

There is also a partial converse of Theorem 4.3, see Theorem 4.5 below. We recall that a set $E \subset \mathbb{R}^n$ is porous, if there exists a constant C such that for every $x \in \mathbb{R}^n$ and $r > 0$ there exists $z \in \mathbb{R}^n$ such that $B(z,Cr) \subset B(x,r)\setminus E$. Porosity can also be characterized using the upper Assouad dimension:

Theorem 4.4 *A set $E \subset \mathbb{R}^n$ is porous if and only if $\overline{\dim}_{\mathrm{A}}(E) < n$.*

For the proof of Theorem 4.4, see for instance [29, Theorem 5.2]. Note that by Theorem 3.5 the conditions in Theorem 4.4 hold if and only if there is $\alpha > 0$ such that $\alpha \in \mathcal{A}(E)$.

Theorem 4.5 *Assume that $E \subset \mathbb{R}^n$ is a non-empty porous set. Let $\alpha \in \mathbb{R}$ and define $w(y) = \operatorname{dist}(y,E)^{-\alpha}$ for every $y \in \mathbb{R}^n$. Then the following assertions hold.*

1. *If $\alpha \ge 0$, $1 \le p < \infty$, and $w \in A_p$, then $\alpha \in \mathcal{A}(E)$.*
2. *If $\alpha < 0$, $1 < p < \infty$, and $w \in A_p$, then $\frac{-\alpha}{p-1} \in \mathcal{A}(E)$.*

Proof In part 1 we may assume that $p > 1$. Let $B_0 = B(x,r)$ be a ball. Since E is porous, there is $z \in B_0$ such that $B(z,Cr) \subset B(x,r)\setminus E$. Then $\operatorname{dist}(y,E) \ge \frac{C}{2}r$

for every $y \in B = B(z, \frac{C}{2}r)$, and since the measures of B_0 and B are comparable, we obtain

$$\begin{aligned}\Big(\fint_{B_0} w(y)^{-\frac{1}{p-1}}\,dy\Big)^{p-1} &\geq C\Big(\fint_{B} w(y)^{-\frac{1}{p-1}}\,dy\Big)^{p-1}\\ &\geq C\Big(\fint_{B} r^{\frac{\alpha}{p-1}}\,dy\Big)^{p-1} \geq Cr^{\alpha}.\end{aligned}$$

On the other hand, the A_p condition (4.1) for $w \in A_p$ gives

$$\Big(\fint_{B_0} w(y)\,dy\Big)\Big(\fint_{B_0} w(y)^{-\frac{1}{p-1}}\,dy\Big)^{p-1} \leq C.$$

By combining the two estimates above we obtain

$$\fint_{B_0} \operatorname{dist}(y, E)^{-\alpha}\,dy = \fint_{B_0} w(y)\,dy \leq C\Big(\fint_{B_0} w(y)^{-\frac{1}{p-1}}\,dy\Big)^{1-p} \leq Cr^{-\alpha},$$

and thus $\alpha \in \mathcal{A}(E)$.

Then we consider part 2. If $w \in A_p$, for $1 < p < \infty$, we have by the A_p duality in (4.3) that

$$\operatorname{dist}(\cdot, E)^{-(\frac{-\alpha}{p-1})} = \operatorname{dist}(\cdot, E)^{\frac{\alpha}{p-1}} = w^{-\frac{1}{p-1}} \in A_{\frac{p}{p-1}}.$$

Hence the claim follows from part 1. □

For porous sets we now have a complete characterization of the A_p properties of the distance functions.

Theorem 4.6 *Let $1 < p < \infty$ and assume that $E \subset \mathbb{R}^n$ is a non-empty porous set. Let $\alpha \in \mathbb{R}$ and define $w(y) = \operatorname{dist}(y, E)^{-\alpha}$ for every $y \in \mathbb{R}^n$. Then the following assertions hold.*

1. *$w \in A_1$ if and only if $0 \leq \alpha < n - \overline{\dim}_{\mathrm{A}}(E)$.*
2. *$w \in A_p$ if and only if*

$$(1-p)\big(n - \overline{\dim}_{\mathrm{A}}(E)\big) < \alpha < n - \overline{\dim}_{\mathrm{A}}(E). \tag{4.5}$$

Proof Since E is porous, $\overline{\dim}_{\mathrm{A}}(E) < n$ by Theorem 4.4.

We consider first part 2. If $0 \leq \alpha < n - \overline{\dim}_{\mathrm{A}}(E)$, we have $\alpha \in \mathcal{A}(E)$ by Lemma 3.3 and thus part 1 of Theorem 4.3 implies $w \in A_p$. On the other hand, if

$$(1-p)\big(n - \overline{\dim}_{\mathrm{A}}(E)\big) < \alpha < 0,$$

then

$$0 < \frac{-\alpha}{p-1} < n - \overline{\dim}_{\mathrm{A}}(E).$$

From Lemma 3.3 we obtain $\frac{-\alpha}{p-1} \in \mathcal{A}(E)$ and hence $w \in A_p$ by part 2 of Theorem 4.3.

Conversely, assume that $w \in A_p$. If $\alpha > 0$, part 1 of Theorem 4.5 implies $\alpha \in \mathcal{A}(E)$, and so $\alpha < n - \overline{\dim}_{\mathrm{A}}(E)$ by Theorem 3.5. If $\alpha = 0$, then (4.5) holds since $\overline{\dim}_{\mathrm{A}}(E) < n$ by porosity. Finally, if $\alpha < 0$, then $\frac{-\alpha}{p-1} \in \mathcal{A}(E)$ by part 2 of Theorem 4.5. Theorem 3.5 gives

$$0 < \frac{-\alpha}{p-1} < n - \overline{\dim}_{\mathrm{A}}(E),$$

showing that (4.5) holds also in this case. The proof of part 2 is complete.

Consider then part 1. If $0 \le \alpha < n - \overline{\dim}_{\mathrm{A}}(E)$, the claim $w \in A_1$ follows from Lemma 3.3 and part 1 of Theorem 4.3 just as in part 2. Conversely, if $w \in A_1$ and $\alpha > 0$, then $\alpha < n - \overline{\dim}_{\mathrm{A}}(E)$ by part 1 of Theorem 4.5 and Theorem 3.5. If $\alpha = 0$, then $0 \le \alpha < n - \overline{\dim}_{\mathrm{A}}(E)$ holds since $\overline{\dim}_{\mathrm{A}}(E) < n$ by porosity. Finally, it is easy to see that $\alpha \ge 0$ is necessary in part 1, and this completes the proof. □

Remark 4.7 The fact that the A_p properties of the weights $w(y) = \operatorname{dist}(y, E)^{-\alpha}$ depend on the dimension(s) of $E \subset \mathbb{R}^n$ has certainly been part of the mathematical folklore, at least for suitably nice sets E. Aikawa [1, 2] mentions explicitly the connections between the Aikawa condition and A_p weights. Horiuchi [15, 16] used a different dimensional condition, called property $P(s)$, in the study of A_p properties of distance weights and in particular distance weighted Sobolev-type embeddings. It was shown in [27, Theorem 3.4] that also this property $P(s)$ can be characterized using the upper Assouad dimension. A sufficient condition in the spirit of Theorem 4.3 was given in [7, Lemma 3.3] for subsets of λ-regular sets of $\mathbb{R}^n$.

Theorem 4.6 was formulated in [8], where corresponding results were also obtained in metric spaces in terms of the so-called lower Assouad codimension. Metric space results of this type were considered in [3], as well, but using a completely different approach and under the stronger assumption that both the space X and the set $E \subset X$ satisfy Ahlfors–David regularity conditions; see [3, Theorems 6 and 7].

5 Hardy–Sobolev Inequalities

Hardy–Sobolev inequalities interpolate between the Sobolev inequality and the p-Hardy inequality. Indeed, for $q = p^* = \frac{np}{n-p}$ inequality (5.1) is the Sobolev inequality, while for $q = p$ we recover the p-Hardy inequality.

Definition 5.1 Let $1 < p \leq q \leq \frac{np}{n-p} < \infty$ and let $\Omega \subsetneq \mathbb{R}^n$ be an open set. We say that the (q, p)-Hardy–Sobolev inequality holds in Ω if there is a constant C such that

$$\left(\int_\Omega |u(x)|^q \operatorname{dist}(x, \Omega^c)^{\frac{q}{p}(n-p)-n}\, dx\right)^{\frac{1}{q}} \leq C\left(\int_\Omega |\nabla u(x)|^p\, dx\right)^{\frac{1}{p}} \tag{5.1}$$

for every $u \in C_0^\infty(\Omega)$.

We also consider weighted versions of these inequalities and say that the (q, p, β)-Hardy–Sobolev inequality holds in Ω, for $\beta \in \mathbb{R}$, if there is a constant C such that

$$\left(\int_\Omega |u(x)|^q \operatorname{dist}(x, \Omega^c)^{\frac{q}{p}(n-p+\beta)-n}\, dx\right)^{\frac{1}{q}} \leq C\left(\int_\Omega |\nabla u(x)|^p \operatorname{dist}(x, \Omega^c)^\beta\, dx\right)^{\frac{1}{p}} \tag{5.2}$$

for every $u \in C_0^\infty(\Omega)$.

For $q = p$, the inequality in (5.2) is called the (p, β)-Hardy inequality.

In this final section we formulate (without proofs) sufficient and necessary conditions for Hardy–Sobolev inequalities in $\Omega \subset \mathbb{R}^n$, given in terms of the upper and lower Assouad dimensions (and also other dimensions) of the complement $\Omega^c = \mathbb{R}^n \setminus \Omega$. It has been understood already for a long time that sufficient conditions for these inequalities naturally split into two cases: either the complement Ω^c has to be sufficiently "thick" or sufficiently "thin". The thickness has been formulated, for instance, using capacity density or Hausdorff content density, and thinness using the Aikawa condition. With Assouad dimensions this duality becomes more transparent: thickness means that Ω^c has large lower Assouad dimension, while thinness means that Ω^c has small upper Assouad dimension.

It can also be shown that suitable combinations of such thick and thin parts give sufficient conditions for Hardy–Sobolev inequalities, as well, but these cases will not be discussed in this work; see e.g. [25, Section 7] for details.

In the case of thin complements, the Hardy–Sobolev inequalities can be obtained by using the following general two weight embedding result together with the Aikawa condition and the knowledge about the A_p properties of the distance functions.

Theorem 5.2 *Let $1 < p \leq q < \infty$ and let (w, v) be a pair of weights such that $w \in A_\infty$ and $\sigma = v^{-\frac{1}{p-1}} \in A_\infty$. Assume that there exists a constant C_1 such that*

$$\left(\frac{1}{|B|^{1-\frac{1}{n}}}\right)^p w(B)^{\frac{p}{q}} \sigma(B)^{p-1} \leq C_1 \tag{5.3}$$

for all balls $B \subset \mathbb{R}^n$. Then there exists a constant C such that

$$\left(\int_{\mathbb{R}^n} |u(x)|^q w(x)\,dx\right)^{\frac{1}{q}} \le C\left(\int_{\mathbb{R}^n} |\nabla u(x)|^p v(x)\,dx\right)^{\frac{1}{p}}$$

for every $u \in C_0^\infty(\mathbb{R}^n)$.

Theorem 5.2 can be proved using the mapping properties of Riesz potentials and maximal operators. Muckenhoupt and Wheeden [31, Theorem 1] gave a single weight control for the Riesz potential I_1 in terms of a fractional maximal operator, and Pérez [32, Theorem 1.1] proved a two weight L^p–L^q control for such maximal operators under the assumption in (5.3). The claim of Theorem 5.2 then follows from the pointwise estimate $|u(x)| \le C I_1|\nabla u|(x)$ for the Riesz potential and the boundedness properties of the maximal operator. See also [33] and [8] for discussion and generalizations of these results to metric spaces.

From Theorem 5.2 we obtain the following weighted global Hardy–Sobolev inequalities where the integrals can be taken over the whole $\mathbb{R}^n$. This is possible since $\overline{\dim}_A(E) < n$ by the assumptions, and consequently $|E| = 0$.

Theorem 5.3 *Let $E \subset \mathbb{R}^n$ be a non-empty closed set and assume that*

$$1 < p \le q \le \frac{np}{n-p} < \infty$$

and

$$\overline{\dim}_A(E) < \min\left\{\frac{q}{p}(n-p+\beta),\, n - \frac{\beta}{p-1}\right\}.$$

Then the inequality

$$\left(\int_{\mathbb{R}^n} |u(x)|^q \operatorname{dist}(x,E)^{\frac{q}{p}(n-p+\beta)-n}\,dx\right)^{\frac{1}{q}} \le C\left(\int_{\mathbb{R}^n} |\nabla u(x)|^p \operatorname{dist}(x,E)^{\beta}\,dx\right)^{\frac{1}{p}} \tag{5.4}$$

holds for every $u \in C_0^\infty(\mathbb{R}^n)$.

In particular, if $E = \Omega^c$ satisfies the assumptions in Theorem 5.3, then the (q, p, β)-Hardy inequality holds in Ω. The dimensional condition in Theorem 5.3, together with Theorem 4.3, implies that the weights in (5.4) satisfy the A_∞ conditions in Theorem 5.2, and then (5.3) for these weights can be checked with the help of the Aikawa condition; see [8, Section 4] for the computations (in the metric setting).

Actually, by the results of Horiuchi [15] (see also [16] and [27, Section 5]) the bound $\overline{\dim}_A(E) < n - \frac{\beta}{p-1}$ can be removed if $\overline{\dim}_A(E) < n-1$, while by [27, Example 4.4] this bound is really needed when $\overline{\dim}_A(E) \ge n-1$ and also sharp at least when $\overline{\dim}_A(E) = n-1$. The proofs in [15] for the case $\overline{\dim}_A(E) < n-1$

however require a completely different approach based on relative isoperimetric inequalities.

On the other hand, it is not hard to show that for $\beta \geq 0$ the bound $\overline{\dim}_{\mathrm{A}}(E) < \frac{q}{p}(n-p+\beta)$ is also necessary for the global Hardy–Sobolev inequality to hold with respect to E (see [27, Theorem 6.1]). Thus we have the following characterization in the case $\beta = 0$.

Theorem 5.4 *Let $1 < p \leq q < \frac{np}{n-p} < \infty$ and assume that $E \subset \mathbb{R}^n$ is a non-empty closed set. Then there exists a constant C such that*

$$\left(\int_{\mathbb{R}^n} |u(x)|^q \operatorname{dist}(x, E)^{\frac{q}{p}(n-p)-n}\, dx\right)^{\frac{1}{q}} \leq C\left(\int_{\mathbb{R}^n} |\nabla u(x)|^p\, dx\right)^{\frac{1}{p}}, \tag{5.5}$$

for every $u \in C_0^\infty(\mathbb{R}^n)$, if and only if

$$\overline{\dim}_{\mathrm{A}}(E) < \frac{q}{p}(n-p).$$

Under some additional conditions the bound $\overline{\dim}_{\mathrm{A}}(E) < \frac{q}{p}(n-p+\beta)$ is necessary also for $\beta < 0$, see [27, Theorem 6.2] and compare also to Theorem 5.7 below.

We now turn to the case of thick complements. A well-known sufficient condition for the unweighted p-Hardy inequality in $\Omega \subset \mathbb{R}^n$ is the uniform p-fatness of the complement Ω^c, see e.g. [28, 34]. Uniform fatness is a density condition for the variational p-capacity, but in fact Ω^c is uniformly p-fat if and only if Ω^c is unbounded and satisfies the λ-Hausdorff density condition in Definition 2.5 for some $\lambda > n-p$; see [19, Section 2.4] for a discussion.

The Hausdorff content density condition is a natural assumption also for weighted Hardy inequalities, but for $\beta \geq p-1$ an additional accessibility condition for the boundary $\partial\Omega$ is needed. This leads to the following theorem. We omit the details and refer to [19] and [24] for the definitions and proofs; see also [5] for recent progress concerning such accessibility conditions.

Theorem 5.5 *Let $1 < p < \infty$, $\lambda \geq 0$, and $\beta \in \mathbb{R}$ be such that $\lambda > n-p+\beta$. Assume that $\Omega \subset \mathbb{R}^n$ is an open set such that Ω^c is unbounded and satisfies the λ-Hausdorff content density condition. Moreover, if $\beta \geq p-1$, we assume an additional accessibility condition for $\partial\Omega$. Then the (p,β)-Hardy inequality holds in Ω.*

Combining this with Theorem 2.7 and an interpolation result in [27, Theorem 2.1], we obtain the corresponding Hardy–Sobolev inequalities under an assumption for the lower Assouad dimension of the complement.

Theorem 5.6 *Let $1 < p \leq q \leq \frac{np}{n-p} < \infty$ and $\beta \in \mathbb{R}$ and assume that $\Omega \subset \mathbb{R}^n$ is an open set such that Ω^c is unbounded and $\underline{\dim}_A(\Omega^c) > n - p + \beta$. Moreover, if $\beta \geq p - 1$, we assume an additional accessibility condition for $\partial\Omega$. Then the (q, p, β)-Hardy–Sobolev inequality holds in Ω.*

Proof Let $\lambda \geq 0$ be such that $\underline{\dim}_A(\Omega^c) > \lambda > n - p + \beta$. By Theorem 2.7 the complement Ω^c satisfies the λ-Hausdorff content density condition (2.4) and thus the (p, β)-Hardy inequality holds in Ω by Theorem 5.5. The (q, p, β)-Hardy–Sobolev inequality then follows from [27, Theorem 2.1]. □

We have seen in Theorems 5.3 and 5.6 that the "dual" conditions

$$\overline{\dim}_A(\Omega^c) < \frac{q}{p}(n - p + \beta) \quad \text{and} \quad \underline{\dim}_A(\Omega^c) > n - p + \beta,$$

possibly together with some additional requirements, are sufficient for the (q, p, β)-Hardy–Sobolev inequality in $\Omega \subset \mathbb{R}^n$. As was already mentioned, also suitable combinations of these conditions suffice for Hardy–Sobolev inequalities, and this rules out the possibility that the conditions above could characterize the validity of Hardy–Sobolev inequalities. Nevertheless, these conditions are not that far from being also necessary, and at least the dimensional bounds $\frac{q}{p}(n-p+\beta)$ and $n-p+\beta$ are optimal. This can be seen from the following result, which is [27, Theorem 4.6]. Interestingly, also the Hausdorff dimension and the (lower) Minkowski dimension are needed here, and they can not be changed to $\underline{\dim}_A(\Omega^c)$ in the respective bounds. However, in the case $q = p$ the inequalities in these dimensional lower bounds can be made strict, see [22]. From this it follows that if

$$\dim_H(\Omega^c) \leq n - p + \beta \leq \overline{\dim}_A(\Omega^c),$$

then the (p, β)-Hardy inequality can not hold in Ω.

Theorem 5.7 *Let $1 < p \leq q < \frac{np}{n-p} < \infty$ and $\beta \in \mathbb{R}$, and assume that the (q, p, β)-Hardy–Sobolev inequality (5.2) holds in an open set $\Omega \subset \mathbb{R}^n$.*

1. *If $\beta \geq 0$ and $\frac{q}{p}(n - p + \beta) \neq n$, then either*

$$\overline{\dim}_A(\Omega^c) < \frac{q}{p}(n - p + \beta) \quad \text{or} \quad \dim_H(\Omega^c) \geq n - p + \beta.$$

2. *If $\beta < 0$ and Ω^c is compact and porous, then either*

$$\overline{\dim}_A(\Omega^c) < \frac{q}{p}(n - p + \beta) \quad \text{or} \quad \underline{\dim}_M(\Omega^c) \geq n - p + \beta.$$

Examples in [27] show that for $\beta < 0$ the compactness assumption can not be completely omitted. However, compactness can be relaxed to the condition that $x \mapsto \operatorname{dist}(x, \Omega^c)^\beta$ is locally integrable, which in turn holds, for instance, if we assume that $\overline{\dim}_{\mathrm{M}}(\mathbf{\Omega}^c \cap B) < n+\beta$ for all balls B centered at $\mathbf{\Omega}^c$. It is not known if the porosity assumption is necessary or if the lower Minkowski dimension (instead of the Hausdorff dimension) is really needed in the case $\beta < 0$.

In conclusion, the moral of this final section is not so much in the actual formulations of all these conditions for Hardy–Sobolev inequalities, but rather in the fact that all five notions of dimensions mentioned in this article (Hausdorff, upper and lower Assouad, and upper and lower Minkowski) have made an appearance. Moreover, in the light of examples at least three of these (Hausdorff, upper and lower Assouad) are certainly needed in order to state the optimal conditions for the validity of Hardy–Sobolev inequalities in a somewhat uniform and condensed way.

Acknowledgments The author is grateful to Steffen Winter for making the writing of this article possible. He also wants to thank Juha Kinnunen and Antti Vähäkangas for numerous discussions related to the topics of this survey, which have helped to shape up this material.

References

1. Aikawa, H.: Quasiadditivity of Riesz capacity. Math. Scand. **69**, 15–30 (1991)
2. Aikawa, H., Essén, M.: Potential theory—selected topics. Lecture Notes in Mathematics, vol. 1633. Springer, Berlin (1996)
3. Aimar, H., Carena, M., Durán, R., Toschi, M.: Powers of distances to lower dimensional sets as Muckenhoupt weights. Acta Math. Hungar. **143**, 119–137 (2014)
4. Assouad, P.: Plongements lipschitziens dans $\mathbf{R}^n$. Bull. Soc. Math. France **111**, 429–448 (1983)
5. Azzam, J.: Accessible parts of the boundary for domains with lower content regular complements. Ann. Acad. Sci. Fenn. Math. **44**, 889–901 (2019)
6. Björn, A., Björn, J.: Nonlinear potential theory on metric spaces. EMS Tracts in Mathematics, vol. 17. European Mathematical Society (EMS), Zürich (2011)
7. Durán, R.G., López García, F.: Solutions of the divergence and analysis of the Stokes equations in planar Hölder-α domains. Math. Models Methods Appl. Sci. **20**, 95–120 (2010)
8. Dyda, B., Ihnatsyeva, L., Lehrbäck, J., Tuominen, H., Vähäkangas, A.V.: Muckenhoupt A_p-properties of distance functions and applications to Hardy–Sobolev-type inequalities. Potential Anal. **50**, 83–105 (2019)
9. Falconer, K.: Fractal Geometry. Mathematical Foundations and Applications. Wiley, Chichester (1990)
10. Fraser, J.M.: Assouad type dimensions and homogeneity of fractals. Trans. Am. Math. Soc. **366**, 6687–6733 (2014)
11. Fraser, J.M.: Interpolating between dimensions. In: Freiberg, U., Hambly, B., Hinz, M., Winter, S. (eds.) Fractal Geometry and Stochastics VI. Progress in Probability, vol. 76, pp. 3–24. Springer, Cham (2020). https://doi.org/10.1007/978-3-030-59649-1_1
12. García-Cuerva, J., Rubio de Francia, J.L.: Weighted Norm Inequalities and Related Topics. North-Holland, Amsterdam (1985)
13. Gehring, F.W.: The L^p-integrability of the partial derivatives of a quasiconformal mapping. Acta Math. **130**, 265–277 (1973)
14. Heinonen, J., Koskela, P., Shanmugalingam, N., Tyson, J.T.: Sobolev spaces on metric measure spaces. An approach based on upper gradients. New Mathematical Monographs, vol. 27. Cambridge University Press, Cambridge (2015)
15. Horiuchi, T.: The imbedding theorems for weighted Sobolev spaces. J. Math. Kyoto Univ. **29**, 365–403 (1989)

16. Horiuchi, T.: The imbedding theorems for weighted Sobolev spaces. II. Bull. Fac. Sci. Ibaraki Univ. Ser. A **23**, 11–37 (1991)
17. Järvi, P., Vuorinen, M.: Uniformly perfect sets and quasiregular mappings. J. Lond. Math. Soc. (2) **54**, 515–529 (1996)
18. Käenmäki, A., Lehrbäck, J., Vuorinen, M.: Dimensions, Whitney covers, and tubular neighborhoods. Indiana Univ. Math. J. **62**, 1861–1889 (2013)
19. Koskela, P., Lehrbäck, J.: Weighted pointwise Hardy inequalities. J. Lond. Math. Soc. (2) **79**, 757–779 (2009)
20. Koskela, P., Zhong, X.: Hardy's inequality and the boundary size. Proc. Am. Math. Soc. **131**, 1151–1158 (2003)
21. Larman, D.G.: A new theory of dimension. Proc. Lond. Math. Soc. (3) **17**, 178–192 (1967)
22. Lehrbäck, J.: Weighted Hardy inequalities and the size of the boundary. Manuscripta Math. **127**, 249–273 (2008)
23. Lehrbäck, J.: Necessary conditions for weighted pointwise Hardy inequalities. Ann. Acad. Sci. Fenn. Math. **34**, 437–446 (2009)
24. Lehrbäck, J.: Weighted Hardy inequalities beyond Lipschitz domains. Proc. Am. Math. Soc. **142**, 1705–1715 (2014)
25. Lehrbäck, J.: Hardy inequalities and Assouad dimensions. J. Anal. Math. **131**, 367–398 (2017)
26. Lehrbäck, J., Tuominen, H.: A note on the dimensions of Assouad and Aikawa. J. Math. Soc. Jpn. **65**, 343–356 (2013)
27. Lehrbäck, J., Vähäkangas, A.V.: In Between the inequalities of Sobolev and Hardy. J. Funct. Anal. **271**, 330–364 (2016)
28. Lewis, J.L.: Uniformly fat sets. Trans. Am. Math. Soc. **308**, 177–196 (1988)
29. Luukkainen, J.: Assouad dimension: antifractal metrization, porous sets, and homogeneous measures. J. Korean Math. Soc. **35**, 23–76 (1998)
30. Mattila, P.: Geometry of sets and measures in Euclidean spaces. Fractals and rectifiability. Cambridge Studies in Advanced Mathematics, vol. 44. Cambridge University Press, Cambridge (1995)
31. Muckenhoupt, B., Wheeden, R.: Weighted norm inequalities for fractional integrals. Trans. Am. Math. Soc. **192**, 261–274 (1974)
32. Pérez, C.: Two weighted norm inequalities for Riesz potentials and uniform L^p-weighted Sobolev inequalities. Indiana Univ. Math. J. **39**, 31–44 (1990)
33. Pérez, C., Wheeden, R.: Potential operators, maximal functions, and generalizations of A_∞. Potential Anal. **19**, 1–33 (2003)
34. Wannebo, A.: Hardy inequalities. Proc. Am. Math. Soc. **109**, 85–95 (1990)

A Survey on Prescription of Multifractal Behavior

Stéphane Seuret

Dedicated to Alain Arnéodo, pioneer in the development of wavelet tools for data analysis.

Abstract Multifractal behavior has been identified and mathematically established for large classes of functions, stochastic processes and measures. Multifractality has also been observed on many data coming from Geophysics, turbulence, Physics, Biology, to name a few. Developing mathematical models whose scaling and multifractal properties fit those measured on data is thus an important issue. This raises several still unsolved theoretical questions about the prescription of multifractality (i.e. how to build mathematical models with a singularity spectrum known in advance), typical behavior in function spaces, and existence of solutions to PDEs or SPDEs with possible multifractal behavior. In this survey, we gather some of the latest results in this area.

Keywords Hausdorff measure and dimension · Fractals and multifractals · Hölder · Sobolev and Besov spaces · Wavelets · Baire category and spaces

Mathematics Subject Classifications (2010) 11K55, 26A21, 28AXX, 42C40, 46E35

1 Multifractality Between Pure and Applied Mathematics

The Supp notion of multifractal functions and measures can be traced back to the interest of physicists in the Hölder singularities structure in fully developed turbulence, which is described in terms of large deviations for the distribution at small scales of Mandelbrot random multiplicative cascades in [35], and in

S. Seuret (✉)
Université Paris-Est, LAMA (UMR 8050), UPEMLV, UPEC, CNRS, Créteil, France
e-mail: seuret@u-pec.fr

U. Freiberg et al. (eds.), *Fractal Geometry and Stochastics VI*,
Progress in Probability 76, https://doi.org/10.1007/978-3-030-59649-1_3

a geometric setting in the version of the so-called multifractal formalism for functions proposed by Frisch and Parisi [23], see Sect. 4. Another source leading to multifractal ideas is provided by the works of Henschel and Procaccia [26] and Halsey et al. [25]. Since then, multifractal analysis was further developed in dynamical systems theory and geometric measure theory, and has become a standard tool to describe the fine geometric structure of objects possessing nice invariance properties, such as self-similar and self-affine measures and functions, many classes of stochastic processes such as Lévy processes and more general Markov processes, as well as random measures emerging from multiplicative chaos theory.

Let us recall the notion of singularity spectrum of a function, leading to multifractals.

Let $d \geq 1$ be an integer. Given a real function $f \in L^{\infty}_{loc}(\mathbb{R}^d)$ and $x_0 \in \mathbb{R}^d$, f is said to belong to $C^H(x_0)$, for some $H \geq 0$, if there exists a polynomial P of degree at most $\lfloor H \rfloor$ and a constant $C > 0$ such that

$$\text{for } x \text{ close to } x_0, \quad |f(x) - P(x - x_0)| \leq C|x - x_0|^H.$$

Definition 1.1 The *pointwise Hölder exponent* of $f \in L^{\infty}_{loc}(\mathbb{R}^d)$ at x_0 is

$$h_f(x_0) = \sup\left\{H \geq 0 : \ f \in C^H(x_0)\right\},$$

and f is said to have a Hölder singularity of order $h_f(x_0)$ at x_0.

The *singularity spectrum* D_f of f is the map:

$$D_f : H \in [0, \infty] \longmapsto \dim E_f(H), \quad \text{where } E_f(H) := \{x_0 \in \mathbb{R}^d : h_F(x_0) = H\}.$$

The notation dim stands for the Hausdorff dimension, and by convention $\dim \emptyset = -\infty$.

The multifractal spectrum D_f encapsulates key information on a given function f, in particular it carries a description of the distribution of the singularities of f. But the computation of D_f often raises deep mathematical questions (for instance, it took almost 130 years to find the multifractal spectrum of the famous Riemann series $\displaystyle\sum_{n=1}^{+\infty} \frac{\sin(n^2 \pi x)}{n^2}$), and in most cases the exact value of D_f happens to be not directly accessible, neither theoretically nor numerically.

Fortunately, the notion of multifractal formalism furnishes a clever way to circumvent this difficulty and to compute the explicit value of the spectrum of large classes of measures and functions. Also, multifractal formalism provides ideas to develop numerical algorithms able to estimate D_f on real-life data. The main idea is that for very large classes of functions f (and also for other mathematical objects like measures, stochastic processes—such examples will be given in this paper), D_f is equal to the Legendre transform of the so-called L^q-spectrum τ_f of f: this L^q-spectrum is computed directly using the values of f, and is numerically accessible.

When these two quantities (D_f and the Legendre transform of τ_f) coincide, it is said that f satisfies the multifractal formalism. Examples of L^q-spectra for functions (and measures) based on increments, wavelet coefficients or wavelet leaders, are given in the upcoming sections (see (4.1), (4.2), (4.8) or (4.9)). The intuition that a multifractal formalism should hold is due to U. Frisch and G. Parisi, we refer the reader to Sect. 4 for an account on the ideas leading to this formula.

The multifractal formalism, and its validity for many mathematical models, explains the success of the multifractal approach used as classification tool in signal and image processing. Indeed, algorithms have been developed (mainly based on wavelet theory, see [38] for the original WTMM method and more recently [2] for a mathematical study of the wavelet leaders algorithm and the latest developments and algorithms based on wavelet leaders) to estimate numerically L^q-scaling functions, the stability and efficiency of these algorithms being mathematically grounded. Using these algorithms, it is now established that many data coming from Geophysics, turbulence, Physics, Biology, exhibit non-linear L^q-scaling functions, which for a given function f is interpreted thanks to the Frisch–Parisi heuristics as a non-trivial singularity spectrum D_f of f. Examples of data and estimated singularity spectra are plotted in Figs. 1 and 2.

Resuming the above, we have on one side many mathematical objects f with non-linear L^q-scaling functions and a non-trivial singularity spectrum D_f, and on the other side an impressive quantity of signals, images and multivariate, multi-dimensional data whose estimated L^q-spectra and singularity spectra are non-trivial. It is worth asking which mathematical objects are indeed the most relevant to model the observed data, and how to create models with any reasonable multifractal behavior.

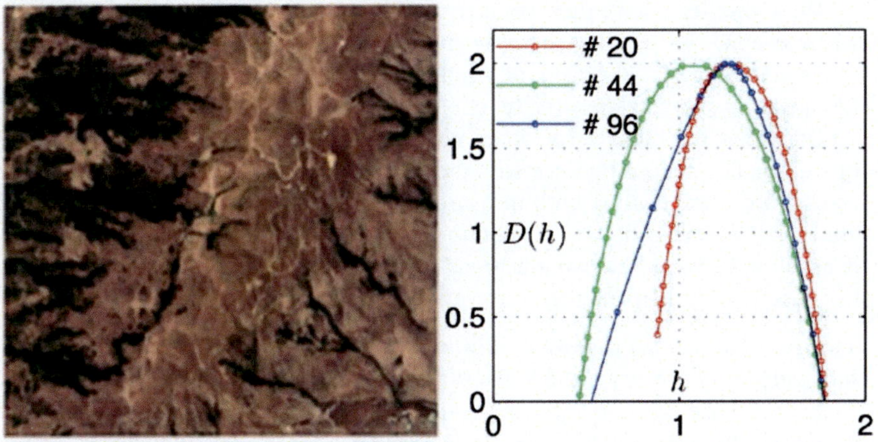

Fig. 1 Image and estimated multifractal spectrum of different color levels of a satellite image. *Courtesy H. Wendt*

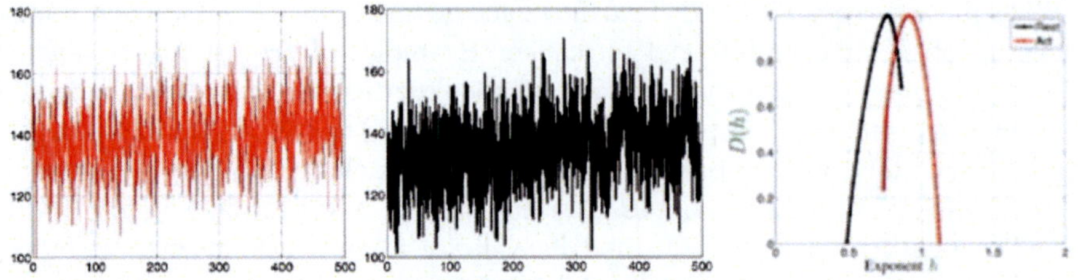

Fig. 2 Two FMRI signals of a resting (in black) and acting (in red) patient. Comparison between their estimated multifractal spectrum. *Courtesy H. Wendt*

This general problematics can be understood in various ways, and raises several theoretical questions, most of them still being open:

1. What are the mappings $\sigma : \mathbb{R}^+ \to [0, d] \cup \{-\infty\}$ that are admissible to be a multifractal spectrum, i.e. there exists a function $f : \mathbb{R}^d \to \mathbb{R}$ such that $D_f = \sigma$?
2. What are the mappings $\sigma : \mathbb{R}^+ \to [0, d] \cup \{-\infty\}$ that are admissible to be a *homogeneous* multifractal spectrum, i.e. there exists a function $f : \mathbb{R}^d \to \mathbb{R}$ such for every cube $I \subset \mathbb{R}^d$ with non-empty interior, $D_{f_I} = \sigma$ where f_I stands for the restriction of f on I?
3. Given an admissible (homogeneous or not) singularity spectrum $\sigma : \mathbb{R}^+ \to [0, d] \cup \{-\infty\}$, is there a functional space in which Baire typical functions have σ as singularity spectrum? Do typical functions satisfy a multifractal formalism?
4. Given an admissible (homogeneous or not) singularity spectrum $\sigma : \mathbb{R}^+ \to [0, d] \cup \{-\infty\}$, is there a differential equation, a PDE or a stochastic (P)DE whose solution has σ as singularity spectrum?

These problems have their counterpart in terms of L^q-spectra: replacing everywhere $\sigma : \mathbb{R}^+ \to [0, d] \cup \{-\infty\}$ by $\tau : \mathbb{R} \to \mathbb{R}$, one may ask for the admissible τ that can be the L^q-spectrum of a function (homogeneous or not), and if such an L^q-spectrum is typical in some functional space.

The same questions arise when considering probability measures instead of functions. The main difference with the function setting is that there are additional constraints when dealing with measures, see Sects. 2 and 4.1. Although the tools used in the two contexts (functions and measures) are of different nature, a connection between the two situations is provided by the following theorem from [8], based on wavelet analysis.

Theorem 1.2 *Let μ be a probability measure on $\mathbb{R}^d$ such that there exist $\alpha, C > 0$ satisfying that for every $x \in \mathbb{R}^d$ and $0 \le r \le 1$, $\mu(B(x, r)) \le Cr^\alpha$.*

Consider the function $F_\mu : \mathbb{R}^d \to \mathbb{R}$ whose wavelet coefficients are given by $d_\lambda = \mu(\lambda)$ for every dyadic cube $\lambda \in \Lambda$ (see Sect. 4.2 for definitions).

Then the multifractal spectra of μ and F_μ coincide.

Our purpose here is to provide a survey on recent results and on some open problems related to these various research directions, which combine many ideas coming from (and having applications to) geometric measure theory, functional and harmonic analysis, and real analysis, as well as ergodic theory and dynamical systems.

2 Prescription of Exponents and Local Dimensions

For a given mapping $f : \mathbb{R}^d \to \mathbb{R}$ belonging to $L^\infty_{loc}(\mathbb{R}^d)$, its associated pointwise Hölder exponent mapping $h_f : x \mapsto h_f(x)$ may be very erratic, changing violently from one point to the other. Nevertheless h_f (viewed as a function) is quite well understood, as confirmed by the following theorem by S. Jaffard which provides a full characterization of h_f [27, 29]. Recall that $C^{\log}(\mathbb{R}^d)$ is the space of those functions $f : \mathbb{R}^d \to \mathbb{R}$ satisfying that there exists $C > 0$ such that for every $x, y \in \mathbb{R}^d$ with $|x - y| \le 1/2$, $|f(x) - f(y)| \le C|\log|x - y||^{-1}$.

Theorem 2.1 *When $f \in C^{\log}(\mathbb{R}^d)$, the mapping h_f is a liminf of a sequence of continuous functions.*

Conversely, let $H : \mathbb{R}^d \to \mathbb{R}^+ \cup \{+\infty\}$ be a liminf of a sequence of continuous functions. There exists a function $f : \mathbb{R}^d \to \mathbb{R}$, $f \in C^{\log}(\mathbb{R}^d)$, such that for every $x \in \mathbb{R}^d$, $h_f(x) = H(x)$.

Let us also mention that in [5] the authors build a continuous nowhere differentiable stochastic process $(M_x)_{x\ge 0}$ whose pointwise Hölder exponents have the most general form, i.e. the mapping $x \mapsto h_M(x) \in (0, 1)$ can be any liminf of a sequence of continuous functions.

It is a natural question to investigate the same issues for local dimensions for measures.

Definition 2.2 Let $\mathcal{M}(K)$ be the set of Borel probability measures on a Borel set $K \subset \mathbb{R}^d$.

For $\mu \in \mathcal{M}(\mathbb{R}^d)$, the support of μ is the set

$$\text{Supp}(\mu) = \{x \in \mathbb{R}^d : \mu(B(x, r)) > 0 \text{ for every } r > 0\}.$$

The (lower) local dimension of μ at $x \in \text{Supp}(\mu)$ is

$$h_\mu(x) = \liminf_{r \to 0^+} \frac{\log \mu(B(x, r))}{\log r} \tag{2.1}$$

and the singularity spectrum of μ is defined for $H \in \mathbb{R} \cup \{+\infty\}$ by

$$D_\mu(H) = \dim E_\mu(H) \quad \text{where } E_\mu(H) = \{x \in \text{Supp}(\mu) : h_\mu(x) = H\}.$$

It is common (and in many situations, relevant and important) to look at points x at which (2.1) turns out to be a limit (and not only a liminf). Nevertheless, in this article only lower local dimensions are considered (we will forget the term "lower" in the following), since we are interested in quantities defined for all $x \in \text{Supp}(\mu)$.

Definition 2.3 A function f (resp. a measure μ) on $\mathbb{R}^d$ is called homogeneous (in short: HM) if the restriction of f (resp. μ) on any finite subcube $I \subset \mathbb{R}^d$ has the same singularity spectrum as f (resp. μ).

The same definition applies to a function or measure when $\mathbb{R}^d$ is replaced by $[0, 1]^d$.

One could expect that an analog of Theorem 2.1 should hold for local dimensions of measures. Unfortunately, the situation is not as clear, as proved by the next lemma [17].

Lemma 2.4 *Let $\mu \in \mathcal{M}(\mathbb{R}^d)$ with a support containing a cube $U \subset \mathbb{R}^d$. If the mapping $x \mapsto h_\mu(x)$ is continuous on U, then h_μ is locally constant and equal to d on U.*

Last lemma leads to the two following open problems: What are the admissible mappings $H : \mathbb{R}^d \to \mathbb{R}^+$ satisfying $H = h_\mu$ for some probability measure μ? Given an admissible mapping H, can one explicitly build a measure $\mu \in \mathcal{M}(\mathbb{R}^d)$ such that $h_\mu = H$?

Even if all these questions are mathematically relevant and raise delicate questions (in geometric measure theory for instance), in many situations it is even more important to construct functions with prescribed singularity spectrum. This is the case in particular when trying to model real-life data, for which essentially only global quantities (like the L^q-spectrum) are accessible.

3 Prescription of Multifractal Behavior

As expected, the prescription of singularity spectrum for functions or measures is more involved than that of exponents. Indeed, there is no obvious characterization for the admissible singularity spectrum for functions. Yet, using wavelet techniques, S. Jaffard was able to prove the following theorem [28]. Let

$$\mathcal{R} = \left\{ \sigma : \mathbb{R}^+ \to [0, d] \cup \{-\infty\} : \left| \begin{array}{l} \exists \text{ bounded interval } I \subset \mathbb{R}^+ \text{ and } \alpha \in [0, d] \\ \text{such that } \sigma = \alpha \mathbf{1}_I + (-\infty)\mathbf{1}_{\mathbb{R}^+ \setminus I} \end{array} \right. \right\}.$$

Theorem 3.1 *Let $\sigma : \mathbb{R}^+ \to [0, d] \cup \{-\infty\}$ be the supremum of a countable sequence of functions $(\sigma_n)_{n \geq 1} \in \mathcal{R}$. Then there exists a continuous function $f : \mathbb{R}^d \to \mathbb{R}$ such that $D_f = \sigma$.*

Although probably not optimal, this theorem already covers a large class of singularity spectra, certainly sufficient to mimic precisely all the singularity spectra that can be estimated on real data.

In particular, any concave mapping $\sigma : \mathbb{R}^+ \to [0, d] \cup \{-\infty\}$ can be written as $\sup_{n\in\mathbb{N}} \sigma_n$ for some well chosen functions $\sigma_n \in \mathcal{R}$, hence it is possible to build a function $f : \mathbb{R}^d \to \mathbb{R}$ such that $D_f = \sigma$.

The same questions were addressed for measures first in [17] and then in [7]. The admissible singularity spectra for measures are not characterized either, but when compared to spectra of functions, there are additional constraints: if $d_\mu = \sigma$ for some $\mu \in \mathcal{M}(\mathbb{R}^d)$, then $\sigma(h) \leq \min(h, d)$ (see [13, 39]).

Another surprising constraint obtained in [17] is that the support of the singularity spectrum of a 1-dimensional HM measure contains an interval. We call Supp(σ) the support of a function $\sigma : \mathbb{R}^d \to \mathbb{R}$, and by abuse of notation, if $\sigma : \mathbb{R} \to \mathbb{R}^+ \cup \{-\infty\}$, $\text{Supp}(\sigma) = \{H : \sigma(H) \geq 0\}$.

Proposition 3.2 *For any non-atomic HM probability measure $\mu \in \mathcal{M}(\mathbb{R})$, the set Supp($D_\mu$) $\cap$ $[0, 1]$ is necessarily an interval of the form $[\alpha, 1]$, where $0 \leq \alpha \leq 1$.*

This proposition leads to the following notation: for $\sigma : \mathbb{R}^+ \to [0, 1] \cup \{-\infty\}$, consider the mapping

$$\sigma^\dagger(H) = \max\big(\sigma(H), 0 \cdot \mathbf{1}_{[\inf(\text{Supp}(\sigma)), \sup(\text{Supp}(\sigma))]}(H)\big).$$

Essentially, $\sigma^\dagger$ fills the gaps in the support of σ by replacing the value $-\infty$ by 0.

The result concerning the prescription of singularity spectrum of measures obtained in [17] is the following.

Theorem 3.3 *Let $\sigma : \mathbb{R}^+ \to [0, 1] \cup \{-\infty\}$ be the supremum of a countable sequence of functions $(\sigma_n)_{n\geq 1} \in \mathcal{R}$ satisfying in addition that for every $n \geq 1$, calling I_n the interval on which σ_n is not $-\infty$,*

- *$I_n \subset [0, 1]$,*
- *I_n is closed,*
- *$\sigma_n(x) \leq x$ for $x \in I_n$.*

Then:

1. *There exists $\mu \in \mathcal{M}(\mathbb{R})$ such that $D_\mu = \sigma$.*
2. *There exists a HM measure $\mu \in \mathcal{M}(\mathbb{R})$ with support equal to $[0, 1]$ such that $D_\mu = \sigma^\dagger$, and $D_\mu(1) = 1$.*

Observe that although the class of singularity spectra obtained here is quite large, only local dimensions less than 1 are dealt with, and only the one-dimensional case is covered. Theorem 3.3 is completed by the result by Barral [7].

Theorem 3.4 *Let $\sigma : \mathbb{R}^+ \to [0, d] \cup \{-\infty\}$ be an upper semi-continuous function with support included in $[\alpha, \beta]$ for some $0 < \alpha < \beta < +\infty$, satisfying $\sigma(h) \leq h$ for every $h \in [\alpha, \beta]$, and such that $\sigma(h) = h$ for some h. Then there exists $\mu \in \mathcal{M}(\mathbb{R}^d)$ such that $D_\mu = \sigma$.*

In the last theorem, Barral was also able to build measures that were "homogeneous" in the sense that the restriction of μ to any bounded cube $I \subset \mathbb{R}^d$ such that

$\mu(I) \neq 0$ has the same singularity spectrum as μ itself. A comparison between Theorems 3.3 and 3.4 yields that (at least) in dimension 1, the measures constructed by Barral are necessarily not supported by a full interval (their support is a Cantor-like set), otherwise σ should be replaced by $\sigma^\dagger$.

Theorems 3.1, 3.3, and 3.4 are not entirely satisfying. Indeed,

- The construction used in Theorem 3.1 does not guarantee that the corresponding spectrum is homogeneous. Homogeneous spectra are yet very common (for instance, trajectories of stationary processes usually exhibit homogeneous spectra).
- In the three previous theorems, even if the prescribed spectrum is concave, the corresponding function or measure a priori does not satisfy a multifractal formalism.
- The functions and measures built along the proofs of Theorems 3.1 and 3.3 are not "typical" in any sense, and may essentially appear, from the modeling standpoint, as mathematical extreme toy examples.

These issues will be addressed in the next sections.

4 Prescription of Multifractal Formalisms

Let us very quickly recall the intuition by Frisch and Parisi [23], who studied the velocity v of a turbulent fluid in a bounded domain $\Omega \subset \mathbb{R}^3$. More precisely, inspired by the seminal works by Kolmogorov on turbulent fluids and the study of the local fluctuations of their velocity, Frisch and Parisi were interested in the moments of the increments of v defined by

$$\text{for every } q \in \mathbb{R}, \quad S_v(q,l) = \int_{\Omega} |v(x+l) - v(x)|^q dx. \tag{4.1}$$

For real data, q being fixed, it has been observed that when $|l|$ becomes small, $S_v(q,l)$ obeys a scaling law:

$$S_v(q,l) \sim |l|^{\zeta_v(q)} \text{ for some exponent } \zeta_v(q) \in \mathbb{R}.$$

The mapping $q \mapsto \zeta_v(q)$ is called the scaling function of the velocity of the fluid. It can be seen that if v is modeled at small scales by a fractional Brownian motion of index H_0 (as did Kolmogorov for instance), then $\zeta_v(q)$ is linear with slope H_0. However, in the 1980s, numerical experiments for the velocity show that $\zeta_v(q)$ is non-linear and concave. The seminal idea by Frisch and Parisi consists in interpreting this non-linearity in terms of multifractality of v, via the following heuristic argument.

Replacing Hausdorff by box dimension, and making all kind of rough approximations (i.e. assuming that limits exist, etc), for all points $x \in \mathbb{R}^3$ at which $h_v(x) = H$, one has $|v(x + l) - v(x)| \sim |l|^H$ for small l. Since $\dim E_v(H) = D_v(H)$, there should exist approximately $|l|^{-D_v(H)}$ cubes of size l in the domain $\boldsymbol{\Omega}$ containing points x which are singularities of order H for the velocity v. All these intuitions lead to the estimates

$$S(q,l) = \int_{\boldsymbol{\Omega}} |v(x+l) - v(x)|^q dx \sim \sum_H |l|^{qH} |l|^{-D_v(H)} |l|^3 \sim \sum_H |l|^{qH - D_v(H)+3}.$$

When $|l| \to 0$, the greatest contribution is obtained for the smallest exponent:

$$\zeta_v(q) = \inf_H (qH - D_v(H) + 3).$$

The corresponding mapping $q \mapsto \zeta_v(q)$ is called the L^q-spectrum or the scaling function of v—soon we will see more relevant formulas for $\zeta_v(q)$ and how to define it for measures. By inverse Legendre transform, one deduces that

$$D_v(H) = \inf_{q \in \mathbb{R}} (qH - \zeta_v(q) + 3)$$

which justifies that D_v has a concave shape.

It is striking that despite the series of crude approximations, this intuition has proved to hold true in many (if not most of) situations, after some renormalization and suitable choices for the scaling functions.

Definition 4.1 We call multifractal formalism any formula relating the singularity spectrum of a function (or a measure) to a scaling function via a Legendre transform.

For almost 30 years now, many efforts have been made to prove the validity of multifractal formalism(s) in various functional spaces, for many mathematical objects (self-similar or self-affine functions and measures) including random processes (Mandelbrot cascades, Gaussian multiplicative chaos, Lévy processes). This line of research was constantly followed and fostered by applications which gave mathematicians lots of signals and physical phenomena to study and work on, see Figs. 1 and 2. In particular, stable algorithms to estimate L^q-spectra of data have been developed, furnishing to the scientific community many robustly analyzed sets of data [2].

A remaining question though lies in the existence of a functional setting in which a given multifractal behavior would be "generic". This is known after [30] as the *Frisch–Parisi conjecture*, which can be formulated as follows:

Conjecture 4.2 Given any admissible concave mapping $\sigma : \mathbb{R}^+ \to [0, d] \cup \{-\infty\}$, is there a functional space in which typical functions have σ as singularity spectrum and satisfy a multifractal formalism?

Notice that ideas leading to a multifractal formalism can also be found in thermodynamics (see [25, 26] and the large literature around thermodynamical formalism). This outlines the universality of the approach consisting in describing local fluctuations via the (Legendre transform of) global statistical quantities computed directly on the object (function, measure, random process) under consideration.

From now on, and without loss of generality, we restrict our statements to measures and functions supported in the cube $[0, 1]^d$.

4.1 Prescription of Multifractal Formalism for Measures

In case of measures $\mu \in \mathcal{M}([0, 1]^d)$, the formula for the L^q-spectrum is quite standard and given by

$$\tau_\mu(q) = \liminf_{j\to+\infty} \frac{1}{-j} \log_2 \sum_{\lambda\in\mathcal{D}_j:\,\mu(\lambda)\neq 0} \mu(\lambda)^q, \tag{4.2}$$

where $\mathcal{D}_j$ stands for the set of dyadic cubes $\lambda_{j,k} = 2^{-j}k + [0, 2^{-j}]^d$, $k \in \mathbb{Z}^d$, of generation $j \in \mathbb{Z}$ (i.e. dyadic cubes with side-length equal to 2^{-j}). It is easily seen that τ_μ is always concave, non-decreasing, and that $-d \le \tau_\mu(0^+) \le \tau_\mu(1) = 0$. In addition, the support of τ_μ is equal to $\mathbb{R}$ when $\limsup_{r\to 0^+} \frac{\log(\inf\{\mu(B(x,r)):\, x\in \mathrm{Supp}(\mu)\})}{\log r} < +\infty$, and it is $[0, +\infty)$ when the same quantity is infinite [7].

Recall that the Legendre transform of a mapping $\tau : \mathbb{R} \to \mathbb{R}$ (used in the previous section) is defined for $H \ge 0$ as

$$\tau^*(H) := \inf_{q\in\mathbb{R}} (qH - \tau(q)).$$

Barral solved in [7] the following inverse problem.

Theorem 4.3 *Let $\tau : \mathbb{R} \to \mathbb{R}$ be concave, non-decreasing, with $-d \le \tau(0^+) \le \tau(1) = 0$. There exists a probability measure $\mu \in \mathcal{M}([0, 1]^d)$ compactly supported, such that $\tau_\mu = \tau$ and μ satisfies the multifractal formalism, i.e. $D_\mu = \tau^*$.*

See Fig. 3 for an illustration.

The drawback of this first important step is that the measure constructed by Barral in [7] has again a Cantor-like set as support (so it is not fully supported on $[0, 1]^d$), hence is not suitable to model any real-life signal supported by, say, an interval. The result is reinforced in the upcoming paper [9], in which we build fully supported measures satisfying a prescribed multifractal formalism, which in addition are almost-doubling in the following sense.

A Borel set function is a mapping μ associating with every Borel set $B \subset [0, 1]^d$ a positive real number $\mu(B) \in [0, +\infty]$. A Borel set function μ is *almost-doubling* when there exists a non increasing function $\theta : (0, 1] \to \mathbb{R}^+ \setminus \{0\}$ such that :

- $\theta(1) = 0$ and $\lim_{r\to 0^+} \frac{\theta(r)}{\log(r)} = 0$

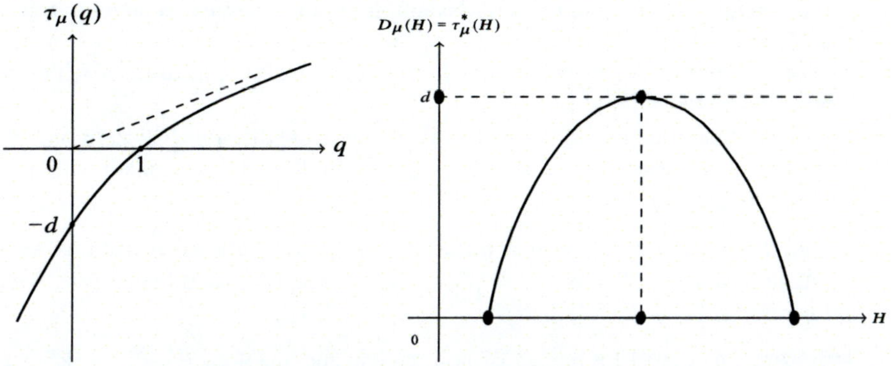

Fig. 3 Left: L^q-spectrum of a measure μ on [0, 1]. Right: The corresponding singularity spectrum of μ when it satisfies a multifractal formalism

- there is a constant $C \geq 1$ such that for all $x \in [0, 1]^d$ and $r \in (0, 1]$ one has

$$C^{-1}e^{-\theta(r)}\mu(B(x, r)) \leq \mu(B(x, 2r)) \leq Ce^{\theta(r)}\mu(B(x, r)). \tag{4.3}$$

When $\theta \equiv 0$, then μ is said to be *doubling*.

Doubling and almost-doubling measures occupy a special place in geometric measure theory since they are easier to deal with in many situations—such properties guarantee a certain stability of the values of μ in the sense that $\mu(B)$ and $\mu(B')$ have comparable values as soon as B and B' are two balls of comparable radii that are close to each other. It is thus important to investigate the possible combination of these properties with the multifractal ones, as done in the following theorem proved in [9].

Theorem 4.4 *Let $\tau : \mathbb{R} \to \mathbb{R}$ be concave, non-decreasing, with $-d = \tau(0^+) \leq \tau(1) = 0$.*

Then there exists an HM almost doubling measure $\mu \in \mathcal{M}([0, 1]^d)$ with full support in $[0, 1]^d$ such that $\tau_\mu = \tau$ and μ satisfies the multifractal formalism, i.e. $D_\mu = \tau^*$.

Although Gibbs measures associated with Hölder regular potentials and smooth maps provide examples of doubling measures with non-trivial multifractal behavior, it may seem surprising that the almost doubling property (which, as said above, limits the local variations of a measure) does not constitute a constraint from the multifractal formalism standpoint: every (admissible) concave mapping can be obtained as the singularity spectrum of a compactly supported probability measure satisfying the multifractal formalism.

Theorem 4.4 leaves open interesting questions in ergodic theory and dynamical systems, and geometric measure theory, which to the best of our knowledge are not completely addressed yet:

1. Can the almost doubling property be simplified in a "simple" doubling property in Theorem 4.4?
2. Given an almost doubling measure μ, is there a doubling measure $\widetilde{\mu}$ with same multifractal behavior as μ?
3. Is it possible to find a Hölder potential on a suitable dynamical system such that the associated invariant measure satisfies the multifractal formalism with a L^q-spectrum given in advance?

Remark 4.5 In Theorem 4.4, it is possible to impose additional conditions on the measures μ so that the same result ($D_\mu = \tau^*$) holds. One useful condition, which will be used later, is the following.

Definition 4.6 Let $\boldsymbol{\Theta}$ be the set of non decreasing functions $\theta : \mathbb{N} \to \mathbb{R}_+^*$ such that:

1. $\theta(j) = o(j)$ as $j \to \infty$
2. $\theta(0) = 0$
3. for all $\varepsilon > 0$, there exists $j_\varepsilon \in \mathbb{N}$ such that for all $j' \geq j \geq j_\varepsilon, \theta(j') - \theta(j) \leq \varepsilon(j' - j)$.

A measure $\mu \in \mathcal{M}([0,1]^d)$ (or $\mu \in \mathcal{M}(\mathbb{R}^d)$) satisfies property (P) if there exist $C, s_1, s_2 > 0$ such that:

(P1) for all $j \in \mathbb{N}$ and $\lambda \in \mathcal{D}_j$, if $\mu(\langle) \neq 0$, then

$$C^{-1} 2^{-js_2} \leq \mu(\lambda) \leq C 2^{-js_1}. \tag{4.4}$$

(P2) There exists $\theta \in \boldsymbol{\Theta}$ such that for all $j, j' \in \mathbb{N}$ with $j' \geq j$, and all $\lambda, \widetilde{\lambda} \in \mathcal{D}_j$ such that $\mu(\langle) \neq 0$, $\mu(\widetilde{\langle}) \neq 0$, $\partial\lambda \cap \partial\widetilde{\lambda} \neq \emptyset$, and $\lambda' \in \mathcal{D}_{j'}$ such that $\lambda' \subset \lambda$:

$$C^{-1} 2^{-\theta(j)} 2^{(j'-j)s_1} \mu(\lambda') \leq \mu(\widetilde{\lambda}) \leq C 2^{\theta(j)} 2^{(j'-j)s_2} \mu(\lambda'). \tag{4.5}$$

Heuristically, this last condition yields for every dyadic cube $\langle \in \mathcal{D}_j$ a control of the μ-mass of the cubes $\widetilde{\langle} \in \mathcal{D}_{\widetilde{j}}$ with $\widetilde{j} \geq j$ and $\widetilde{\langle} \subset 3\langle$. It is easily checked on self-similar measures satisfying an open-set condition for instance.

In [9], it is proved that there exist measures satisfying (P) for which the conclusion of Theorem 4.4 holds.

4.2 Prescription of Multifractal Formalism for Functions

While the definition of the L^q-spectrum for measures is quite standard and intuitive, finding a suitable formula for the L^q-spectrum of functions is not straightforward. Indeed, one easily sees that Eq. (4.1) does not allow one to catch and describe the local regularity characteristics of smooth functions (with pointwise exponents greater than 1). Many alternative formulas have been proposed, and most of them are based on wavelets. It is thus useful at this point to set the notation concerning wavelets coefficients and wavelet leaders.

Let $\Phi : \mathbb{R}^d \to \mathbb{R}$ be a scaling function and consider an associated family of smooth wavelets $\Psi = \{\psi^{(i)}\}_{i=1,\dots,2^d-1}$ belonging to $C^r(\mathbb{R}^d)$, with $r \in \mathbb{N}^*$ (for a general construction, see [37, Ch. 3]). For simplicity, we assume that Φ and the wavelets Ψ are compactly supported [19]. For every $j \in \mathbb{Z}$, recall that $\mathcal{D}_j$ is the set of dyadic cubes of generation j, i.e. if $k = (k_1, \dots, k_d) \in \mathbb{Z}^d$ and

$$\lambda_{j,k} := \prod_{i=1,\dots,d} [k_i 2^{-j}, (k_i + 1)2^{-j}) \subset \mathbb{R}^d$$

then $\mathcal{D}_j = \{\lambda_{j,k} : k \in \mathbb{Z}^d\}$ (see the beginning of Sect. 4.1). Further we consider the set

$$\Lambda_j = \{\lambda = (i, j, k) : k \in \mathbb{Z}^d,\ i \in \{1, \dots, 2^d - 1\}\},$$

and $\Lambda = \bigcup_{j\in\mathbb{Z}} \Lambda_j$. By abuse of notation, $\lambda \in \Lambda_j$ will still be called a dyadic cube of generation j and identified with $\lambda = \lambda_{j,k} \in \mathcal{D}_j$.

For every cube $\lambda = (i, j, k) \in \Lambda$, we denote by ψ_λ the function $x \mapsto \psi^{(i)}(2^j x - k)$. The set of functions $2^{dj/2}\psi_\lambda$, $j \in \mathbb{Z}$, $\lambda \in \Lambda_j$, forms a Hilbert basis of $L^2(\mathbb{R}^d)$, so that every $f \in L^2(\mathbb{R}^d)$ can be expanded as

$$f = \sum_{j\in\mathbb{Z}} \sum_{\lambda\in\Lambda} d_\lambda \psi_\lambda, \quad \text{with } d_\lambda = \int_{\mathbb{R}^d} 2^{dj} \psi_\lambda(x) f(x)\, dx,$$

where equality holds in L^2 (we will work with smooth functions, so equality will also hold pointwise). Observe that we choose an L^∞ normalization for the so-called *wavelet coefficients* $(d_\lambda)_{\lambda\in\Lambda}$ of $f \in L^2(\mathbb{R}^d)$ (more generally, of $f \in L^p(\mathbb{R}^d)$ for some $p \in [1, \infty]$). For $f \in L^2(\mathbb{R}^d)$, define also for $k \in \mathbb{Z}^d$

$$\beta(k) = \int_{\mathbb{R}^d} f(x)\Phi(x - k)\, dx. \tag{4.6}$$

Finally, for a function $f \in L^p(\mathbb{R}^d)$ with $p \in [1, \infty]$ whose wavelet coefficients are denoted by $(d_\lambda)_{\lambda\in\Lambda}$, the wavelet leader associated with $\lambda \in \mathcal{D}_j$ is

$$d_\lambda^L = \sup_{\lambda' \in \Lambda,\, \lambda' \subset 3\lambda} |d_{\lambda'}|,$$

where for $\lambda \in \mathcal{D}_j$, 3λ stands for the cube with same center as λ and radius $\frac{3}{2}2^{-j}$ (it is the cube that contains λ as well as its $2^d - 1$ neighbors in $\mathcal{D}_j$). While wavelet coefficients are usually sparse (only a few coefficients carry the important information about f), wavelet leaders possess a strong hierarchical structure since $0 \le d_{\lambda'}^L \le d_\lambda^L$ when $\lambda' \subset \lambda$.

Remark 4.7 Although the notations for wavelet coefficients and wavelet leaders do not mention the function f, they highly depend on f and we should never forget about it!

Wavelet coefficients and wavelet leaders characterize the pointwise Hölder exponents: indeed, if $f \in C^\epsilon(\mathbb{R}^d)$ for some $\epsilon > 0$, then for every $x_0 \in [0,1]^d$ one has

$$h_f(x_0) = \liminf_{j\to\infty} \frac{\log d^L_{\lambda_j(x_0)}}{\log(2^{-j})}, \tag{4.7}$$

where $\lambda_j(x_0)$ is the unique cube $\lambda \in \mathcal{D}_j$ that contains x_0 (see [31]).

It was quite clear from the beginning that a formula based on increments like (4.1) was not stable neither mathematically nor numerically. To circumvent this difficulty, the idea of introducing wavelets (whose computation requires local means, bringing simultaneously a numerical stability crucial for applications and a natural connection with characterizations of standard functional spaces, see Sect. 5) was introduced by Alain Arnéodo and his collaborators. Two formulations are nowadays recognized to be the most relevant:

- Formula based on wavelets:

$$T_f(q,j) = \sum_{\lambda\in\Lambda_j : d_(\neq 0} |d_\lambda|^q \longrightarrow \eta_f(q) = \liminf_{j\to+\infty} \frac{\log_2 T_f(q,j)}{-j}. \tag{4.8}$$

- Formula based on wavelet leaders:

$$L_f(q,j) = \sum_{\lambda\in\mathcal{D}_j : d^L_(\neq 0} |d^L_\lambda|^q \longrightarrow L_f(q) = \liminf_{j\to+\infty} \frac{\log_2 L_f(q,j)}{-j}. \tag{4.9}$$

Even if wavelets brought some stability in the computations, wavelet leaders are now recognized as the most efficient, relevant and numerically exploitable measurements of local and global regularity. In particular, the hierarchical structure of wavelet leaders (i.e. $0 \leq d_\lambda \leq d_{\lambda'}$ as soon as $\lambda \subset \lambda'$) makes all computations easier and more stable [2].

Definition 4.8 The wavelet multifractal formalism WMF (*resp.* wavelet leader multifractal formalism WLMF) is satisfied for a function f on an interval $J \subset \mathbb{R}^+$ when $D_f(H) = (\eta_f)^*(H)$ (*resp.* $D_f(H) = (L_f)^*(H)$) for every $H \in J$.

We also say that a function f satisfies the weak wavelet leader multifractal formalism (weak-WLMF) on an interval $J \subset \mathbb{R}^+$ when there exists an increasing sequence $(j_n)_{n\geq 1}$ of integers such that if $\widetilde{L}_f(q) = \liminf_{n\to+\infty} \frac{\log_2 L_f(q,j_n)}{-j_n}$, then $D_f(H) = (\widetilde{L}_f)^*(H)$ for every $H \in J$.

Remark 4.9 The above definition of formalisms depends a priori on the chosen wavelets Ψ. Actually it does not depend on Ψ in the increasing part of the multifractal spectrum [31], but it does in the decreasing part. For simplicity, we do not mention this dependence in the notations.

Let $\mathcal{S}_d$ be the set of admissible singularity spectra for functions satisfying a multifractal formalism, i.e.

$$\mathcal{S}_d = \left\{ \sigma : \mathbb{R}^+ \to [0, d] \cup \{-\infty\} : \left\{ \begin{array}{l} \sigma \text{ is compactly supported in } (0, +\infty), \\ \text{concave, with maximum equal to } d. \end{array} \right\} \right. . \tag{4.10}$$

We are now able to state the result on multifractal formalism prescription for functions.

Theorem 4.10 *For every mapping $\sigma \in \mathcal{S}_d$, there exists a function $f \in L^2(\mathbb{R}^d)$ satisfying the WLMF and whose singularity spectrum is equal to σ.*

Proof Observe that if a function f has its wavelet coefficients d_λ given by $\mu(\lambda)$ for some probability measure $\mu \in \mathcal{M}([0, 1]^d)$, then for every choice of $\alpha, \beta > 0$, the function $f_{\alpha,\beta}$ whose wavelet coefficients are $\tilde{d}_\lambda := d_\lambda^\alpha 2^{-j\beta}$ satisfies

$$\text{for every } H \geq 0, \quad D_{f_{\alpha,\beta}}(H) = D_f\left(\frac{H-\beta}{\alpha}\right).$$

This simply follows from (4.7) and the fact that $h_{f_{\alpha,\beta}}(x_0) = \alpha h_f(x_0) + \beta$ for all x_0.

Let $\sigma : \mathbb{R}^+ \to [0, d] \cup \{-\infty\} \in \mathcal{S}_d$ be a mapping satisfying the conditions to be a singularity spectrum of a function satisfying a multifractal formalism.

Let α, β be two strictly positive real numbers such that the mapping $\sigma_{\alpha,\beta}(H) = \sigma(\alpha H + \beta)$ satisfies $\sigma_{\alpha,\beta}(H) \leq H$ and there exists $H_0 > 0$ such that $\sigma_{\alpha,\beta}(H_0) = H_0$. The existence of (α, β) is an exercise (notice that (α, β) need not be unique).

Theorem 4.4 provides us with a measure μ satisfying the multifractal formalism for measures and $D_\mu = \sigma_{\alpha,\beta}$.

Then, Theorem 1.2 yields that the function F_μ whose wavelet coefficients are given by $d_\lambda = \mu(\lambda)$ has the same singularity spectrum as μ. In addition, comparing (4.2) with (4.9), and using the hierarchical structure of the measure (i.e. $\mu(\lambda') \leq \mu(\lambda)$ whenever $\lambda' \subset \lambda$), one sees that $\tau_\mu(q) = L_{F_\mu}(q)$ for every $q \in \mathbb{R}$, hence F_μ satisfies the WLMF.

Finally, using the first remark of this proof, the function F whose wavelet coefficients equal $\mu(\lambda)^\alpha 2^{-j\beta}$ has its singularity spectrum equal to σ and satisfies the WLMF.

We thus have a complete answer for the prescription of multifractal formalism for functions. But at this point, one may have the feeling that the functions we built are mathematical toy examples. The purpose of the last sections is to explain that for any choice of concave admissible mapping σ, there are natural functional spaces in which typical functions have exactly σ as singularity spectrum. This confirms and strengthens the overall presence of multifractals in most of science fields, and reinforces the position of multifractal machinery as legitimate tool in signal processing and data analysis.

5 Typical Multifractal Behavior in Classical Functional Spaces

As emphasized above, it is possible to find mathematical models that mimic large classes of multifractal behavior, in particular including all concave singularity spectra. This last part of the results is key, since for real-life data (multi-dimensional and/or multivariate signals, images, ...) only estimates for the L^q-spectrum are numerically accessible (based on log-log plots on a well-chosen range of scales). Indeed, the standard paradigm is to assume that the discrete data f (say, a signal) is obtained from discrete samples of a mathematical model obeying a multifractal formalism, and to consider that the Legendre transform of the estimated L^q-spectrum contains relevant information regarding the distribution of the singularities of f (somehow extrapolating on Frisch–Parisi heuristics). This Legendre transform is thus viewed as an "approximation" of the singularity spectrum of the data, although the meaning of the singularity spectrum of a discretized signal is not made precise. The obtained estimated singularity spectrum of the data f possesses various characteristics (values of the largest and the smallest exponents, locations of the maximum, curvature of the concave spectrum at its maximum,...) which are then used as classification tools between numerous samples of a physical, medical,...phenomenon. This has proven to be relevant in various fields going from medicine (heart-beat rate and X-ray analysis) and turbulence [32] to, recently, more surprising areas (paintings analysis [1], text analysis [33]).

Inspired by these applications, it is thus key to investigate whether the mathematical objects we regularly meet satisfy a multifractal formalism (so that all these heuristics described above lie on solid mathematical grounds). In this survey, we focus on "typical" objects in the sense of Baire: in a Baire space E, a property $\mathcal{P}$ of elements $x \in E$ is *typical* or *generic* when the set $\{x \in E : x \text{ satisfies } \mathcal{P}\}$ is a residual set, i.e. its complement is included in a first Baire category set (a union of countably many nowhere dense sets in E).

Regularity properties of typical functions have been explored since the pioneer works of Banach [6] or Mazurkiewicz [36] for instance. The seminal result concerning multifractal properties of typical functions is due to Buczolich and Nagy, who proved the following [14].

Theorem 5.1 *Let $\mathcal{M}on([0,1])$ be the set of continuous monotone functions $f : [0,1] \to \mathbb{R}$ equipped with the supremum norm of functions. Typical functions in $\mathcal{M}on([0,1])$ are multifractal with singularity spectrum equal to $D_f(H) = H \cdot \mathbf{1}_{[0,1]}(H) + (-\infty) \cdot \mathbf{1}_{(1,+\infty]}(H)$.*

Theorem 5.1 was the starting point of an abundant literature on the subject, examples of which are given in the following. The method consists first in finding an upper bound for the singularity spectrum of all functions in $\mathcal{M}on([0,1])$ (here, the diagonal $\sigma(H) = H$)), then an explicit function F_{typ} whose local behavior is the one suspected to be typical, and finally to construct a countable sequence $(A_n)_{n\geq 1}$ of sets of functions, dense in $\mathcal{M}on([0,1])$, which are for a given n, really close to F_{typ} at a given scale (depending on n). If the parameters are correctly settled, the intersection of the $(A_n)_{n\geq 1}$ will be the set of typical functions with multifractal behavior similar to that of F_{typ}.

The proof is based on a careful analysis on local oscillations of functions, and simultaneous constructions of Cantor-like sets $E_f(H)$ carrying the sets of points with pointwise Hölder exponent equal to H, for every $f \in \bigcap_{n\geq 1} A_n$.

After Theorem 5.1, the first direction consisted in exploring the typical behavior in other standard functional spaces. The first, spectacular, results were obtained by Jaffard [30], who implemented the same strategy as [14] but added wavelet tools to deal with the important examples of Hölder and Besov spaces.

Theorem 5.2

(1) Let $\alpha > 0$ and consider the space $C^\alpha([0,1]^d)$ of α-Hölder functions on $[0,1]^d$. Typical functions in $C^\alpha([0,1]^d)$ are monofractal and satisfy

$$D_f(H) = d \cdot \mathbf{1}_{\{\alpha\}}(H) + (-\infty) \cdot \mathbf{1}_{[0,+\infty]\setminus\{\alpha\}}(H).$$

(2) Let $p \geq 1$ and $s > d/p$, and consider the Besov space $B_q^{s,p}([0,1]^d)$. Typical functions in $B_q^{s,p}([0,1]^d)$ are multifractal and satisfy

$$D_f(H) = p(H - (s - d/p)) \cdot \mathbf{1}_{[s-d/p,s]}(H) + (-\infty) \cdot \mathbf{1}_{[0,+\infty]\setminus[s-d/p,s]}(H).$$

In addition, typical functions satisfy the WLMF.

See Fig. 4 for an illustration.

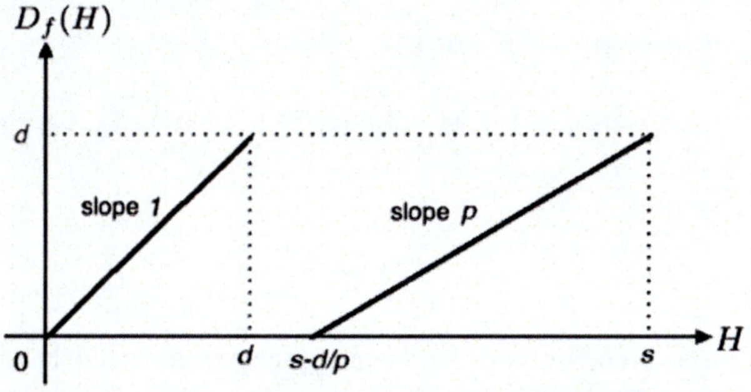

Fig. 4 Typical singularity spectra of measures supported on $[0,1]^d$ (Left) and of functions in $B_q^{s,p}(\mathbb{R}^d)$ (Right)

Theorems 5.1 and 5.2 are striking since they underline the preeminence of multifractal properties for "everyday" functions. Jaffard also described the multifractal behavior of typical functions belonging to countable intersections of Besov spaces, leading to a first answer to the Frisch–Parisi conjecture. Although these results were a giant step in the domain, only increasing singularity spectra with restricted shapes can be obtained and the typical functions do not obey a satisfactory multifractal formalism. Let us also mention that Besov spaces with indices $s < d/p$ were also considered in [30].

Other directions have been investigated. The most natural one concerns probability measures: typical multifractal properties were explored in [15] for measures supported on $[0, 1]^d$ and these results were extended by Bayart [11] for measures supported on general compact sets.

Theorem 5.3 *Let $K \subset \mathbb{R}^d$ be a compact set, and let $\mathcal{M}(K)$ be the set of probability measures on K.*

A typical measure $\mu \in \mathcal{M}(K)$ satisfies for any $H \in [0, \dim(K))$, $D_\mu(H) = H$.

In addition, when the $\dim(K)$-Hausdorff measure of K is strictly positive, then typical measures satisfy $D_\mu(\dim(K)) = \dim(K)$ and obey the multifractal formalism.

Another extension of typical monotone functions is provided by the set of monotone increasing in several variables: A function $f : [0, 1]^d \to \mathbb{R}$ is continuous monotone increasing in several variables (in short: MISV) if for all $i \in \{1, \dots, d\}$, the coordinate functions

$$f^{(i)}(t) = f(x_1, \dots, x_{i-1}, t, x_{i+1}, \dots, x_d)$$

are continuous monotone increasing. The set of MISV functions is denoted by MISV^d. With Z. Buczolich, we also investigated the set CC^d of continuous convex functions $f : [0, 1]^d \to \mathbb{R}$. Equipped with the supremum norm $\| \cdot \|$, CC^d and MISV^d are separable complete metric spaces. In [16] and [18] we obtained the following results.

Theorem 5.4

(1) Typical functions in $MISV^d$ satisfy

$$D_f(H) = (d - 1 + H) \cdot \mathbf{1}_{[0,1]}(H) + (-\infty) \cdot \mathbf{1}_{[0,+\infty]\setminus[0,1]}(H).$$

(2) Typical functions $f \in CC^d$ satisfy

$$D_f(H) = (d-1)\cdot\mathbf{1}_{\{0\}}(H)+(d-2+H)\cdot\mathbf{1}_{[1,2]}(H)+(-\infty)\cdot\mathbf{1}_{[0,+\infty]\setminus[1,2]\cup\{0\}}(H).$$

See Fig. 5 for a comparison between typical multifractal behavior in various functional spaces. This shall also be compared to Fig. 3. It appears clearly that in all the previous situations, the singularity spectra of typical functions have the same shape: it is an affine, increasing, mapping, with no decreasing part.

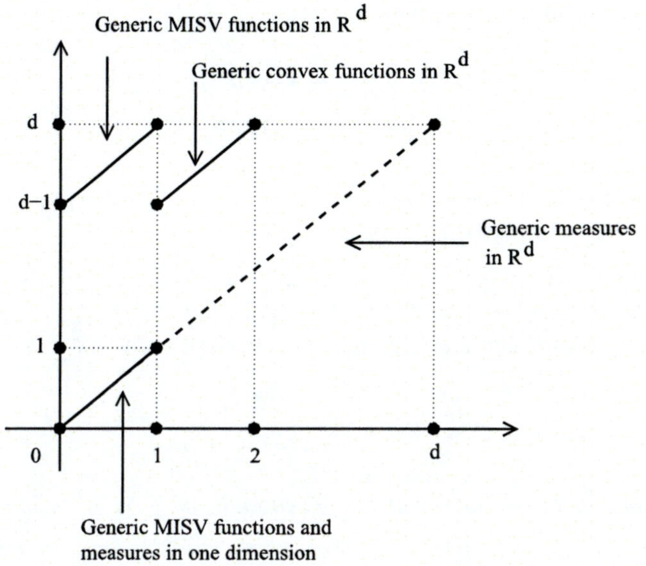

Fig. 5 Typical singularity spectra for measures, MISV and convex functions

Other functional spaces, called S_ν spaces were built in [3], in which typical functions all exhibit a singularity spectrum which is *visibly increasing* in the sense of [34], enlarging the class of possible typical multifractal behavior in functional spaces. In addition, these typical functions do not satisfy a multifractal formalism in the sense of Definition 4.1.

In order to break this limitation (no decreasing part in the singularity spectrum), new (and natural) functional spaces have been introduced in [9].

6 Besov Spaces in Multifractal Environment

Since standard functional spaces do not fulfill our requirements (i.e. typical functions in such spaces do not exhibit concave singularity spectra), it is natural to ask whether there are other functional spaces in which typical functions have any singularity spectrum given in advance, and satisfy a multifractal formalism. This solves the Frisch–Parisi conjecture as stated in Conjecture 4.2.

Let $\mathcal{B}(\mathbb{R}^d)$ be the Borel sets included in $\mathbb{R}^d$, and let us introduce the set of Hölder set functions

$$C(\mathbb{R}^d) := \left\{ \mu : \mathcal{B}(\mathbb{R}^d) \to \mathbb{R}^+ \text{ such that } \begin{Bmatrix} \exists s_1, s_2 \geq 0, \ \forall I \subset \mathbb{R}^d \text{ with } |I| \leq 1, \\ |I|^{s_2} \leq \mu(I) \leq |I|^{s_1} \end{Bmatrix} \right\}. \tag{6.1}$$

For $\mu \in C(\mathbb{R}^d)$ and $s \in \mathbb{R}$, we write

$$\mu^s(I) = \mu(I)^s,$$
$$\mu^{(s)}(I) = \mu(I)|I|^s.$$

We will use the following notation: for $x, y \in \mathbb{R}^d$, $B[x, y]$ is the smallest Euclidean ball that contains x and y.

Definition 6.1 Let $h \in \mathbb{R}^d$, $f : \mathbb{R}^d \to \mathbb{R}$, and consider the finite difference operator $\Delta_h f : x \mapsto f(x+h) - f(x)$. Define for $n \geq 2$ by iteration $\Delta_h^n f := \Delta_h(\Delta_h^{n-1} f)$.

For every set function $\mu \in C(\mathbb{R}^d)$, let us introduce for $n \geq 2$

$$\Delta_h^{\mu,n} f(x) = \frac{\Delta_h^n f(x)}{\mu(B[x, x+nh])}. \tag{6.2}$$

The *μ-adapted n-th order modulus of continuity of f* on $\mathbb{R}^d$ is defined for $t > 0$ by

$$\omega_n^{\mu}(f, t)_p = \sup_{t/2 \leq |h| \leq t} \|\Delta_h^{\mu,n} f\|_{L^p(\mathbb{R}^d)}. \tag{6.3}$$

It is trivial to check that that when $\mu(I) = 1$ for every set I, then $\omega_n^{\mu}(f, t)_p$ coincides with the so-called homogeneous n-th order modulus of continuity of f

$$\omega_n(f, t)_p = \sup_{t/2 \leq |h| \leq t} \|\Delta_h^n f\|_{L^p(\mathbb{R}^d)}.$$

Definition 6.2 Let $\mu \in C(\mathbb{R}^d)$ associated with exponents $0 < s_1 \leq s_2$ in (6.1).

Let $n \geq s_2$. For $1 \leq p, q \leq +\infty$, the Besov space in μ-environment $B_q^{\mu,p}(\mathbb{R}^d)$ is the space of those functions $f : \mathbb{R}^d \to \mathbb{R}$ such that $\|f\|_{L^p(\mathbb{R}^d)} < +\infty$ and

$$|f|_{B_q^{\mu,p}} = \|2^{jd/p}(\omega_n^{\mu}(f, 2^{-j})_p)_{j\geq 1}\|_{\ell^q(\mathbb{N})} < +\infty. \tag{6.4}$$

Finally, let us introduce the spaces

$$\widetilde{B}_q^{\mu,p}(\mathbb{R}^d) = \bigcap_{0<\varepsilon<s_1/2} B_q^{\mu^{(-\varepsilon)},p}(\mathbb{R}^d). \tag{6.5}$$

The reader can check that $B_q^{\mu,p}(\mathbb{R}^d)$, when endowed with the topology induced by the norm $\|f\|_{B_q^{\mu p}} = \|(\beta(k))_{k\in\mathbb{Z}^d}\|_p + |f|_{B_q^{\mu,p}}$, forms a Banach space (recall (4.6) for the definition of $\beta(k)$).

The intuition behind Definition 6.2 consists in introducing some space-dependent constraints that will create heterogeneity at all scales. Indeed, when a function f belongs to $B_q^{\mu,p}(\mathbb{R}^d)$, its oscillations $\Delta_h^n f(x)$ must be very small in certain regions (around points x where $\mu(B(x,r)) \sim r^\alpha$ with α large), while in other regions (where $\mu(B(x,r)) \sim r^\alpha$ with α small) the control of the oscillations can be relaxed.

In [9], a wavelet characterization of $B^{\mu,p}_q(\mathbb{R}^d)$ and $\widetilde{B}^{\mu,p}_q(\mathbb{R}^d)$ is proved when μ is an almost-doubling set function satisfying condition (P) (recall Eq. (4.3) and Remark 4.5). Observe indeed that Definition 4.6 of the condition (P) for measures can easily be extended for set functions $\mu \in C(\mathbb{R}^d)$.

For this, let us introduce a second semi-norm for $f \in L^p(\mathbb{R}^d)$: we set

$$|f|_{p,q,\mu} = \|(A_j)_{l\geq 1}\|_{\ell^q(\mathbb{N})}, \text{ where } A_j = \left(\sum_{\lambda\in\Lambda_j} \left|\frac{d_\lambda}{\mu(\lambda)}\right|^p\right)^{1/p}.$$

The following inequalities are proved in [9].

Theorem 6.3 *Let $\mu \in C(\mathbb{R}^d)$ be an almost doubling set function satisfying condition (P), and let Φ be a scaling function associated with wavelets Ψ (see Sect. 4.2).*

Let $p \geq 1$, and $q \in [1, +\infty]$.

Assume that the wavelets Ψ are compactly supported, belong to the standard Besov space $B^{s,p}_q(\mathbb{R}^d)$ for some $s > d/p+s_2$, and possess at least $\lfloor s_2\rfloor+1$ vanishing moments (s_1 and s_2 are the exponents associated with μ in (6.1)).

For every $0 < \varepsilon < s_1$, there exists a constant $C > 1$ (not depending on f) such that

$$\|f\|_{L^p} + |f|_{B^{\mu,p}_q} \leq C(\|f\|_{L^p} + |f|_{\mu^{(+\varepsilon)},p,q}) \tag{6.6}$$

$$\|f\|_{L^p} + |f|_{\mu,p,q} \leq C(\|f\|_{L^p} + |f|_{B^{\mu^{(+\varepsilon)},p}_q}). \tag{6.7}$$

Moreover, when μ is doubling, (6.6) and (6.7) hold for $\varepsilon = 0$, and the norms $\|f\|_{L^p} + |f|_{p,q,\mu}$ and $\|f\|_{L^p} + |f|_{B^{\mu,p}_q}$ are equivalent.

Last theorem supports the idea that $\widetilde{B}^{\mu,p}_q(\mathbb{R}^d)$ is the right space to work with, since it is characterized by wavelet coefficients, while the spaces $B^{\mu,p}_q(\mathbb{R}^d)$ are not (unless μ is doubling).

The main theorem in [9] is the following.

Theorem 6.4 *Let $\sigma \in \mathcal{S}_d$ be an admissible singularity spectrum (recall (4.10)). Call H_s the smallest value at which $\sigma(H) = d$.*

There exists an almost doubling set function $\mu \in C(\mathbb{R}^d)$ satisfying condition (P) and $p \in [1, +\infty]$ such that for every $q \in [1, +\infty]$, typical functions $f \in \widetilde{B}^{\mu,p}_q(\mathbb{R}^d)$ possess the following properties:

- $D_f = \sigma$
- *f satisfies the WLMF for every $H \leq H_s$.*
- *f satisfies the weak-WLMF for every $H > H_s$.*

In addition:

- *when $p = +\infty$, typical functions in $\widetilde{B}^{\mu,p}_q(\mathbb{R}^d)$ satisfy $D_f = D_\mu$.*
- *when μ is doubling, the same holds for $B^{\mu,p}_q(\mathbb{R}^d)$ instead of $\widetilde{B}^{\mu,p}_q(\mathbb{R}^d)$.*

Also, given $\sigma \in \mathcal{S}_d$, from the proof in [9] it can be checked that the couple (μ, p) in Theorem 6.4 is not unique.

Theorem 6.4 brings a solution to the Frisch–Parisi conjecture (Conjecture 4.2). The fact that the (strong) multifractal formalism holds only for the increasing part of the singularity spectrum (when $H \le H_s$) seems to be unavoidable. A heuristic explanation of the weak validity of the multifractal formalism in the decreasing part of the spectrum (and not the full validity) is that functions have usually sparse wavelet representations, generating very large values for negative values of q for $L_f(q, j)$ on some values of j.

Let us conclude this section by mentioning that a deeper study of the $B^{\mu,p}_q$ and $\widetilde{B}^{\mu,p}_q$ spaces is performed in [9], leading to results that have their own interest. More precisely, a uniform upper bound for the singularity spectrum of all functions in $B^{\mu,p}_q$ and $\widetilde{B}^{\mu,p}_q$ is found, as well as the singularity spectrum of typical functions in these spaces for large classes of almost-doubling measures μ. Without giving details on the results, it appears that the singularity spectra D_f of typical functions f may have very different shapes depending on the initial measure μ, and the proofs involve many arguments coming from geometric measure theory, ergodic theory and harmonic analysis.

7 Perspectives

First of all, we are far from being exclusive on generic dimensional results in analysis (see for instance [22, 24]), and many other regularity properties shall definitely be studied from the Baire genericity standpoint.

In this survey we focused on the notion of Baire genericity—the same issues can (and must) be addressed in the *prevalence* sense. Many results regarding prevalent multifractal properties have been obtained, see [4, 20, 21, 40, 41] amongst many references, and asking whether prevalent properties coincide with generic ones can sometimes bring some surprises (when they do not coincide).

Finally, one challenging research direction consists in establishing multifractal properties for (classes of) solutions to ordinary or partial differential equations, as well as for the stochastic counterparts. Indeed, multifractal ideas originate from the study of turbulence and other physical phenomena that are ruled by ODEs, SDEs or (S)PDEs, and it would be a fair return to demonstrate the multifractality of (some of) those functions that are solutions to such equations. A few examples already exist (i.e., Burgers equation with a Brownian motion as initial condition [12] and large classes of stochastic jump diffusions [10, 42]), but they are only a first step toward a systematic multifractal analysis of solutions to (partial) differential equations, which will certainly require the development of new techniques and approaches.

Acknowledgments The author wishes to thank the organizers of the fantastic conference "Fractal Geometry and Stochastics 6", and the referee for all her/his relevant and useful comments on this text. The research was partly supported by the grant ANR MULTIFRACS.

References

1. Abry, P., Wendt, H., Jaffard, S.: When Van Gogh meets Mandelbrot: multifractal classification of painting's texture. Signal Process. **93**(3), 554–572 (2013)
2. Abry, P., Jaffard, S., Wendt, H.: Irregularities and scaling in signal and image processing: Multifractal analysis. In: Ed, M.F. (ed.) Benoit Mandelbrot: A Life in Many Dimensions, pp. 31–116. World Scientific, New York (2015)
3. Aubry, J.M., Bastin, F., Dispa, S., Jaffard, S.: The spaces s_ν: new spaces defined with wavelet coefficients and related to multifractal analysis. Int. J. Appl. Math. Statist. **7**, 82–95 (2007)
4. Aubry, J.M., Maman, D., Seuret, S.: Local behavior of traces of besov functions: prevalent results. J. Func. Anal. **264**(3), 631–660 (2013)
5. Ayache, A., Jaffard, S., Taqqu, M.S.: Wavelet construction of generalized multifractional processes. Rev. Mat. Iberoamericana **23**(1), 327–370 (2007)
6. Banach, S.: Über die Baire'sche kategorie gewisser funktionenmengen. Studia Math. **3**, 174–179 (1931)
7. Barral, J.: Inverse problems in multifractal analysis of measures. Ann. Ec. Norm. Sup **48**(6), 1457–1510 (2015)
8. Barral, J., Seuret, S.: From multifractal measures to multifractal wavelet series. J. Fourier Anal. Appl. **11**(5), 589–614 (2005)
9. Barral, J., Seuret., S.: Besov spaces in multifractal environment and the Frisch-Parisi conjecture (2020). Preprint. https://hal-cnrs.archives-ouvertes.fr/hal-02899957
10. Barral, J., Fournier, N., Jaffard, S., Seuret, S.: A pure jump Markov process with a random singularity spectrum. Ann. Probab. **38**(5), 1924–1946 (2010). http://dx.doi.org/10.1214/10-AOP533
11. Bayart, F.: Multifractal spectra of typical and prevalent measures. Nonlinearity **26**, 353–367 (2013)
12. Bertoin, J., Jaffard, S.: Solutions multifractales de l'équation de burgers. Matapli **52**, 19–28 (1997)
13. Brown, G., Michon, G., Peyrière, J.: On the multifractal analysis of measures. J. Stat. Phys. **66**, 775–790 (1992)
14. Buczolich, Z., Nagy, J.: Hölder spectrum of typical monotone continuous functions. Real Anal. Exchange **26**(1), 133–156 (1999)
15. Buczolich, Z., Seuret, S.: Typical borel measures on $[0, 1]^d$ satisfy a multifractal formalism. Nonlinearity **23**(11), 7–13 (2010)
16. Buczolich, Z., Seuret, S.: Hölder spectrum of functions monotone in several variables. J. Math. Anal. Appl. **1**, 110–126 (2011)
17. Buczolich, Z., Seuret, S.: Measures and functions with prescribed singularity spectrum. J. Fractal Geometry **1**(3), 295–333 (2014)
18. Buczolich, Z., Seuret, S.: Multifractal properties of typical convex functions. Monatshefte fur Mathematik **187**(1), 59–78 (2019)
19. Daubechies, I.: Ten Lectures on Wavelets. SIAM, Philadelphia (1992)
20. Fraysse, A.: Regularity criteria of almost every function in a sobolev space. J. Func. Anal. **258**(6), 1806–1821 (2010)
21. Fraysse, A., Jaffard, S.: How smooth is almost every function in a sobolev space? Rev. Mat. Iberoamericana **22**, 663–682 (2006)
22. Fraysse, A., Jaffard, S., Kahane, J.P.: Quelques propriétés génériques en analyse. Notes aux CRAS Série I Math. **340**, 645–651 (2005)

23. Frisch, U., Parisi, D.: Fully developed turbulence and intermittency in turbulence, and predictability in geophysical fluid dynamics and climate dynamics. In: Ghil, M. (ed.) International School of Physics "Enrico Fermi", course 88, pp. 84–88. North Holland, New York (1985)
24. Gruslys, V., Jonusas, V., Mijović, V., Ng, O., Olsen, L., Petrykiewicz, I.: Dimensions of prevalent continuous functions. Monatshefte für Mathematik **166**(2), 153–180 (2012)
25. Halsey, T., Jensen, M., Kadanoff, L., Procaccia, I., Shraiman, B.: Fractal measures and their singularities: the characterisation of strange sets. Phys. Rev. A **33**, 1141–1151 (1986)
26. Hentschel, H., Procaccia, I.: The infinite number of generalized dimensions of fractals and strange attractors. Physica D **8**(3), 435–444 (1983)
27. Jaffard, S.: Exposants de Hölder en des points donnés et coefficients d'ondelettes. C. R. Acad. Sci. Paris Série I **308**, 79–81 (1989)
28. Jaffard, S.: Construction de fonctions multifractales ayant un spectre de singularités prescrit. C.R.A.S. **315**(1), 19–24 (1992)
29. Jaffard, S.: Functions with prescribed Hölder exponent. Appl. Comput. Harmon. Anal. **2**, 400–401 (1995)
30. Jaffard, S.: On the Frisch-Parisi conjecture. J. Math. Pures Appl. **79**(6), 525–552 (2000)
31. Jaffard, S.: Wavelet techniques in multifractal analysis. In: Fractal Geometry and Applications: A Jubilee of Benoit Mandelbrot, Proceedings of the Symposia in Pure Mathematic. AMS, Providence (2004)
32. Lashermes, B., Roux, S., Jaffard, S., Abry, P.: Comprehensive multifractal analysis of turbulent velocity using wavelet leaders. Eur. Phys. J. B. **61**(2), 201–215 (2008)
33. Leonarduzzi, R., Abry, P., Jaffard, S., Wendt, H., Gournay, L., Kyriacopoulou, T., Martineau, C., Martinez, C.: p-leader multifractal analysis for text type identification. In: IEEE International Conference on Acoustics, Speech and Signal Processing (ICASSP). New Orleans, USA (2017)
34. Maman, D., Seuret, S.: Fixed points for the multifractal spectrum map. Constr. Approx. **43**(3), 337–356 (2016)
35. Mandelbrot, B.B.: Multiplications aléatoires itérées et distributions invariantes par moyennes pondérées. C. R. Acad. Sci. Paris **278**, 289–292 and 355–358 (1974)
36. Mazurkiewicz, S.: Sur les fonctions non dérivables. Studia Math. **3**, 92–94 (1931)
37. Meyer, Y.: Ondelettes et opérateurs I. In: Hermann (1990)
38. Muzy, J.F., Bacry, E., Arnéodo, A.: Multifractal formalism for fractal signals: the structure-function approach versus the wavelet-transform modulus-maxima method. Phys. Rev. E **47**, 875–884 (1993)
39. Olsen, L.: A multifractal formalism. Adv. Math. **116**, 92–195 (1995)
40. Olsen, L.: Fractal and multifractal dimensions of prevalent measures. Indiana Univ. Math. J. **59**(2), 661–690 (2010)
41. Olsen, L.: Prevalent L^q-dimensions of measures. Math. Proc. Cambridge Philos. Soc **149**(3), 553–571 (2010)
42. Yang, X.: Multifractality of jump diffusion processes. Ann. Inst. Henri Poincaré Probab. Stat. **54**(4), 2042–2074 (2018). http://arxiv.org/abs/1502.03938

Renewal Theorems and Their Application in Fractal Geometry

Sabrina Kombrink

Abstract A selection of probabilistic renewal theorems and renewal theorems in symbolic dynamics are presented. The selected renewal theorems have been widely applied. Here, we will show how they can be utilised to solve problems in fractal geometry with particular focus on counting problems and the question of Minkowski measurability. The fractal sets we consider include self-similar and self-conformal sets as well as limit sets of graph-directed systems consisting of similarities and conformal mappings.

Keywords Renewal theorem · Dependent interarrival times · Symbolic dynamics · Minkowski content · Counting problems in fractal geometry · Ruelle Perron-Frobenius theory

Mathematics Subject Classifications (2010) Primary: 60K05, 60K15; Secondary: 28A80, 28A75

1 Introduction

Renewal theorems have found wide applicability in various areas of mathematics (such as fractal and hyperbolic geometry), economics (such as queuing, insurance and ruin problems) and biology (such as population dynamics). Classically, they describe the asymptotic behaviour of waiting times in-between occurrences of a repetitive pattern connected with repeated trials of a stochastic experiment. These probabilistic renewal theorems have been extended and generalised in several ways, resulting in an even broader applicability.

The purpose of this article is to provide an overview of a selection of renewal theorems and to highlight in which situation which renewal theorem is natural to

S. Kombrink (✉)
University of Birmingham, School of Mathematics, Birmingham, UK
e-mail: s.kombrink@bham.ac.uk

U. Freiberg et al. (eds.), *Fractal Geometry and Stochastics VI*,
Progress in Probability 76, https://doi.org/10.1007/978-3-030-59649-1_4

be applied. This will be done by considering two questions in fractal geometry in different settings. These motivating questions will be stated in Sect. 2. Subsequently, a selection of probabilistic renewal theorems is introduced in Sect. 3 and applied to obtain answers to the previously raised questions in the setting of similarities. In Sect. 4 renewal theorems in symbolic dynamics are presented and applied to solve the questions raised in Sect. 2 in more general settings. Additionally, in an Appendix we provide background material and address the relationships between the mentioned renewal theorems.

2 Some Questions in Fractal Geometry

In fractal geometry various notions of dimension such as Minkowski-, Hausdorff- and packing dimension are well-established tools to describe the fractal nature of a given set. Characterising sets beyond their dimension is one of the many applications of renewal theorems. In Sect. 2.1 we raise two questions which we answer by means of renewal theory in Sects. 3 and 4 for the classes of sets that we introduce in Sect. 2.2.

2.1 Characterising Sets Beyond Dimension

Our first question relates to counting problems. The most basic counting problems associated with fractal sets E arise in the situation when E is a subset of $[0, 1]$. Letting $\{I_\ell\}_{\ell\in L}$ denote the family of connected components of $[0, 1] \setminus E$ a natural question is:

Question 2.1 What is the asymptotic behaviour as $r \to 0$ of the number of intervals I_ℓ whose lengths lie in the interval $[r, rh)$ for some $h > 1$, i. e. of

$$N_{\log h}(r) := \#\{\ell \in L \mid rh > |I_\ell| \geq r\}?$$

Here # denotes cardinality and $|I_\ell|$ denotes the length of the interval I_ℓ.

An example of a more advanced counting problem is to count the number of closed geodesics on manifolds related to Schottky groups that do not exceed a given length. This problem can also be treated by means of renewal theory. We refer the interested reader to [28].

Before addressing how the answer to Question 2.1 helps our understanding of the fine geometric structure of a set in Remark 2.3 we turn to the second question, which relates to the asymptotic behaviour of the volume function.

For arbitrary $d \in \mathbb{N}$ the d-dimensional Lebesgue measure shall be denoted by λ_d. Further, for $r > 0$ we let $E_r := \{x \in \mathbb{R}^d \mid \mathrm{dist}(x, E) \leq r\}$ denote the *r-parallel set* of $E \subset \mathbb{R}^d$, where $\operatorname{dist}(x, E) := \inf_{y \in E}|x - y|$ denotes the distance of x to E with respect to the euclidean metric $|\cdot|$ on $\mathbb{R}^d$.

Question 2.2 What is the asymptotic behaviour as $r \to 0$ of the volume of the r-parallel set of E, i.e. of

$$\lambda_d(E_r) \quad \text{as } r \to 0?$$

Supposing that the *Minkowski dimension* $\dim_M(E) := d - \lim_{r \to 0} \frac{\log \lambda_d(E_r)}{\log r}$ of E exists, the above question can be reformulated as follows. How does the function $f\colon (0, \infty) \to \mathbb{R}$, $f(r) := r^{\dim_M(E)-d} \lambda_d(E_r)$ behave as $r \to 0$? If $\lim_{r \to 0} f(r)$ exists, we call the limit the *Minkowski content* of E and denote it by $\mathcal{M}(E)$. If $\lim_{r \to 0} f(r)$ exists, is positive and finite, then we say that E is *Minkowski measurable*. In recent years the question of Minkowski measurability of a given set has attracted much attention and is for instance related to the question 'Can you hear the shape of a drum with fractal boundary?', see for instance [29].

Remark 2.3 Knowledge of the asymptotic behaviour of $N_{\log h}(r)$ and $\lambda_d(E_r)$ as $r \to 0$ provides insight to the fine structure of E and can for instance be used to describe the lacunarity of E. The word lacunarity originates from the Latin word lacuna which means gap. According to [33] 'a fractal is to be called *lacunar* if its gaps tend to be large, in the sense that they include large intervals (discs or balls)'. A nice exposition of lacunarity, its geometric meaning and its relationship to the above introduced counting function and asymptotic behaviour of $\lambda_d(E_r)$ is provided in [33], see also [24]. We will provide further insight to the geometric meaning of the Minkowski content in Remark 3.6.

2.2 *Classes of Fractal Sets*

We address the above questions for the following classes of fractal sets.

2.2.1 Self-Similar and Self-Conformal Sets

Let $\Phi := \{\phi_1, \ldots, \phi_M\}$ denote an IFS of $M \geq 2$ contracting maps $\phi_i\colon X \to X$ acting on a compact subset X of $\mathbb{R}^d$. The famous Hutchinson-Hata Theorem states that there exists a unique, non-empty and compact subset $J \subset X$, which is invariant under Φ, that is $J = \bigcup_{i=1}^{M} \phi_i(J)$. If all the maps ϕ_i are similarities, i.e. there exist $r_i \in (0, 1)$ such that $|\phi_i(x) - \phi_i(y)| = r_i|x - y|$ for any $x, y \in X$, then the invariant set J is called *self-similar*. If all the maps ϕ_i extend to *conformal maps* on an open neighbourhood U of X, i.e. $\phi_i\colon U \to U$ is a C^1-diffeomorphism whose total derivative at every point is a similarity, then the invariant set J is called *self-conformal*. For background we refer the reader to [15].

Below, self-similar and self-conformal sets appear as special cases if $A_{i,j} = 1$ for all $i, j \in \Sigma$ and all ϕ_i are similarities resp. extend to conformal maps.

2.2.2 Limit Sets of Graph-Directed Systems

Here, we restrict to a special class of graph-directed systems, namely those which arise from iterated function systems (IFS) by forbidding certain transitions. However, the results presented below are not limited to this special class but also hold for general graph-directed systems as defined in [36]. We will provide references at the appropriate places.

Let $\Phi := \{\phi_1, \ldots, \phi_M\}$ denote an IFS of finitely many contracting maps $\phi_i : X \to X$ acting on a compact subset X of $\mathbb{R}^d$. Further, let A be an *irreducible* $M \times M$ matrix of zeros and ones, i. e. for each pair $i, j \in \Sigma := \{1, \ldots, M\}$ there exists $n \in \mathbb{N}$ such that $(A^n)_{i,j} > 0$. We allow to concatenate $\phi_i \circ \phi_j$ if and only if $A_{i,j} = 1$. Let $\Sigma_A^n := \{(\omega_1, \ldots, \omega_n) \in \Sigma^n \mid A_{\omega_i, \omega_{i+1}} = 1 \; \forall i \in \{1, \ldots, n-1\}\}$. The *limit set* of this type of *graph-directed system* is defined to be

$$J := \bigcap_{n \in \mathbb{N}} \bigcup_{\omega \in \Sigma_A^n} \phi_\omega(X),$$

where $\phi_\omega := \phi_{\omega_1} \circ \cdots \circ \phi_{\omega_n}$ for $\omega = (\omega_1, \ldots, \omega_n)$. We in particular study the cases in which all the maps ϕ_i are *similarities*, and in which all the maps ϕ_i extend to *conformal maps* on an open neighbourhood U of X.

3 Probabilistic Renewal Theorems and Their Applications to Questions 2.1 and 2.2 for Self-Similar Sets and Limit Sets of Graph-Directed Systems of Similarities

Probabilistic renewal theory is concerned with waiting times in-between occurrences of a repetitive pattern connected with repeated trials of a stochastic experiment. In classical renewal theory, it is assumed that after each occurrence of the pattern, the trials start from scratch. This means that the trials which follow an occurrence of the pattern form a replica of the whole stochastic experiment. In other words, the *waiting times* in-between successive occurrences of the pattern, also called *inter-arrival times*, are assumed to be mutually independent random variables with the same distribution (see [16, Ch. XIII] and [17]). The classical renewal theorems have been extended in various ways and to various different settings. One such extension, which we focus on here is given by Markov renewal theory, where the independence condition is weakened. The literature on classical and Markov renewal theory is vast. Therefore, we abstain from presenting a complete list of references but instead refer the reader to the following monographs and fundamental articles, where further references can be found: [1, 2, 10, 16, 17, 34].

The aim of this section is to present the afore-mentioned renewal theorems and to demonstrate to which question in which setting the respective renewal theorems are natural to apply. We will present a solution to a selection of the problems, focus on the ideas and provide references for the details. More precisely, we study the fundamental setting of renewal theory in Sects. 3.1 and 3.3 and show how its results can be utilised to answer Questions 2.1 and 2.2 for self-similar sets in Sects. 3.2 and 3.4. Subsequently, in Sect. 3.5 we turn to Markov renewal theory and apply Markov renewal theorems to answer Questions 2.1 and 2.2 for limit sets of graph-directed systems of similarities in Sect. 3.6.

3.1 Expected Number of Renewals—Blackwell's Renewal Theorem

In the afore-mentioned setting it is of interest how many occurrences of the pattern (renewals) are expected in a given time interval, if the process has been going on for a long time.

Let $W, W_1, W_2, \ldots$ denote independent identically distributed (i. i. d.) non-negative random variables on a common probability space $(\Omega, \mathcal{A}, \mathbb{P})$. We interpret W_i as the waiting time between the $(i-1)$-st and the i-th occurrence of the pattern and set $W_0 := 0$. For $n \in \mathbb{N}_0 := \mathbb{N} \cup \{0\}$ define $S_n := \sum_{i=0}^{n} W_i$, which is the epoch of the $(n+1)$-*st occurrence* of a renewal, the origin counting as a renewal epoch. Further, introduce the *renewal function* $N\colon [0,\infty) \times (0,\infty) \to \mathbb{R} \cup \{\infty\}$ by

$$N(t,h) := \mathbb{E}\left(\sum_{n=0}^{\infty} \mathbb{1}_{\{t-h<S_n\leq t\}}\right) = \mathbb{E}\left(\sum_{n=0}^{\infty} \mathbb{1}_{[0,h)}(t-S_n)\right), \tag{3.1}$$

where $\mathbb{E}$ denotes expectation. Thus, $N(t,h)$ gives the expected number of renewals in the time interval $(t-h,t]$.

The asymptotic behaviour of $N(t,h)$ as $t \to \infty$ depends on whether the common distribution F of the W_i is lattice or non-lattice. Recall that a distribution function is called *lattice* if its set of discontinuities lies in a discrete subgroup of $\mathbb{R}$, i. e. in $a\mathbb{Z}$ for some $a > 0$. If a is maximal as such, we say that the distribution is *a-lattice*. If no such a exists, then it is called *non-lattice*.

Intuitively, in the non-lattice situation we would expect h renewals in a time interval of length $h\mathbb{E}(W)$ if the process has been going for a long while. Thus, in a time interval of length h intuition yields $h/\mathbb{E}(W)$ to be the expected number of renewals. In the a-lattice situation the same is plausible with h replaced by a.

This intuition was made rigorous in a series of publications, in which different situations were covered, see [5, 6, 12, 17, 23] and references therein, resulting in the following renewal theorem, which sometimes is referred to as *Blackwell's renewal theorem*.

We say that $f, g\colon \mathbb{R} \to (0, \infty)$ are *asymptotic* and write $f(t) \sim g(t)$ as $t \to \infty$ if $\lim_{t\to\infty} f(t)/g(t) = 1$.

Theorem 3.1 *Suppose the setting of the current subsection. In particular, assume that F is supported on* $[0, \infty)$. *Further, interpret* $\mathbb{E}(W)^{-1}$ *as 0 if* $\mathbb{E}(W) = \infty$.

(i) *If F is a-lattice then*

$$N(t, a) \sim a\mathbb{E}(W)^{-1} \quad \text{as } t \to \infty.$$

(ii) *If F is non-lattice then*

$$N(t, h) \sim h\mathbb{E}(W)^{-1} \quad \text{as } t \to \infty$$

for any $h > 0$.

3.2 *Question 2.1 for Self-Similar Sets—Application of Blackwell's Renewal Theorem*

We fix the following notation. $\Phi := \{\phi_1, \ldots, \phi_M\}$ shall denote an IFS of finitely many contracting similarities ϕ_i with similarity ratio r_i acting on $[0, 1]$ with invariant set $E \subset [0, 1]$. For ease of exposition, we assume that $\{0, 1\} \subset E$ and that $\phi_i([0, 1]) \cap \phi_j([0, 1]) = \varnothing$ for distinct i, j, however, note that the open set condition is sufficient. (If we assume the milder open set condition to be satisfied with a bounded feasible open set O, i. e. $\phi_i O \cap \phi_j O = \varnothing$ for $i \neq j$ and $\phi_i O \subseteq O$ for all i, then we would consider the connected components of $O \setminus \bigcup_{i=1}^{M} \phi_i O$ below, of which there might be infinitely many.)

Let $G_1, \ldots, G_q$ denote the connected components of $[0, 1] \setminus \bigcup_{i=1}^{M} \phi_i([0, 1])$. Then the connected components of $[0, 1] \setminus E$ are precisely the intervals $\phi_\omega(G_j)$. Recall, $\phi_\omega := \phi_{\omega_1} \circ \cdots \circ \phi_{\omega_n}$ and $r_\omega := r_{\omega_1} \cdots r_{\omega_n}$ for $\omega = (\omega_1, \ldots, \omega_n)$. Thus,

$$N_{\log h}(r) = \sum_{j=1}^{q} \sum_{n=0}^{\infty} \#\{\omega \in \Sigma^n \mid hr > r_\omega |G_j| \geq r\} = \sum_{j=1}^{q} M_{\log h}\left(\tfrac{r}{|G_j|}\right), \tag{3.2}$$

where $\Sigma := \{1, \ldots, M\}$, $M_{\log h}(r) := \sum_{n=0}^{\infty} \#\{\omega \in \Sigma^n \mid hr > r_\omega \geq r\}$ and $\Sigma^0 := \{\varnothing\}$, with $\varnothing$ denoting the empty word and $r_\varnothing := 1$. For applying Blackwell's renewal theorem, we introduce random variables W_i in the following. By the Moran-Hutchinson formula, $1 = \sum_{i=1}^{M} r_i^D$ where D is the Hausdorff dimension of E. Thus, $\mathbb{P}(W = -\log r_i) = r_i^D$ for $i \in \Sigma$ defines the distribution of a non-negative random variable W. With $W, W_1, W_2, \ldots$ being i.i.d. the distribution of $S_n := W_1 + \cdots + W_n$

is given by $\mathbb{P}(S_n = -\log t) = \sum_{\omega\in\Sigma^n : t=r_\omega} r_\omega^D$ for $t > 0$. With this notation

$$e^{-Dt} M_{\log h}(e^{-t}) = \mathbb{E}\left(\sum_{n=0}^{\infty} z(t - S_n)\right), \tag{3.3}$$

where $z\colon \mathbb{R} \to \mathbb{R}$, $z(t) := \mathbb{1}_{[0,\log h)}(t)e^{-Dt}$.

3.2.1 The Lattice Case

If $-\log r_1, \ldots, -\log r_M$ lie in the discrete subgroup $a\mathbb{Z}$ of $\mathbb{R}$ with $a > 0$ maximal as such, then W is a-lattice. As $-\log r_\omega \in a\mathbb{Z}$ for each ω, it follows that $t - a < -\log r_\omega \le t$ is equivalent to $-\log r_\omega/a = \lfloor t/a \rfloor := \max\{k \in \mathbb{Z} \mid k \le t/a\}$. Whence, Theorem 3.1 implies for $t \to \infty$

$$\begin{aligned} M_a(e^{-t})e^{-aD\lfloor t/a \rfloor} &= \sum_{n=0}^{\infty} \sum_{\omega\in\Sigma^n} r_\omega^D \mathbb{1}_{(t-a,t]}(-\log r_\omega) \\ &= \mathbb{E}\Big(\sum_{n=0}^{\infty} \mathbb{1}_{(t-a,t]}(S_n)\Big) \sim \frac{a}{\mathbb{E}(W)}, \end{aligned}$$

yielding

$$N_a(e^{-t}) \sim \frac{a}{-\sum_{i=1}^{M} r_i^D \log r_i} \sum_{j=1}^{q} e^{aD\lfloor (t+\log|G_j|)/a \rfloor} \quad \text{as} \quad t \to \infty.$$

3.2.2 The Non-lattice Case

If $-\log r_1, \ldots, -\log r_M$ do not generate a discrete subgroup of $\mathbb{R}$ then W is non-lattice. Let $h > 0$ be arbitrary. Theorem 3.1 implies for $t \to \infty$

$$M_{\log h}(e^{-t})e^{-Dt} \le \sum_{n=0}^{\infty} \sum_{\omega\in\Sigma^n} r_\omega^D \mathbb{1}_{(t-h,t]}(-\log r_\omega) = \mathbb{E}\Big(\sum_{n=0}^{\infty} \mathbb{1}_{(t-h,t]}(S_n)\Big) \sim \frac{h}{\mathbb{E}(W)},$$

$$M_{\log h}(e^{-t})e^{-Dt} > e^{-hD} \sum_{n=0}^{\infty} \sum_{\omega\in\Sigma^n} r_\omega^D \mathbb{1}_{(t-h,t]}(-\log r_\omega) \sim \frac{he^{-hD}}{\mathbb{E}(W)}.$$

We abstain from gaining the precise asymptotics here as these can be easily deduced from the key renewal theorem, see Remark 3.8.

3.3 The Key Renewal Theorem

The considerations of Sect. 3.1 are intimately related to the asymptotic behaviour of the solution $Z\colon \mathbb{R} \to \mathbb{R}$ of the *renewal equation*

$$Z(t) = z(t) + \int_{-\infty}^{\infty} Z(t-y)\,\mathrm{d}F(y) = (z + F \star Z)(t) \tag{3.4}$$

with given $z\colon \mathbb{R} \to \mathbb{R}$, where $\star$ denotes convolution and F is a distribution on $\mathbb{R}$.

For obtaining statements on the uniqueness and on the asymptotic behaviour of $Z(t)$ as $t \to \infty$ it is required that z be directly Riemann integrable.

Definition 3.2 For a function $f\colon \mathbb{R} \to \mathbb{R}$, $h > 0$ and $k \in \mathbb{Z}$ set

$$\underline{m}_k(f,h) := \inf\{f(t) \mid (k-1)h \le t < kh\} \qquad \text{and}$$
$$\overline{m}_k(f,h) := \sup\{f(t) \mid (k-1)h \le t < kh\}.$$

The function f is called *directly Riemann integrable (d. R. i.)* if for some sufficiently small $h > 0$

$$\underline{R}(f,h) := \sum_{k\in\mathbb{Z}} h \cdot \underline{m}_k(f,h) \qquad \text{and} \qquad \overline{R}(f,h) := \sum_{k\in\mathbb{Z}} h \cdot \overline{m}_k(f,h)$$

are finite and tend to the same limit, denoted by $\int f(T)\,\mathrm{d}T$, as $h \to 0$.

Direct Riemann integrability excludes wild oscillations of the function at infinity and is stronger than Riemann integrability. For further insights into this notion we refer the reader to [17, Ch. XI] and [2, Ch. B.V].

As before, $W, W_1, W_2, \ldots$ shall denote i. i. d. random variables with common distribution F. Note that here the W_i are not necessary non-negative. Recall that $S_n := \sum_{i=0}^{n} W_i$ with $W_0 := 0$.

Lemma 3.3 ([1, Ch. 3.2]) *If z is d. R. i. then the function $Z\colon \mathbb{R} \to \mathbb{R}$ given by*

$$Z(t) := \mathbb{E}\left(\sum_{n=0}^{\infty} z(t - S_n)\right) \tag{3.5}$$

is the unique solution of the renewal equation (3.4) that satisfies $\lim_{t\to-\infty} Z(t) = 0$.

Being a solution of the renewal equation (3.4), Z from (3.5) is called *renewal function*. Setting $z = \mathbb{1}_{[0,h)}$ and assuming that F is concentrated on $[0,\infty)$ we recover the renewal function $N(\cdot, h)$ of Sect. 3.1, see Eq. (3.1). Thus, it is apparent that the present setting is much more general than that of Sect. 3.1.

Theorem 3.4 ([1, Satz 3.2.2]) *Denote by $z\colon \mathbb{R} \to \mathbb{R}$ a d. R. i. function. Let F be a distribution supported on $\mathbb{R}$ with positive mean and let Z be the unique* *solution* (3.5) *of the renewal equation* (3.4) *which satisfies* $\lim_{t\to-\infty} Z(t) = 0$. *Then the following hold.*

(i) If F is non-lattice, then as t $\to \infty$

$$Z(t) \sim \mathbb{E}(W)^{-1} \int_{-\infty}^{\infty} z(T)\,\mathrm{d}T.$$

(ii) If F is a-lattice, then as t $\to \infty$

$$Z(t) \sim a\mathbb{E}(W)^{-1} \sum_{\ell=-\infty}^{\infty} z(a\ell + t).$$

Notice, direct Riemann integrability of z ensures convergence of the series $\sum_{\ell=-\infty}^{\infty} z(a\ell + t)$ in the above theorem, which can be seen as follows. If $m \in \mathbb{N}$ is minimal such that $\overline{R}(z, a/m) < \infty$ then $\overline{R}(z, a) \leq m\overline{R}(z, a/m) < \infty$. Thus, $m = 1$ and we are done.

Remark 3.5 A nice exposition of the key renewal theorem tailored to fractal geometry can be found in [14, Ch. 7], where it is applied to obtain results on the asymptotic behaviour of the covering number of a self-similar subset of $\mathbb{R}^d$, and to Questions 2.1 and 2.2 for self-similar subsets of $\mathbb{R}$.

3.4 *Questions 2.1 and 2.2 for Self-Similar Sets—Application of the Key Renewal Theorem*

In the setting of self-similar sets both Questions 2.1 and 2.2 can be solved by means of the key renewal theorem and the ideas below stem from [40]. We focus on the solution to Question 2.2 and briefly discuss Question 2.1 in Remark 3.8. We fix the following notation. $\Phi := \{\phi_1, \ldots, \phi_M\}$ shall denote an IFS of finitely many contracting similarities ϕ_i with similarity ratio r_i acting on $X \subset \mathbb{R}^d$ with invariant set E. We suppose that the open set condition (OSC) is satisfied and that O is a feasible open set for Φ, i. e. $\phi_i(O) \subset O$ and $\phi_i(O) \cap \phi_j(O) = \varnothing$ for $i \neq j$. Assume w. l. o. g. that O is bounded.

Often, depending on the shape of O, the expression $\lambda_d(E_r \setminus O)$ is very easy to determine. For the Sierpiński carpet E for instance, (i.e. for the invariant set E of the IFS $\{x \mapsto x/3 + (i/3, j/3)\}_{i,j\in\{0,1,2\}^2\setminus\{(1,1)\}}$ acting on $X = [0, 1]^2$) one can choose $O = (0, 1)^2$, giving $\lambda_d(E_r \setminus O) = 4r + \pi r^2$. Moreover, it is known that $\lambda_d(E_r \setminus O) = \mathfrak{o}(r^{d-\dim_M(E)})$ as $r \to 0$ for general self-similar sets E under the OSC with the little Landau symbol $\mathfrak{o}$, see [40]. (For functions $f, g\colon \mathbb{R} \to \mathbb{R}$ we write $f = \mathfrak{o}(g)$ as $t \to \infty$ if $\lim_{t\to\infty} f(t)/g(t) = 0$.) Therefore, we consider $\lambda_d(E_r \cap O)$ and let $\Gamma := O \setminus \Phi(O)$, where the action of Φ on a subset U of X is defined via $\Phi(U) := \bigcup_{i=1}^{M} \phi_i(U)$. Then O can be decomposed as

$$O = \dot{\bigcup_{n=0}^{\infty}} \bigcup_{\omega \in \Sigma^n} \phi_\omega \Gamma \cup \bigcap_{n=0}^{\infty} \Phi^n O,$$

where the unions are disjoint. We have $\Phi\left(\overline{\bigcap_{n=0}^{\infty} \Phi^n O}\right) = \overline{\bigcap_{n=0}^{\infty} \Phi^n O}$. Thus, $\overline{\bigcap_{n=0}^{\infty} \Phi^n O}$ is either empty or coincides with E by uniqueness of the self-similar set. Therefore, $\lambda_d\left(\bigcap_{n=0}^{\infty} \Phi^n O\right) \le \lambda_d(E)$. Let D denote the Minkowski dimension of E. If $D < d$ then $\lambda_d(E) = 0$ and whence $\lambda_d\left(\bigcap_{n=0}^{\infty} \Phi^n O\right) = 0$. Suppose that O is chosen in such a way that $E_r \cap \phi_\omega \Gamma = (\phi_\omega E)_r \cap \phi_\omega \Gamma$ for each ω. This condition is known as the *locality property* and it is shown in [40] that to each IFS of similarities satisfying the OSC there is a feasible open set O which satisfies the locality property, namely the central open set as introduced in [3]. Thus,

$$\begin{aligned}
\lambda_d(E_{\mathrm{e}^{-t}} \cap O) &= \sum_{n=0}^{\infty} \sum_{\omega \in \Sigma^n} \lambda_d(E_{\mathrm{e}^{-t}} \cap \phi_\omega \Gamma) \\
&= \sum_{n=0}^{\infty} \sum_{\omega \in \Sigma^n} \lambda_d((\phi_\omega E)_{\mathrm{e}^{-t}} \cap \phi_\omega \Gamma) \\
&= \sum_{n=0}^{\infty} \sum_{\omega \in \Sigma^n} r_\omega^d \lambda_d(E_{\mathrm{e}^{-t-\log r_\omega}} \cap \Gamma).
\end{aligned}$$

Let $W, W_1, W_2, \ldots$ denote i. i. d. random variables with common distribution given by $\mathbb{P}(W = -\log r_i) = r_i^D$ as in Sect. 3.1. In [40] it is shown that $t \mapsto z(t) := \mathrm{e}^{-t(D-d)} \lambda_d(E_{\mathrm{e}^{-t}} \cap \Gamma)$ is d. R. i., which allows to apply the key renewal theorem to

$$Z(t) := \mathrm{e}^{-t(D-d)} \lambda_d(E_{\mathrm{e}^{-t}} \cap O) = \sum_{n=0}^{\infty} \sum_{\omega \in \Sigma^n} r_\omega^D z(t + \log r_\omega) = \mathbb{E}\Big(\sum_{n=0}^{\infty} z(t - S_n)\Big).$$

3.4.1 The Lattice Case

If $-\log r_1, \ldots, -\log r_M$ lie in the discrete subgroup $a\mathbb{Z}$ of $\mathbb{R}$ with $a > 0$ maximal as such, then W is a-lattice. Thus, Theorem 3.4 yields for $t \to \infty$

$$\begin{aligned}
Z(t) &\sim a \mathbb{E}(W)^{-1} \sum_{\ell=-\infty}^{\infty} z(a\ell + t) \\
&= \frac{-a}{\sum_{i=1}^{M} r_i^D \log r_i} \sum_{\ell=-\infty}^{\infty} \mathrm{e}^{-(a\ell+t)(D-d)} \lambda_d(E_{\mathrm{e}^{-a\ell-t}} \cap \Gamma) =: g(t).
\end{aligned} \tag{3.6}$$

Note that $g(t)$ is periodic in t with period a. In general it is not known whether g is strictly periodic (implying that E is not Minkowski-measurable) or constant (implying Minkowski-measurability of E). For self-similar subsets of $\mathbb{R}$ arising from lattice IFS E being not Minkowski measurable has been shown in [26], building on [20, 27, 30]. In the higher dimensional setting the analogue statement has been verified under further assumptions in various works, see [27] and references therein.

3.4.2 The Non-lattice Case

If $-\log r_1, \ldots, -\log r_M$ do not generate a discrete subgroup of $\mathbb{R}$ then W is non-lattice and Theorem 3.4 gives for $t \to \infty$

$$\begin{aligned} Z(t) &\sim \mathbb{E}(W)^{-1} \int_{-\infty}^{\infty} z(T)\, \mathrm{d}T \\ &= \frac{-1}{\sum_{i=1}^{M} r_i^D \log r_i} \int_{-\infty}^{\infty} \mathrm{e}^{-T(D-d)} \lambda_d(E_{\mathrm{e}^{-T}} \cap \Gamma)\, \mathrm{d}T. \end{aligned} \tag{3.7}$$

Thus, (3.7) implies that E is Minkowski measurable in the non-lattice setting. Furthermore, the Minkowski content of E is given by the right hand side of (3.7).

Remark 3.6 Just like there is a variety of sets of the same topological dimension, e.g. 3-dimensional balls and cubes, there are various distinct fractal sets of the same Minkowski dimension. The formula in (3.7) shows that we can use the Minkowski content to distinguish between such sets. The value that the Minkowski content takes highly depends on the geometric structure of Γ. Equation (3.7) shows that if Γ includes large intervals (discs or balls), i.e. is highly lacunar, then $\mathcal{M}(E)$ will be relatively small, compared to the case when Γ is made up of several connected components of smaller size. We refer the interested reader to [24, 33] for further details.

Remark 3.7 In the setting of self-similar sets, Question 2.2 has been studied by various authors. References include [11, 13, 18, 29, 31, 32, 40] and several related articles by the same authors. Related to the Minkowski measurability question is the question of existence of fractal curvature measures, see e.g. [8, 37, 41].

Remark 3.8 Combining the methods presented above with those of Sect. 3.2 leads to an answer of Question 2.1 in the setting of Sect. 3.2: combining (3.2) with (3.3) gives

$$\mathrm{e}^{-Dt} N_{\log h}(\mathrm{e}^{-t}) = \sum_{j=1}^{q} |G_j|^D \mathbb{E}\left(\sum_{n=0}^{\infty} z(t + \log|G_j| - S_n) \right)$$

with $z\colon \mathbb{R} \to \mathbb{R}$, $z(t) := \mathbb{1}_{[0,\log h)}(t)\mathrm{e}^{-Dt}$. In the a-lattice situation an application of the key renewal theorem leads to

$$\mathrm{e}^{-Dt} N_a(\mathrm{e}^{-t}) \sim \frac{a}{-\sum_{i=1}^{M} r_i^D \log r_i} \sum_{j=1}^{q} |G_j|^D \mathrm{e}^{aD\{(t+\log|G_j|)/a\}},$$

where $\{x\} := x - \lfloor x \rfloor \in [0, 1)$ for $x \in \mathbb{R}$. In the non-lattice situation an application of the key renewal theorem yields

$$\mathrm{e}^{-Dt} N_{\log h}(\mathrm{e}^{-t}) \sim \frac{1 - h^{-D}}{D \sum_{i=1}^{M} r_i^D \log r_i} \sum_{j=1}^{q} |G_j|^D.$$

3.5 *Markov Renewal Theory*

In Markov renewal theory one is concerned with the asymptotic behaviour of solutions of the Markov renewal equation, which is a system of coupled renewal equations that we will introduce momentarily. Before, let us allude to the stochastic motivation.

By a *Markov random walk*, we understand a point process for which the inter-arrival times $W_0, W_1, \ldots$ are not necessarily i. i. d. (as in the preceding subsections), but *Markov dependent* on a Markov chain $(X_n)_{n\in\mathbb{N}_0}$ with finite or countable state space Σ. This means that W_n is sampled according to the current and proximate values X_n, X_{n+1} but is independent of the past values $X_{n-1}, \ldots, X_0$ of the underlying Markov chain. Thus, $(X_{n+1}, W_n)_{n\in\mathbb{N}_0}$ has an interpretation as a stochastic process with state space $\Sigma \times \mathbb{R}$ and transition kernel $U\colon \Sigma \times (\mathcal{P}(\Sigma) \otimes \mathfrak{B}(\mathbb{R})) \to \mathbb{R}$ given by

$$U(i, \{j\} \times (-\infty, t]) := \mathbb{P}(X_{n+1} = j, W_n \le t \mid X_n = i) =: F_{i,j}(t). \tag{3.8}$$

Here $\mathcal{P}(\Sigma)$ denotes the power set of Σ and $\mathfrak{B}(\mathbb{R})$ denotes the Borel σ-algebra on $\mathbb{R}$ and $F_{i,j}$ defines a distribution function of a finite measure with total mass $\|F_{i,j}\|_\infty := \mathbb{P}(X_1 = j \mid X_0 = i)$ for given $i, j \in \Sigma$.

The system of equations

$$N(t, i) = f_i(t) + \sum_{j\in\Sigma} \int_{-\infty}^{\infty} N(t - u, j) F_{i,j}(\mathrm{d}u), \tag{3.9}$$

for varying $i \in \Sigma$ and given $f_i : \mathbb{R} \to \mathbb{R}$ is called a *Markov renewal equation*, *multivariate renewal equation* or *system of coupled renewal equations*. This system of equations is a direct analogue of (3.4) to the current setting, taking the Markov dependence into account.

The Laplace transform of $F_{i,j}$ at $s \in \mathbb{R}$ is given by

$$B_{i,j}(s) := (\mathcal{L}F_{i,j})(s) := \int_{-\infty}^{\infty} \mathrm{e}^{-sT} \mathrm{d}F_{i,j}(T).$$

Setting $B(s) := (B_{ij}(s))_{i,j \in \Sigma}$, and assuming that Σ is of finite cardinality, the Perron-Frobenius theorem for matrices yields a unique s for which $B(s)$ has spectral radius one.

Theorem 3.9 (A Markov Renewal Theorem) *Let $M \geq 2$ be an integer and assume that $\Sigma = \{1, \dots, M\}$. For $i, j \in \Sigma$ let $F_{i,j}(t)$ be as in (3.8) and suppose that $F := (\|F_{i,j}\|_\infty)_{i,j \in \Sigma}$ is irreducible. Let $\delta > 0$ denote the unique positive real number for which the matrix $B(\delta)$ given by $B_{i,j}(\delta) := \int \mathrm{e}^{-\delta u} F_{i,j}(\mathrm{d}u)$ has spectral radius one. For $i \in \Sigma$ let $f_i : \mathbb{R} \to \mathbb{R}$ denote d. R. i. functions. Suppose that there exist $C, s > 0$ such that $\mathrm{e}^{-\delta t}|f_i(t)| \leq C\mathrm{e}^{st}$ for $t < 0$ and $i \in \Sigma$. Choose vectors ν, h with $\nu B(\delta) = \nu$, $B(\delta)h = h$ and $\nu_i, h_i > 0$ for $i \in \Sigma$. Let $N(t, i)$ for $i \in \Sigma$ solve the Markov renewal equation (3.9).*

(i) If $F_{i,j}$ is non-lattice for some $(i, j) \in \Sigma^2$, then

$$\mathrm{e}^{-\delta t} N(t, i) \sim \frac{h_i \sum_{j=1}^{M} \nu_j \int \mathrm{e}^{-\delta T} f_j(T) \mathrm{d}T}{\sum_{k,j=1}^{M} \nu_k h_j \int T \mathrm{e}^{-\delta T} F_{k,j}(\mathrm{d}T)} =: G(i).$$

(ii) We always have

$$\lim_{t \to \infty} t^{-1} \int_0^t \mathrm{e}^{-T\delta} N(T, i) \mathrm{d}T = G(i).$$

A statement for the lattice situation, i. e. when all $F_{i,j}$ are lattice, can be deduced from Theorem 4.2.

Remark 3.10 The above theorem is presented in a similar form in [2, VII. Thm. 4.6]. More general versions of Markov renewal theorems can be found in the literature (see e.g. [1]). The precise version of Theorem 3.9 is a direct consequence of the more general Renewal Theorem 4.2, which we present in the next section. In Appendix B.2 we allude to how Theorem 3.9 can be deduced from Theorem 4.2.

3.6 *Questions 2.1 and 2.2 for Limit Sets of Graph-Directed Systems of Similarities—Application of Markov Renewal Theory*

We demonstrate how to apply Markov renewal theory by considering the following example. Let $X := [0,1]$ and let $\phi_1, \phi_2, \phi_3 \colon X \to X$ be given by

$$\phi_1(x) = \frac{x}{4}, \quad \phi_2(x) = \frac{x}{6} + \frac{5}{12}, \quad \phi_3(x) = \frac{x}{4} + \frac{3}{4} \qquad \text{and} \qquad A = \begin{pmatrix} 1 & 1 & 1 \\ 1 & 0 & 1 \\ 1 & 1 & 1 \end{pmatrix}.$$

Further, let J denote the limit set of $\Phi := \{\phi_1, \phi_2, \phi_3\}$ associated with A. The first two steps in the construction of J are depicted in Fig. 1.

For a given $h > 1$ let $N_1(r)$, $N_2(r)$, $N_3(r)$ respectively denote the number of connected components of $[0, 1/4] \setminus J$, $[5/12, 7/12] \setminus J$, $[3/4, 1] \setminus J$ of lengths between r and rh. We have that

$$N_1(r) = N_3(r) = 2 \cdot \mathbb{1}_{(\frac{1}{24h}, \frac{1}{24}]}(r) + N_1(4r) + N_2(4r) + N_3(4r),$$
$$N_2(r) = 2 \cdot \mathbb{1}_{(\frac{1}{12h}, \frac{1}{12}]}(r) + N_1(6r) + N_3(6r).$$

Setting

$$N(t,i) := N_i(\mathrm{e}^{-t}),$$
$$f_1 := f_3 := 2 \cdot \mathbb{1}_{(\log(24), \log(24h)]}, \quad f_2 := 2 \cdot \mathbb{1}_{(\log(12), \log(12h)]}, \quad \text{and}$$
$$F_{i,j} = \begin{cases} \mathbb{1}_{[\log 4, \infty)} & : (i,j) \in \{1,3\} \times \{1,2,3\} \\ 0 & : (i,j) = (2,2) \\ \mathbb{1}_{[\log 6, \infty)} & : (i,j) \in \{2\} \times \{1,3\} \end{cases}$$

we see that $N(t,i) = f_i(t) + \sum_{j \in \Sigma} \int_{-\infty}^{\infty} N(t-u, j) F_{i,j}(\mathrm{d}u)$ for $i \in \Sigma$. Thus, the system of coupled renewal equations (3.9) is satisfied. As we are in the non-lattice situation and all hypotheses of Theorem 3.9 are clearly satisfied, Theorem 3.9

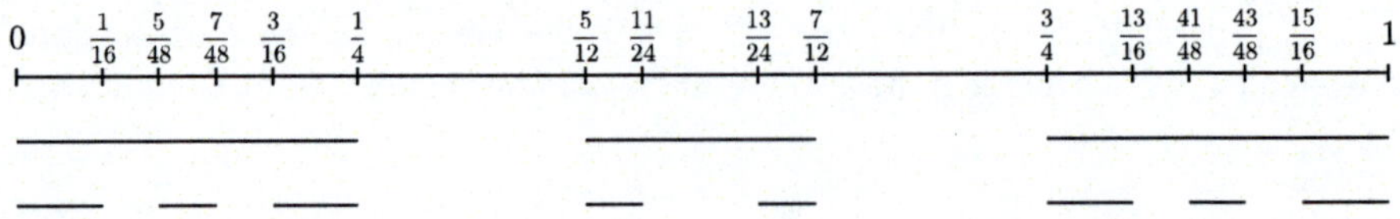

Fig. 1 First two steps in the construction of the limit set of the graph-directed system studied in Sect. 3.6

thus yields

$$e^{-\delta t}N(t,1) = e^{-\delta t}N(t,3) \sim \frac{(1-h^{-\delta})(1+2^{-\delta})}{2\delta[\log 4(1+6^{\delta}) + \log 6]} =: G \quad \text{and}$$

$$e^{-\delta t}N(t,2) \sim 6^{-\delta}G$$

as $t \to \infty$, where $\delta \approx 0.6853$. Now the asymptotics of the total number of complementary intervals of lengths between e^{-t} and he^{-t} can be obtained from the above through evaluating

$$N(t,1) + N(t,2) + N(t,3) + 2 \cdot \mathbb{1}_{[\log 6,\infty)}(t).$$

There is nothing particular about this example and the general setting, assuming $\phi_i(\text{int}(X)) \cap \phi_j(\text{int}(X)) = \varnothing$ for $i \neq j$, can be treated analogously. Here, $\text{int}(X)$ denotes the topological interior of X.

The author is not aware that this approach has been carried out in the literature. However, general results in the current setting were obtained in [20] for the more general class of limit sets of conformal graph-directed systems, by means of the renewal theorems that we turn to in the following section.

4 Renewal Theory in Symbolic Dynamics

The renewal theorem which is presented in the current section was developed in [25] and extended to the setting of infinite state space in [21]. Here, the focus lies on the situation of finite state space.

Now, the assumption of the previous section that $(X_n)_{n\in\mathbb{N}_0}$ is a Markov chain and that W_n is Markov dependent on $(X_n)_{n\in\mathbb{N}_0}$ is dropped. Instead, we consider a time-homogeneous (i.e. stationary increments) stochastic process $(X_n)_{n\in\mathbb{Z}}$ with finite state space $\Sigma = \{1,\ldots,M\}$ and time-set $\mathbb{Z}$ and extend to the setting that W_n may depend on the current values X_{n+1}, X_n as well as on the whole past $X_{n-1}, X_{n-2},\ldots$ of the stochastic process $(X_n)_{n\in\mathbb{Z}}$.

In this situation it is of interest to study the limiting behaviour as $t \to \infty$ of the renewal function $N\colon \mathbb{R} \times \Delta \to \mathbb{R}$ given by

$$N(t,x) := \mathbb{E}_x\left[\sum_{n=0}^{\infty} f_{X_n\cdots X_1 x}\left(t - \sum_{k=0}^{n-1} W_k\right)\right], \tag{4.1}$$

where $\{f_y\colon \mathbb{R} \to \mathbb{R} \mid y \in \Delta\}$ is a family of functions, $\mathbb{E}_x$ is the conditional expectation given $X_0X_{-1}\cdots = x$, for $n = 0$ we interpret $f_{X_n\cdots X_1 x}(t - \sum_{k=0}^{n-1} W_k)$ to be $f_x(t)$, and Δ is a subset of $\Sigma^{\mathbb{N}}$. For instance, if $f_y = \mathbb{1}_{[0,\infty)}$, then $N(t,x)$ gives the expected number of renewals in the time-interval $(0,t]$ given $X_0X_{-1}\cdots = x$.

In view of the questions in fractal geometry that we raised in Sect. 2 we impose some assumptions, which turn the renewal function from (4.1) into a deterministic one. For this, well-known terminology and theorems from symbolic dynamics are used. For convenience, these are introduced in Appendix A and referred to at the appropriate places.

4.1 Setting

The admissible transitions of the stochastic process $(X_n)_{n\in\mathbb{N}_0}$ are assumed to be governed by an irreducible $(M \times M)$- incidence matrix A of zeros and ones. Hence infinite paths of the process are encoded by elements of the code space $\Sigma_A := \{x \in \Sigma^{\mathbb{N}} \mid A_{x_k,x_{k+1}} = 1\, \forall k \in \mathbb{N}\}$, see Sect. A.1. Thus, we consider the renewal function $N\colon \mathbb{R}\times\Sigma_A \to \mathbb{R}$ from (4.1) acting on $\mathbb{R}\times\Sigma_A$.

A natural assumption in applications is that the recent history of $(X_n)_{n\in\mathbb{N}_0}$ has more influence on which state will be visited next than the earlier history. This is reflected in the assumption that the function $\eta\colon \Sigma_A \to \mathbb{R}$ given by

$$\eta(ix) := \log \mathbb{P}_x(X_1 = i)$$

belongs to the class $\mathcal{F}_\alpha(\Sigma_A)$ of real-valued α-Hölder continuous functions on Σ_A for some $\alpha \in (0, 1)$, see Sect. A.2. Here, $i \in \Sigma$ and $\mathbb{P}_x$ is the distribution corresponding to $\mathbb{E}_x$. Note that $\mathbb{P}_x(X_1 = i) := \mathbb{P}(X_1 = i \mid X_0X_{-1}\cdots = x) > 0$ if $ix \in \Sigma_A$ by the definition of Σ_A. Similarly, it is assumed that the dependence of W_n on $X_{n+1}, X_n, \ldots$ is described by a Hölder continuous function. That is we assume existence of $\xi \in \mathcal{F}_\alpha(\Sigma_A)$ with

$$W_n = \xi(X_{n+1}X_nX_{n-1}\cdots).$$

This notation allows us to evaluate the conditional expectation and express $N(t, x)$ in a deterministic way. Let σ denote the left-shift on Σ_A and S_n the n-th Birkhoff sum, see Sects. A.1 and A.3. Since $\sum_{k=0}^{n-1} W_k = S_n\xi(X_nX_{n-1}\cdots)$ and, for $x, y \in \Sigma_A$ with $\sigma^n y = x$, we have $\mathbb{P}(X_nX_{n-1}\cdots = y \mid X_0X_{-1}\cdots = x) = \exp(S_n\eta(y))$ it follows that

$$N(t, x) = \sum_{n=0}^{\infty} \sum_{y\in\Sigma_A:\sigma^n y=x} f_y(t - S_n\xi(y))\mathrm{e}^{S_n\eta(y)}. \tag{4.2}$$

From this, one can deduce the renewal-type equation

$$N(t, x) = \sum_{y\in\Sigma_A:\sigma y=x} N(t - \xi(y), y)\mathrm{e}^{\eta(y)} + f_x(t),$$

which justifies calling N a renewal function. Intuitively, inter-arrival times are non-negative and probabilities take values in $[0, 1]$. However, when considering the deterministic form (4.2), ξ is allowed to take negative values, provided there exists $n \in \mathbb{N}$ for which $S_n\xi$ is strictly positive. Note that this condition is equivalent to ξ being co-homologous (see Definition A.2) to a strictly positive function, see [25, Rem. 2.1]. Moreover, η is allowed to be chosen freely from the class $\mathcal{F}_\alpha(\Sigma_A)$.

4.2 The Renewal Theorem

For $y \in \Sigma_A$ and $t \in \mathbb{R}$ write

$$f_y(t) = \chi(y) \cdot g_y(t)$$

with non-negative but not identically zero $\chi \in \mathcal{F}_\alpha(\Sigma_A)$, where $g_y \colon \mathbb{R} \to \mathbb{R}$, for $y \in \Sigma_A$, need to satisfy a regularity condition, which is related to the direct Riemann integrability assumption of the classical key renewal theorem (see Sect. 3.3), and which we introduce next.

Definition 4.1 A family of functions $\{f_x \colon \mathbb{R} \to \mathbb{R} \mid x \in I\}$ with some index set I is called *equi directly Riemann integrable (equi d. R. i.)* if f_x is d. R. i. for all $x \in I$ (see Definition 3.2) and if

$$\sum_{k\in\mathbb{Z}} h \cdot \sup_{x\in I} \left(\underline{m}_k(f_x, h) - \overline{m}_k(f_x, h) \right)$$

tends to zero as $h \to 0$.

For the following, fix ξ and η as in Sect. 4.1 and let $C(\Sigma_A)$ denote the space of real-valued continuous functions on Σ_A, see Sect. A.2.

Theorem 4.2 (Renewal Theorem in Symbolic Dynamics, [25, Thm. 3.1] and [21, Thm. 3.1]) *Let A be irreducible, fix $x \in \Sigma_A$ and take $\alpha \in (0, 1)$. Further, let $\xi, \eta \in \mathcal{F}_\alpha(\Sigma_A)$ be so that $S_n\xi$ is strictly positive on Σ_A for some $n \in \mathbb{N}$. Let $\delta > 0$ denote the unique real for which $P(\eta - \delta\xi) = 0$, where P denotes the topological pressure function (see Sect. A.3). Assume that $x \mapsto g_x(t)$ is α-Hölder continuous for any $t \in \mathbb{R}$, that $\{t \mapsto \mathrm{e}^{-t\delta}|g_x(t)| \mid x \in \Sigma_A\}$ is equi d. R. i. and that there exist $C, s > 0$ such that $\mathrm{e}^{-t\delta}|g_x(t)| \le C\mathrm{e}^{st}$ for $t < 0$ and $x \in \Sigma_A$.*

(i) If ξ is non-lattice (see Definition A.2) then there exists $G(x) \in \mathbb{R}$, explicitly stated in Sect. A.5, such that

$$N(t, x) \sim \mathrm{e}^{t\delta} G(x)$$

as $t \to \infty$, uniformly for $x \in \Sigma_A$.

(ii) *Assume that ξ is lattice (see Definition A.2) and let $\zeta, \psi \in C(\Sigma_A)$ satisfy the relation*

$$\xi - \zeta = \psi - \psi \circ \sigma,$$

where $\zeta(\Sigma_A) \subseteq a\mathbb{Z}$ for some $a > 0$. Suppose that ξ is not co-homologous to any function with values in a proper subgroup of $a\mathbb{Z}$, see Definition A.2. Then

$$N(t, x) \sim \mathrm{e}^{t\delta} \widetilde{G}_x(t)$$

as $t \to \infty$, uniformly for $x \in \Sigma_A$. Here $\widetilde{G}_x$ is periodic with period a and explicitly stated in Sect. A.5.

(iii) *We always have*

$$\lim_{t \to \infty} t^{-1} \int_0^t \mathrm{e}^{-T\delta} N(T, x) \mathrm{d}T = G(x).$$

Remark 4.3

(i) In [25] it is shown that weaker assumptions than $\{t \mapsto \mathrm{e}^{-t\delta}|g_x(t)| \mid x \in \Sigma_A\}$ being equi d.R.i. suffice, see [25, Sec. 3, (A)–(D)].

(ii) In [28] the case that η is the constant zero-function in conjunction with $g_x := \mathbb{1}_{[0,\infty)}$ for every $x \in \Sigma_A$ is addressed. With these restrictions, [25, Sec. 3, (A) and (Da)] are immediate and [25, Sec. 3, (B) and (C)] are shown in [28, Lemma 8.1]. The renewal function from (4.2) becomes

$$N(t, x) := \sum_{n=0}^{\infty} \sum_{y:\sigma^n y = x} \chi(y) \mathbb{1}_{[0,\infty)}(t - S_n\xi(y)),$$

which is a counting function. [25, Thm. 3.1] provides its asymptotic behaviour as $t \to \infty$, recovering [28, Thms. 1 to 3].

(iii) Notice, in [25] the above theorem was obtained under the stronger assumption of A being primitive. This was weakened to A being irreducible in [21], where additionally Theorem 4.2 was extended to the setting of Σ being countably infinite.

In Appendix B we show how versions of the probabilistic renewal theorems, which we stated in Sect. 3, can be deduced from the renewal theorems in symbolic dynamics presented above.

4.3 *Questions 2.1 and 2.2 for Limit Sets of Graph-Directed Systems of Conformal Maps, Including Self-conformal Sets—Application of the Renewal Theorems in Symbolic Dynamics*

Both Questions 2.1 and 2.2 can be solved for limit sets of graph-directed systems of conformal maps by means of the Renewal theorems in symbolic dynamics. We will show how this is done for Question 2.2 below. Since the main ideas are similar we will not execute how to solve Question 2.1 in this setting. Moreover, we will focus on the case of self-conformal sets here and refer to [22] for the graph-directed case, where details are provided.

As in Sect. 3.4 assume that Φ satisfies the OSC with feasible open set O and w. l. o. g. that O is bounded. Recall from Sect. 3.4 that $\Gamma := O \setminus \bigcup_{i=1}^{M} \phi_i O$ and that

$$\lambda_d(O) = \lambda_d\left(\bigcup_{n=0}^{\infty} \bigcup_{u\in\Sigma^n} \phi_u \Gamma\right), \tag{4.3}$$

where the unions are disjoint. In the following we assume that O can be chosen so that $\lambda_d(E_{\mathrm{e}^{-t}} \cap \Gamma) = \mathfrak{o}(\mathrm{e}^{t(D-d)})$ as $t \to \infty$ with the little Landau symbol $\mathfrak{o}$, where E denotes the self-conformal set associated with Φ and D denote its Minkowski dimension. (For functions $f, g\colon \mathbb{R} \to \mathbb{R}$ we write $f = \mathfrak{o}(g)$ as $t \to \infty$ if $\lim_{t\to\infty} f(t)/g(t) = 0$.) This is a mild condition, which is always satisfied for self-similar systems with any feasible open set O, see [40]. For $D < d$ Eq. (4.3) thus gives

$$\begin{aligned}\lambda_d(E_{\mathrm{e}^{-t}} \cap O) &= \sum_{n=0}^{\infty} \sum_{u\in\Sigma^n} \lambda_d(E_{\mathrm{e}^{-t}} \cap \phi_u \Gamma)\\ &= \sum_{\omega\in\Sigma^m} \sum_{n=0}^{\infty} \sum_{u\in\Sigma^n} \lambda_d(E_{\mathrm{e}^{-t}} \cap \phi_u\phi_\omega \Gamma) + \mathfrak{o}(\mathrm{e}^{t(D-d)})\end{aligned}$$

for any $m \in \mathbb{N}$. In the current setting we need to assume that $\lambda_d(E_{\mathrm{e}^{-t}} \cap \phi_u\phi_\omega\Gamma) = \lambda_d((\phi_u E)_{\mathrm{e}^{-t}} \cap \phi_u\phi_\omega\Gamma)$. As conformal maps locally behave like similarities the expression $\lambda_d((\phi_u E)_{\mathrm{e}^{-t}} \cap \phi_u\phi_\omega\Gamma)$ can be approximated by

$$|\phi_u'(\pi\sigma\omega x)|^d \lambda_d(E_{\mathrm{e}^{-t}/|\phi_u'(\pi\sigma\omega x)|} \cap \phi_\omega\Gamma) \tag{4.4}$$

with an arbitrary $x \in \Sigma^{\mathbb{N}}$. Here, $\pi\colon \Sigma^{\mathbb{N}} \to E$ is the *code map* defined by $\{\pi(\omega)\} := \bigcap_{n=0}^{\infty} \phi_{\omega|_n}(X)$. Introducing the *geometric potential function* $\xi\colon \Sigma^{\mathbb{N}} \to \mathbb{R}$ associated with the IFS Φ by

$$\xi(\omega) := -\log|\phi_{\omega_1}'(\pi\sigma\omega)|$$

we obtain $\exp(-S_n\xi(u\omega x)) = |\phi_u'(\pi\sigma\omega x)|$. Thus, $\lambda_d(E_{e^{-t}} \cap O)$ can be approximated by

$$\sum_{\omega\in\Sigma^m}\sum_{n=0}^{\infty}\sum_{u\in\Sigma^n} \mathrm{e}^{-dS_n\xi(u\omega x)}\lambda_d(E_{\mathrm{e}^{-t+S_n\xi(u\omega x)}} \cap \phi_\omega\Gamma) + o(\mathrm{e}^{t(D-d)}).$$

Setting $f_y(t) := \lambda_d(E_{\mathrm{e}^{-t}} \cap \phi_\omega\Gamma)$ for $y \in \Sigma^{\mathbb{N}}$, $\chi := \mathbb{1}_{\Sigma^{\mathbb{N}}}$, $\eta := -d\xi$ and assuming the condition of $\{\mathrm{e}^{-t\delta} f_y \mid y \in \Sigma_A\}$ being equi d.R.i. we can apply Thm. 4.2 and, if ξ is non-lattice, obtain

$$\sum_{n=0}^{\infty}\sum_{u\in\Sigma^n} \mathrm{e}^{-dS_n\xi(u\omega x)}\lambda_d(E_{\mathrm{e}^{-t+S_n\xi(u\omega x)}} \cap \phi_\omega\Gamma) = N(t,\omega x)$$

$$\sim \mathrm{e}^{t\delta}\frac{h_{-(d+\delta)\xi}(\omega x)}{\int \xi \mathrm{d}\mu_{-(d+\delta)\xi}}\int_{-\infty}^{\infty} \mathrm{e}^{-T\delta}\lambda_d(E_{\mathrm{e}^{-T}} \cap \phi_\omega\Gamma)\mathrm{d}T,$$

where $\delta > 0$ is the unique value for which $P(-(d+\delta)\xi) = 0$, see Sect. A.3 and the terms appearing in the fraction are explained in Sect. A.3, see also Sect. A.5. It is proven in [4] that the Minkowski dimension D of E is the unique solution to $P(-D\xi) = 0$ thus, $d + \delta = D$. Using the bounded distortion property [35, Lem. 2.3.1] this shows, in the non-lattice situation, that

$$\lambda_d(E_{\mathrm{e}^{-t}} \cap O) \sim \mathrm{e}^{t(D-d)} \lim_{m\to\infty}\sum_{\omega\in\Sigma^m}\frac{h_{-D\xi}(\omega x)}{\int \xi \mathrm{d}\mu_{-D\xi}}\int_{-\infty}^{\infty} \mathrm{e}^{-T(D-d)}\lambda_d(E_{\mathrm{e}^{-T}} \cap \phi_\omega\Gamma)\mathrm{d}T.$$

The lattice case can be treated similarly.

Remark 4.4 Question 2.2 for self-conformal subsets of $\mathbb{R}$ and limit sets of graph-directed systems in $\mathbb{R}$, including Fuchsian groups of Schottky type, are treated in [19] and [20], where results of [28] were applied. The desire of obtaining an answer to Question 2.2 in the higher dimensional setting of limit sets of conformal graph-directed systems gave the motivation for developing the renewal theorem in symbolic dynamics that we stated in Theorem 4.2 in [25] and its generalisation to the infinite alphabet case in [21]. In [25] and [22] more details on the above can be found. Results on curvature measures are for instance provided in [7].

Acknowledgments The author would like to thank the anonymous referee for their helpful and constructive suggestions.

A Appendix: Symbolic Dynamics

Here, we provide some background from symbolic dynamics which we use in Sect. 4. Good references for the exposition below are [9, 38].

A.1 *Sub-Shifts of Finite Type—Admissible Paths of a Random Walk Through Σ*

Recall the following setting from Sect. 4. $\Sigma = \{1, \dots, M\}$, $M \geq 2$ denotes the state space of the stochastic process $(X_n)_{n\in\mathbb{N}_0}$ and A denotes an *irreducible* $(M \times M)$-incidence matrix of zeros and ones. The set of *one-sided infinite admissible paths* of $(X_n)_{n\in\mathbb{N}_0}$ through Σ consistent with $A = (A_{i,j})_{i,j\in\Sigma}$ is defined by

$$\Sigma_A := \{x \in \Sigma^{\mathbb{N}} \mid A_{x_k,x_{k+1}} = 1\ \forall k \in \mathbb{N}\}.$$

Elements of Σ_A are interpreted as paths which describe the history of the process, supposing that the process has been going on forever.

The path of the process prior to the current time is described by $\sigma(x)$, where $\sigma\colon \Sigma_A \to \Sigma_A$ denotes the (left) shift-map on Σ_A given by $\sigma(\omega_1\omega_2\dots) := \omega_2\omega_3\dots$. The set of *admissible words of length* $n \in \mathbb{N}$ is defined by

$$\Sigma_A^n := \{\omega \in \Sigma^n \mid A_{\omega_k,\omega_{k+1}} = 1 \text{ for } k \leq n-1\}.$$

If ω has infinite length or length $m \geq n$ we define $\omega|_n := \omega_1 \cdots \omega_n$ to be the sub-path of length n. Further, $[\omega] := \{u_1u_2\cdots \in \Sigma_A \mid u_i = \omega_i \text{ for } i \leq n\}$ is the *ω-cylinder set* for $\omega \in \Sigma_A^n$.

A.2 *(Hölder-)Continuous and (Non-)lattice Functions*

Equip $\Sigma^{\mathbb{N}}$ with the product topology of the discrete topologies on Σ and equip $\Sigma_A \subset \Sigma^{\mathbb{N}}$ with the subspace topology, i.e. the weakest topology with respect to which the canonical projections onto the coordinates are continuous. Denote by $C(\Sigma_A)$ the space of continuous real-valued functions on Σ_A. Elements of $C(\Sigma_A)$ are called *potential functions*.

Definition A.1 For $\xi \in C(\Sigma_A)$, $\alpha \in (0,1)$ and $n \in \mathbb{N}_0$ define

$$\mathrm{var}_n(\xi) := \sup\{|\xi(\omega) - \xi(u)| \mid \omega, u \in \Sigma_A \text{ and } \omega_i = u_i \text{ for all } i \in \{1,\dots,n\}\},$$

$$|\xi|_\alpha := \sup_{n\geq 0} \frac{\mathrm{var}_n(\xi)}{\alpha^n} \quad\text{and}$$

$$\mathcal{F}_\alpha(\Sigma_A) := \{\xi \in C(\Sigma_A) \mid |\xi|_\alpha < \infty\}.$$

Elements of $\mathcal{F}_\alpha(\Sigma_A)$ are called *α-Hölder continuous* functions on Σ_A.

Definition A.2 Functions $\xi_1, \xi_2 \in C(\Sigma_A)$ are called *co-homologous*, if there exists $\psi \in C(\Sigma_A)$ such that $\xi_1 - \xi_2 = \psi - \psi \circ \sigma$. A function $\xi \in C(\Sigma_A)$ is said to be *lattice*, if it is co-homologous to a function whose range is contained in a discrete subgroup of $\mathbb{R}$. Otherwise, we say that ξ is *non-lattice*.

A.3 Topological Pressure Function and Gibbs Measures

The *topological pressure function* $P\colon C(\Sigma_A) \to \mathbb{R}$ is given by the well-defined limit

$$P(\xi) := \lim_{n\to\infty} n^{-1} \log \sum_{\omega\in\Sigma_A^n} \exp \sup_{u\in[\omega]} S_n\xi(u). \tag{A.1}$$

Here, $S_n\xi := \sum_{k=0}^{n-1} \xi \circ \sigma^k$ denotes the *n-th Birkhoff sum* of ξ with $n \in \mathbb{N}$ and $S_0\xi := 0$.

Proposition A.3 *Let $\xi, \eta \in C(\Sigma_A)$ be so that $S_n\xi$ is strictly positive on Σ_A, for some $n \in \mathbb{N}$. Then $s \mapsto P(\eta + s\xi)$ is continuous, strictly monotonically increasing and convex with $\lim_{s\to-\infty} P(\eta+s\xi) = -\infty$ and $\lim_{s\to\infty} P(\eta+s\xi) = \infty$. Hence, there is a unique $\delta \in \mathbb{R}$ for which $P(\eta - \delta\xi) = 0$.*

A finite Borel measure μ on Σ_A is said to be a *Gibbs measure* for $\xi \in C(\Sigma_A)$ if there exists a constant $c > 0$ such that

$$c^{-1} \leq \frac{\mu([\omega|_n])}{\exp(S_n\xi(\omega) - n \cdot P(\xi))} \leq c \tag{A.2}$$

for every $\omega \in \Sigma_A$ and $n \in \mathbb{N}$.

A.4 Ruelle's Perron-Frobenius Theorem

The Ruelle-Perron-Frobenius operator to a potential function $\xi \in C(\Sigma_A)$ is defined by $\mathcal{L}_\xi : C(\Sigma_A) \to C(\Sigma_A)$,

$$\mathcal{L}_\xi \chi(x) := \sum_{y\in\Sigma_A:\sigma y=x} \chi(y)\mathrm{e}^{\xi(y)}. \tag{A.3}$$

The dual operator acting on the set of Borel probability measures supported on Σ_A, is denoted by $\mathcal{L}_\xi^*$.

By [39, Thm. 2.16, Cor. 2.17] and [9, Theorem 1.7], for each $\xi \in \mathcal{F}_\alpha(\Sigma_A)$, some $\alpha \in (0, 1)$, there exists a unique Borel probability measure ν_ξ on Σ_A satisfying $\mathcal{L}_\xi^* \nu_\xi = \gamma_\xi \nu_\xi$ for some $\gamma_\xi > 0$. This equation uniquely determines γ_ξ, which satisfies $\gamma_\xi = \exp(P(\xi))$ and which coincides with the spectral radius of $\mathcal{L}_\xi$. Further, there exists a unique strictly positive eigenfunction $h_\xi \in C(\Sigma_A)$ satisfying $\mathcal{L}_\xi h_\xi = \gamma_\xi h_\xi$ and $\int h_\xi \mathrm{d}\nu_\xi = 1$. Define μ_ξ by $\mathrm{d}\mu_\xi / \mathrm{d}\nu_\xi = h_\xi$. This is the unique σ-invariant Gibbs measure for the potential function ξ.

Proposition A.3 and the relation $\gamma_\xi = \exp(P(\xi))$ imply the following.

Proposition A.4 *Let $\xi, \eta \in C(\Sigma_A)$ be such that for some $n \in \mathbb{N}$ the n-th Birkhoff sum $S_n\xi$ of ξ is strictly positive on Σ_A. Then $s \mapsto \gamma_{\eta+s\xi}$ is continuous, strictly monotonically increasing, log-convex in $s \in \mathbb{R}$ with $\lim_{s\to-\infty} \gamma_{\eta+s\xi} = 0$ and satisfies $\lim_{s\to\infty} \gamma_{\eta+s\xi} = \infty$. The unique $\delta \in \mathbb{R}$ from Proposition A.3 is the unique $\delta \in \mathbb{R}$ for which $\gamma_{\eta-\delta\xi} = 1$.*

A.5 *The Constants in Theorem 4.2*

Using the notation from Sect. A.3 we can explicitly state the form of $G(x)$ and $G_x(t)$ occurring in the Renewal Theorem 4.2. For this, write $\lfloor t \rfloor$ for the largest integer $k \in \mathbb{Z}$ satisfying $k \leq t$, where $t \in \mathbb{R}$. Moreover, set $\{t\} := t - \lfloor t \rfloor \in [0, 1)$. Notice, for $t \in \mathbb{R}$ positive, $\lfloor t \rfloor$ is the integer part and $\{t\}$ is the fractional part of t.

$$G(x) = \frac{h_{\eta-\delta\xi}(x)}{\int \xi \mathrm{d}\mu_{\eta-\delta\xi}} \int_{\Sigma_A} \chi(y) \int_{-\infty}^{\infty} \mathrm{e}^{-T\delta} g_y(T) \,\mathrm{d}T \mathrm{d}\nu_{\eta-\delta\xi}(y) \quad \text{and}$$

$$\widetilde{G}_x(t) = \int_{\Sigma_A} \chi(y) \sum_{l=-\infty}^{\infty} \mathrm{e}^{-al\delta} g_y \left(al + a \left\{ \tfrac{t+\psi(x)}{a} \right\} - \psi(y) \right) \mathrm{d}\nu_{\eta-\delta\zeta}(y)$$
$$\times\, \mathrm{e}^{-a\left\{\frac{t+\psi(x)}{a}\right\}\delta} \frac{a \mathrm{e}^{\delta\psi(x)}}{\int \zeta \mathrm{d}\mu_{\eta-\delta\zeta}} \cdot h_{\eta-\delta\zeta}(x).$$

B Appendix: Relation to the Probabilistic Renewal Theorems

The setting of Sect. 4 extends and unifies the setting of established renewal theorems. In brief: in the context of classical renewal theory for finitely supported measures (in particular of the key renewal theorem), η and ξ only depend on the first coordinate. When η and ξ only depend on the first two coordinates, we are in the setting of Markov renewal theory. If η is the constant zero-function and $f_y(t) = \chi(y)\mathbb{1}_{[0,\infty)}(t)$, where $\chi \in \mathcal{F}_\alpha(\Sigma_A)$ is non-negative, we are precisely in the setting of [28], where renewal theorems for counting measures in symbolic dynamics were developed, see Remark 4.3. The results of the infinite alphabet case obtained in [21] even yield the respective cases for general discrete measures.

In the following we expand upon the above and let $N\colon \Sigma_A \times \mathbb{R} \to \mathbb{R}$ denote the renewal function given in (4.2).

B.1 The Key Renewal Theorem for Finitely Supported Measures

The special case of Theorem 4.2 that N is independent of Σ_A gives the classical key renewal theorem for measures on $[0, \infty)$ that are finitely supported:

N being independent of Σ_A can be achieved by the following assumptions. First, $\Sigma_A = \Sigma^{\mathbb{N}}$ (i.e. full shift). Second, $g_x = f$ is independent of $x \in \Sigma^{\mathbb{N}}$ implying that equi d.R.i. of $\{t \mapsto \mathrm{e}^{-t\delta}|g_x(t)| \mid x \in \Sigma_A\}$ is equivalent to $z\colon \mathbb{R} \to \mathbb{R}$ with $z(t) := \mathrm{e}^{-\delta t} f(t)$ being absolutely d.R.i. Third, $\chi = \mathbb{1}_{\Sigma_A}$. Fourth and most importantly, ξ and η are constant on cylinder sets of length one. To emphasise local constancy, write $s_u := S_n\xi(u_1 \cdots u_n\omega)$ and $p_u := \exp[S_n(\eta - \delta\xi)(u_1 \cdots u_n\omega)]$ for $u = u_1 \cdots u_n \in \Sigma^n$ and $\omega = \omega_1\omega_2 \cdots \in \Sigma^{\mathbb{N}}$. Setting $Z(t) := \mathrm{e}^{-\delta t} N(t)$ we obtain that

$$Z(t) = \sum_{n=0}^{\infty} \sum_{\omega \in \Sigma^n} z(t - s_\omega) p_\omega \quad \text{and} \quad Z(t) = \sum_{i=1}^{M} Z(t - s_i) p_i + z(t), \tag{B.1}$$

for $t \in \mathbb{R}$. Notice, the latter equation of (B.1) is the classical renewal equation (3.4). The assumption $S_n\xi > 0$ for some $n \in \mathbb{N}$ implies $s_i > 0$ for all $i \in \Sigma$. Thus, the distribution F which assigns mass p_i to s_i is concentrated on $(0, \infty)$. On the other hand, any vector $(s_1, \ldots, s_M)$ with $s_1, \ldots, s_M > 0$ determines a strictly positive function $\xi \in \mathcal{F}_\alpha(\Sigma^{\mathbb{N}})$ via $\xi(\omega_1\omega_2 \cdots) := s_{\omega_1}$. Furthermore, in the setting of Theorem 4.2, $(p_1, \ldots, p_M)$ is a probability vector with $p_i \in (0, 1)$ since

$$0 = P(\eta - \delta\xi) = \lim_{n \to \infty} n^{-1} \log \left(\sum_{i \in \Sigma} p_i \right)^n = \log \sum_{i \in \Sigma} p_i$$

by Proposition A.3. Thus, F is a probability distribution. On the other hand, any probability vector $(p_1, \ldots, p_M)$ with $p_1, \ldots, p_M \in (0, 1)$ determines $\eta \in \mathcal{F}_\alpha(\Sigma^{\mathbb{N}})$ via $\eta(\omega_1\omega_2 \cdots) := \log(p_{\omega_1} \mathrm{e}^{\delta s_{\omega_1}})$.

Consequently, Theorem 4.2 provides the asymptotic behaviour of Z under the assumptions that $(p_1, \ldots, p_M)$ is a probability vector and that $s_1, \ldots, s_M > 0$. In order to present the asymptotic term in a common form, observe that $\mathcal{L}_{\eta-\delta\xi}\mathbf{1} = \mathbf{1}(x)$ for any $x \in \Sigma^{\mathbb{N}}$, where $\mathbf{1} = \mathbb{1}_{\Sigma^{\mathbb{N}}}$. Thus,

$$h_{\eta-\delta\xi} = \mathbf{1} \qquad \text{and} \qquad \mu_{\eta-\delta\xi}([i]) = \nu_{\eta-\delta\xi}([i]) = p_i,$$

where the last equality follows by considering the dual operator of $\mathcal{L}_{\eta-\delta\xi}$. If ξ is lattice then the range of ξ itself lies in a discrete subgroup of $\mathbb{R}$: If there exist $\zeta, \psi \in C(\Sigma^{\mathbb{N}})$ with $\xi - \zeta = \psi - \psi \circ \sigma$ and $\zeta(\Sigma^{\mathbb{N}}) \subset a\mathbb{Z}$ for some $a > 0$, then ξ and ζ need to coincide on $\{\omega \in \Sigma^{\mathbb{N}} \mid \omega = \sigma\omega\}$. As every cylinder set of length one contains a periodic word of period one the claim follows. Hence, we can choose $\zeta = \xi$ and ψ to be the constant zero-function. We deduced the key renewal theorem, Theorem 3.4 for finitely supported measures on $[0, \infty)$ and $f \geq 0$. In exactly the same way [21, Thm. 3.1] yields the key renewal theorem for discrete measures.

B.2 Relation to Markov Renewal Theorems

Suppose that we are in the setting of Sect. 4.

If we assume that η and ξ are constant on cylinder sets of length two, then the point process with inter-arrival times $W_0, W_1, \ldots$ becomes a Markov random walk: To see this, define $\widetilde{\eta}, \widetilde{\xi} \colon \Sigma_A^2 \to \mathbb{R}$ by $\widetilde{\eta}(ij) := \eta(ij\omega)$ and $\widetilde{\xi}(ij) := \xi(ij\omega)$ for any $\omega \in \Sigma_A$ for which $ij\omega \in \Sigma_A$. Then

$$\mathbb{P}(X_1 = i \mid X_0 X_{-1} \cdots = x) = \mathrm{e}^{\eta(ix)} = \mathrm{e}^{\widetilde{\eta}(ix_1)} = \mathbb{P}(X_1 = i \mid X_0 = x_1).$$

Thus, $(X_n)_{n\in\mathbb{Z}}$ is a Markov chain. Further, $W_n = \xi(X_{n+1}X_nX_{n-1}\cdots) = \widetilde{\xi}(X_{n+1}X_n)$ implies that the inter-arrival times $W_0, W_1, \ldots$ are Markov dependent on $(X_n)_{n\in\mathbb{Z}}$. Applying Theorem 4.2 to such Markov random walks gives the Markov renewal theorem presented in Theorem 3.9. In order to state its conclusions in the form of Theorem 3.9 we present several simplifications and conversions in the following. Set

$$\begin{aligned}\widetilde{F_{i,j}}(t) &:= \mathbb{P}(X_{n+1} = j, W_n \leq t \mid X_n = i)\\ &= \begin{cases} \mathbb{1}_{(-\infty,t]}(\widetilde{\xi}(ji))\mathrm{e}^{\widetilde{\eta}(ji)} & : ji \in \Sigma_A^2 \\ 0 & : \text{otherwise.}\end{cases}\end{aligned}$$

and define $F := (\widetilde{F}_{ij})_{i,j\in\Sigma}$ to be the matrix with entries $F_{ij} := \|\widetilde{F}_{ij}\|_\infty = \exp(\widetilde{\eta}(ji))\mathbb{1}_{\Sigma_A^2}(ji)$. Then, F is irreducible if and only if A is irreducible. Moreover, $\widetilde{F}_{ij}$ is a distribution function of a discrete measure. Thus, ξ is lattice if and only if $\widetilde{F}_{ij}$ is lattice for all i, j. For $s \in \mathbb{R}$ and $i, j \in \Sigma$ we have

$$B_{i,j}(s) := \int \mathrm{e}^{-sT}\, \widetilde{F}_{i,j}(\mathrm{d}T) = \begin{cases} \exp(\widetilde{\eta}(ji) - s\widetilde{\xi}(ji)) & : ji \in \Sigma_A^2 \\ 0 & : \text{otherwise.}\end{cases}$$

Setting $B(s) := (B_{ij}(s))_{i,j\in\Sigma}$ we see that the action of $B(-s)$ on vectors coincides with the action of the Ruelle-Perron-Frobenius operator $\mathcal{L}_{\eta+s\xi}$ on functions

$g \colon \Sigma_A \to \mathbb{R}$ which are constant on cylinder sets of length one. That is, setting $\widetilde{g}_i := g(ix)$, for $x \in \Sigma_A$ with $ix \in \Sigma_A$, gives

$$\mathcal{L}_{\eta+s\xi} g(ix) = \sum_{j\in\Sigma,\, ji\in\Sigma_A^2} \mathrm{e}^{\widetilde{\eta}(ji)+s\widetilde{\xi}(ji)} \widetilde{g}_j = \sum_{j\in\Sigma} B_{ij}(-s)\widetilde{g}_j = (B(-s)\widetilde{g})_i .$$

By the Perron-Frobenius theorem for matrices there is a unique s for which $B(s)$ has spectral radius one. By the above this value coincides with the unique s for which $\mathcal{L}_{\eta-s\xi}$ has spectral radius one, which we denoted by δ in Proposition A.4. Similarly, $h_{\eta-\delta\xi}$ is constant on cylinder sets of length one. Thus, setting $h_i := h_{\eta-\delta\xi}(ix)$ for $x \in \Sigma_A$ with $ix \in \Sigma_A$ we obtain a vector $(h_i)_{i\in\Sigma}$ with strictly positive entries which satisfies $B(\delta)h = h$, since

$$(B(\delta)h)_i = \mathcal{L}_{\eta-\delta\xi} h_{\eta-\delta\xi}(ix) = h_{\eta-\delta\xi}(ix) = h_i .$$

Moreover, the vector ν given by $\nu_i := \nu_{\eta-\delta\xi}([i])$ satisfies $\nu_i > 0$ for all $i \in \Sigma$ and $\nu B(\delta) = \nu$, since $\mathcal{L}^*_{\eta-\delta\xi}\nu_{\eta-\delta\xi} = \nu_{\eta-\delta\xi}$. By the Perron-Frobenius theorem h and ν are unique with these properties. Additionally assuming $\chi = \mathbb{1}_{\Sigma_A}$ and that f_x only depends on the first letter of $x \in \Sigma_A$ it follows that $N(t, x)$ only depends on the first letter of x. Thus, for $i \in \Sigma$ write $N(t, i) := N(t, ix)$ with $x \in \Sigma_A$ for which $ix \in \Sigma_A$. Now, the renewal equation becomes

$$\begin{aligned} N(t, i) &= \sum_{j\in\Sigma,\ ji\in\Sigma_A^2} N(t - \widetilde{\xi}(ji), j)\mathrm{e}^{\widetilde{\eta}(ji)} + f_i(t) \\ &= \sum_{j\in\Sigma} \int_{-\infty}^{\infty} N(t-u, j)\widetilde{F}_{i,j}(\mathrm{d}u) + f_i(t), \end{aligned} \tag{B.2}$$

for $i \in \Sigma$, where $f_i(t) := f_{ix}(t)$ for $x \in \Sigma_A$ with $ix \in \Sigma_A$, compare (3.8). Using the above in conjunction with the constants provided in Sect. A.5 thus yields Theorem 3.9.

References

1. Alsmeyer, G.: Erneuerungstheorie [Renewal Theory]. Analyse stochastischer Regenerationsschemata. [Analysis of stochastic regeneration schemes]. Teubner Skripten zur Mathematischen Stochastik. [Teubner Texts on Mathematical Stochastics]. B.G. Teubner, Stuttgart (1991)
2. Asmussen, S.: Applied Probability and Queues. Applications of Mathematics (New York), vol. 51, 2nd edn. Springer-Verlag, New York (2003). Stochastic Modelling and Applied Probability
3. Bandt, C., Hung, N., Rao, H.: On the open set condition for self-similar fractals. Proc. Am. Math. Soc. **134**(5), 1369–1374 (2006)

4. Bedford, T.: Hausdorff dimension and box dimension in self-similar sets. In: Proceedings of the Conference on Topology and Measure V, Ernst-Moritz-Arndt Universitat Greisfwald, pp. 17–26 (1988)
5. Blackwell, D.: A renewal theorem. Duke Math. J. **15**, 145–150 (1948)
6. Blackwell, D.: Extension of a renewal theorem. Pac. J. Math. **3**, 315–320 (1953)
7. Bohl, T.J.: Fractal curvatures and Minkowski content of self-conformal sets (2012). Preprint. arXiv:1211.3421
8. Bohl, T.J., Zähle, M.: Curvature-direction measures of self-similar sets. Geom. Dedicata **167**, 215–231 (2013)
9. Bowen, R.: Equilibrium states and the ergodic theory of Anosov diffeomorphisms, 2nd revised edn. Lecture Notes in Mathematics, vol. 470. Springer, Berlin (2008)
10. Çinlar, E.: Markov renewal theory: a survey. Manage. Sci. **21**(7), 727–752 (1974/1975)
11. Deniz, A., Koçak, M.Ş., Özdemir, Y., Ratiu, A., Üreyen, A.E.: On the Minkowski measurability of self-similar fractals in $\mathbb{R}^d$. Turk. J. Math. **37**(5), 830–846 (2013)
12. Erdős, P., Feller, W., Pollard, H.: A property of power series with positive coefficients. Bull. Am. Math. Soc. **55**, 201–204 (1949)
13. Falconer, K.J.: On the Minkowski measurability of fractals. Proc. Am. Math. Soc. **123**(4), 1115–1124 (1995)
14. Falconer, K.J.: Techniques in Fractal Geometry. Wiley, Chichester (1997)
15. Falconer, K.J.: Fractal Geometry. Mathematical Foundations and Applications, 2nd edn. Wiley, Chichester (2003)
16. Feller, W.: An Introduction to Probability Theory and Its Applications, vol. I, 3rd edn. Wiley, New York/London/Sydney (1968)
17. Feller, W.: An Introduction to Probability Theory and Its Applications, vol. II, 2nd edn. Wiley, New York/London/Sydney (1971)
18. Gatzouras, D.: Lacunarity of self-similar and stochastically self-similar sets. Trans. Am. Math. Soc. **352**(5), 1953–1983 (2000)
19. Kesseböhmer, M., Kombrink, S.: Fractal curvature measures and Minkowski content for self-conformal subsets of the real line. Adv. Math. **230**(4–6), 2474–2512 (2012)
20. Kesseböhmer, M., Kombrink, S.: Minkowski content and fractal Euler characteristic for conformal graph directed systems. J. Fractal Geom. **2**, 171–227 (2015)
21. Kesseböhmer, M., Kombrink, S.: A complex Ruelle-Perron-Frobenius theorem for infinite Markov shifts with applications to renewal theory. Discrete Contin. Dyn. Syst. Ser. S **10**(2), 335–352 (2017)
22. Kesseböhmer, M., Kombrink, S.: Minkowski measurability of infinite conformal graph directed systems and application to Apollonian packings (2017). Preprint. arXiv:1702.02854
23. Kolmogoroff, A. (1936). Anfangsgründe der Theorie der Markoffschen Ketten mit unendlich vielen möglichen Zuständen. Rec. Math. Moscou, n. Ser. **1**, 607–610 (1936)
24. Kombrink, S.: A survey on Minkowski measurability of self-similar and self-conformal fractals in $\mathbb{R}^d$. In: Fractal Geometry and Dynamical Systems in Pure and Applied Mathematics. I. Fractals in Pure Mathematics. Contemporary Mathematics, vol. 600, pp. 135–159. American Mathematical Society, Providence, RI (2013)
25. Kombrink, S.: Renewal theorems for processes with dependent interarrival times. Adv. Appl. Probab. **50**(4), 1193–1216 (2018)
26. Kombrink, S., Winter, S.: Lattice self-similar sets on the real line are not Minkowski measurable. Ergodic Theory Dynam. Syst. **40**(1), 221–232 (2020)
27. Kombrink, S., Pearse, E.P.J., Winter, S.: Lattice-type self-similar sets with pluriphase generators fail to be Minkowski measurable. Math. Z. **283**(3–4), 1049–1070 (2016)
28. Lalley, S.P.: Renewal theorems in symbolic dynamics, with applications to geodesic flows, noneuclidean tessellations and their fractal limits. Acta Math. **163**(1–2), 1–55 (1989)
29. Lapidus, M.L., Pomerance, C.: The Riemann zeta-function and the one-dimensional Weyl-Berry conjecture for fractal drums. Proc. Lond. Math. Soc. (3) **66**(1), 41–69 (1993)
30. Lapidus, M.L., van Frankenhuysen, M.: Complex Dimensions of Fractal Strings and Zeros of Zeta Functions. Fractal Geometry and Number Theory. Birkhäuser, Boston (2000)

31. Lapidus, M.L., van Frankenhuijsen, M.: Fractal Geometry, Complex Dimensions and Zeta Functions Geometry and Spectra of Fractal Strings. Springer, New York (2006)
32. Lapidus, M.L., Pearse, E.P.J., Winter, S.: Minkowski measurability results for self-similar tilings and fractals with monophase generators. In: Fractal Geometry and Dynamical Systems in Pure and Applied Mathematics I: Fractals in Pure Mathematics., pp. 185–203. American Mathematical Society, Providence, RI (2013)
33. Mandelbrot, B.B.: Measures of fractal lacunarity: Minkowski content and alternatives. In: Fractal Geometry and Stochastics. Birkhäuser Verlag, Basel, Boston, Berlin (1995)
34. Mitov, K.V., Omey, E.: Renewal Processes. SpringerBriefs in Statistics. Springer, Cham (2014)
35. Mauldin, R.D., Urbański, M.: Dimensions and measures in infinite iterated function systems. Proc. Lond. Math. Soc. III. Ser. **73**(1), 105–154 (1996)
36. Mauldin, R.D., Urbański, M.: Graph Directed Markov Systems: Geometry and Dynamics of Limit Sets. Cambridge University Press, Cambridge (2003)
37. Rataj, J., Zähle, M.: Curvature Measures of Singular Sets. Springer, Cham (2019)
38. Walters, P.: An Introduction to Ergodic Theory. Graduate Texts in Mathematics, vol. 79. Springer, New York/Berlin (1982)
39. Walters, P.: Convergence of the Ruelle operator for a function satisfying Bowen's condition. Trans. Am. Math. Soc. **353**(1), 327–347 (2001)
40. Winter, S.: Minkowski content and fractal curvatures of self-similar tilings and generator formulas for self-similar sets. Adv. Math. **274**, 285–322 (2015)
41. Winter, S., Zähle, M.: Fractal curvature measures of self-similar sets. Adv. Geom. **13**(2), 229–244 (2013)

Part II
Random Graphs and Complexes

Fractal Dimension of Discrete Sets and Percolation

Markus Heydenreich

Abstract There are various notions of dimension in fractal geometry to characterise (random and non-random) subsets of $\mathbb{R}^d$. In this expository text, we discuss their analogues for infinite subsets of $\mathbb{Z}^d$ and, more generally, for infinite graphs. We then apply these notions to critical percolation clusters, where the various dimensions have different values.

Keywords Discrete fractal · Fractal dimension · Mass dimension · Spectral dimension · Discrete Hausdorff dimension · Percolation · Incipient infinite cluster

Mathematics Subject Classifications (2010) 28A80, 60K35, 82B43

1 What is the Dimension of a Graph?

Motivation There are various notions of dimension for subsets of $\mathbb{R}^d$, see the classical work of Falconer [22] as well as texts by Fraser and Lehrbäck in this volume [24, 44]. Hausdorff dimension is perhaps the most commonly used, other examples are box dimension and Assouad dimension. Any reasonable notion of dimension yields the same value for strictly self-similar sets, but already for affine self-similar sets these values may differ. All these notions depend on microscopic properties of the set, i.e. local properties.

In statistical physics, many interesting models give rise to (random) subsets of the lattice $\mathbb{Z}^d$ or even general graphs, and therefore "dimension" in this context should describe the macroscopic properties of the set rather than the microscopic ones.

M. Heydenreich (✉)
Mathematisches Institut, Universität München, München, Germany
e-mail: m.heydenreich@lmu.de

U. Freiberg et al. (eds.), *Fractal Geometry and Stochastics VI*,
Progress in Probability 76, https://doi.org/10.1007/978-3-030-59649-1_5

In this expository text, we shall describe and compare three notions of dimension for graphs: *fractal dimension* and *spectral dimension* can be defined for any (connected and locally finite) graph, while the *mass dimension* requires the graph to be embedded in an "external" metric space (for our purpose, we can think of $\mathbb{R}^d$ equipped with the Euclidean norm). In the second part, we investigate these notions for (high-dimensional) critical percolation as a prime example of a rich and interesting subset of $\mathbb{Z}^d$, and we shall see that the three notions of dimension yield different values.

It appears that different mathematical communities use different vocabulary, and it is one of our aims to draw the connection between the various concepts involved.

Preparatory Notions We start by recalling basic notions from graph theory. Let $G = (V, E)$ be a graph with non-empty vertex set V and edge set $E \subset \binom{V}{2}$ and distinguished vertex $\mathbf{0} \in V$ ("the root"). We interpret G as a metric space with *intrinsic metric* (or 'graph metric')

$$d_G(x, y) = \inf\big\{n \in \mathbb{N} \colon \exists v_1, \dots, v_n \in V \text{ such that } \{x, v_1\}, \{v_1, v_2\}, \dots, \{v_{n-1}, v_n = y\} \in E\big\}, \tag{1.1}$$

for the shortest number of edges forming a *path* from x to y (including the case that $d_G(x, y) = \infty$ whenever there is no such path).

We henceforth assume that the graph is *locally finite*, i.e. for all $x \in V$:

$$\deg_G(x) := \sum_{e \in E} \mathbb{1}_{\{x \in e\}} < \infty, \tag{1.2}$$

and *connected*, i.e. $d_G(x, y) < \infty$ for all $x.y \in V$. For $x \in V$ and $n \in \mathbb{N}_0$, we denote by

$$B_x(n) := \big\{y \in V : d_G(x, y) \le n\big\}$$

the ball w.r.t. the intrinsic metric d_G, and abbreviate $B(n) := B_{\mathbf{0}}(n)$ for the ball of the root. We write $\partial B_x(n) := B_x(n) \setminus B_x(n-1)$ for the inner vertex boundary of $B_x(n)$.

Fractal Dimension The first notion of dimension is the *fractal dimension* (or "volume growth dimension") defined as

$$\dim_f(G) := \lim_{n\to\infty} \frac{\log |B(n)|}{\log n} \tag{1.3}$$

whenever the limit exists. More generally, we refer to the upper (resp. lower) fractal dimension as lim sup (resp., lim inf) of (1.3). The fractal dimension appears to be a very natural concept, and it characterises the structure of G viewed as a metric space. In case of existence of the limit (1.3), we can write $|B(n)| = n^{\dim_f(G)+o(1)}$.

Spectral Dimension A second, completely different approach to dimensionality is given through random walks on the graph G. To this end, we define the *(simple) random walk* on the (locally finite) graph G as the (discrete-time) stochastic process with $(S_n)_{n\in\mathbb{N}_0}$ with probability measure P and the property that

- $P(S_0 = \mathbf{0}) = 1$;
- For all $n \in \mathbb{N}$ and $x, y \in V$:

$$P(S_n = x \mid S_{n-1} = y) = \begin{cases} \frac{1}{\deg_G(y)} & \text{if } d_G(x, y) = 1, \\ 0 & \text{otherwise.} \end{cases} \tag{1.4}$$

In words, the random walk starts at the root at time $n = 0$, and in each time step it moves to one of the neighbouring vertices (chosen independently with equal probability).

We are now interested in the event that the random walk returns to the origin after a given number $2n$ of steps. Indeed, we may use the decay rate of this probability to define the *spectral dimension* of the graph G as

$$\dim_s(G) := \lim_{n\to\infty} -2\frac{\log P(S_{2n} = \mathbf{0})}{\log n} \tag{1.5}$$

provided that the limit exists.

Mind that we are interested in returning after an *even* number of steps only, the reason for this is that $P(S_{2n} = \mathbf{0}) > 0$ for all n (e.g. by "reversing" the first n steps). However, on bipartite graphs, the random walk can return to the origin only after an even number of steps, so that automatically $P(S_n = \mathbf{0}) = 0$ whenever n is odd.

Both notions $\dim_f$ and $\dim_s$ use the special vertex $\mathbf{0}$ as 'base point'. However, it might be easily observed that $\mathbf{0}$ is not relevant for the dimension (as long as the graph is connected), and any other vertex of G as base point would lead to the same value of $\dim_f$ and $\dim_s$.

The spectral dimension is closely linked to the concept of recurrence and transience of a graph, which we introduce next. To this end, we investigate the probability that the random walk *always* returns to its starting point or not. We call the graph *recurrent* if this is the case, i.e., if $P(\exists n \in \mathbb{N}\colon S_n = \mathbf{0}) = 1$. Otherwise, we call the graph *transient*.

Lemma 1.1 *The graph G is recurrent if* $\dim_s < 2$, *and it is transient if* $\dim_s > 2$.

Proof A well-known theorem about random walks (e.g. Theorem 5.3.1 in [21]) states that the random walk $(S_n)_n$ is transient if and only if $\sum_{n\in\mathbb{N}} P(S_n = \mathbf{0}) < \infty$. Thus for $\dim_s < 2$, we have

$$\sum_{n\in\mathbb{N}} P(S_n = \mathbf{0}) \geq \sum_{n\in\mathbb{N}} P(S_{2n} = \mathbf{0}) = \sum_{n\in\mathbb{N}} n^{-\dim_s/2+o(1)} = \infty.$$

For an upper bound, we use $P(S_{2n+1} = \mathbf{0}) \le P(S_{2n} = \mathbf{0})$ for all $n \in \mathbb{N}$ (cf. Lemma 4.1 in [7]), and thus

$$\sum_{n\in\mathbb{N}} P(S_n = \mathbf{0}) \le 1 + \sum_{n\in\mathbb{N}} 2P(S_{2n} = \mathbf{0}) = 1 + 2\sum_{n\in\mathbb{N}} n^{-\dim_s/2+o(1)},$$

and this is summable whenever $\dim_s > 2$. □

The "borderline case" $\dim_s = 2$ relies on the finer asymptotics of $P(S_{2n} = \mathbf{0})$, and thus the limit (1.5) is too coarse to give an answer.

The terminology "spectral dimension" suggests a connection with the eigenvalues of the graph Laplacian, see for example Rammal and Toulouse [47] for a discussion in the Physics literature. For Brownian motion on a class of compact fractals, Kigami and Lapidus [39] prove a rigorous correspondence between the fractal dimension and the spectrum of the associated Laplacian.

Examples An important example in the present text is the hypercubic lattice $\mathbb{L}^d = (\mathbb{Z}^d, \mathbb{E}^d)$ with edge set $\mathbb{E}^d = \big\{\{x, y\}: |x - y| = 1\big\}$. It is easily observed that $\dim_f(\mathbb{L}^d) = \dim_s(\mathbb{L}^d) = d$; indeed, this property should hold for any meaningful notion of dimension for discrete sets. A Cayley graph is a graph that encodes the abstract structure of a (usually finitely generated) group. The class of Cayley graphs is very rich, and includes the hypercubic lattice, homogeneous trees, and many other graphs. Gromov [27] proved that the limit (1.3) exists as an integer number for every Cayley graph. Hebisch and Saloff-Coste [31, Thm. 5.1] verified that $\dim_f = \dim_s$ for Cayley graphs. This equality is true for many other classes of graphs.

The Escape Time Exponent A second notion characterising random walks on graphs is the *escape time exponent* β, which is defined as

$$E\big[\inf\{n \in \mathbb{N}: S_n \in \partial B(n)\}\big] = n^{\beta+o(1)}. \tag{1.6}$$

Thus β describes how long it typically takes to reach the boundary of n-balls; by $E[\,\cdot\,]$ we denote expectation w.r.t. the random walk measure P. For the Euclidean lattice $\mathbb{L}^d$ we have $\beta = 2$. If $\beta > 2$, we speak of *anomalous diffusion*, which relates to the fact that the random walk moves on average much slower than in Euclidean space: after n steps, the random walk is typically at distance $n^{1/\beta}$ from its starting point. An example for anomalous diffusion is random walk on the Sierpinski gasket, for which Barlow and Perkins [9] proved that $\beta = \log 5/\log 2$. The exponent β is closely linked to $\dim_f$ and $\dim_s$. Indeed, Barlow and Bass [8] prove that $\beta = 2\dim_f/\dim_s$ for any generalized Sierpinski carpet. However, all values of β in the interval $[2, \dim_f + 1]$ are possible, as pointed out by Barlow [6].

Mass Dimension The graph notions described above are rather versatile tools for abstract graphs. We shall now consider graphs that are embedded into Euclidean space $\mathbb{R}^d$ (by this we mean that $V \subset \mathbb{R}^d$). For our purpose we can be more restrictive and require that $V \subset \mathbb{Z}^d$. We denote by

$$Q(n) := [-n, n]^d \cap \mathbb{Z}^d \tag{1.7}$$

the ball of radius n with respect to the supremum-metric on $\mathbb{Z}^d$. The *mass dimension* of a graph $G = (V, E)$ is then defined via

$$\dim_m(G) := \lim_{n\to\infty} \frac{\log |V \cap Q(n)|}{\log n}. \tag{1.8}$$

Mind the difference between $\dim_f$ and $\dim_m$: while the former identifies the growth exponent of balls w.r.t. the *intrinsic* (graph) metric, the latter measures balls w.r.t. the *extrinsic* (Euclidean) metric. This makes no difference for $G = \mathbb{L}^d$, but we will encounter examples, where this is indeed very different. The use of the supremum metric in (1.7) might appear arbitrary, but since all metrics on $\mathbb{Z}^d$ are equivalent, they will all lead to the same value of $\dim_m$.

Other Notions of Dimension In this exposition we focus on the formerly defined dimensions. However, there are various other notions of dimensions for subsets of $\mathbb{Z}^d$ (mostly graph analogues of "continuum dimensions" for subsets of $\mathbb{R}^d$). We explain two of these notions, which were introduced by Barlow and Taylor [10, 11].

The first definition is the *discrete Hausdorff dimension* $\dim_H$, which is defined for subsets of $\mathbb{Z}^d$ as follows. We say that a set $A \subset \mathbb{Z}^d$ is a *finite cube* if there exists $x \in \mathbb{Z}^d$ and $r \in \mathbb{N}$ such that $A = Q(r) + \{x\}$ (where $+$ is the Minkowski sum). For a finite set $A \subset \mathbb{Z}^d$, we denote by

$$\mathcal{R}(A) = \min\{r : A \subset Q(r) + \{x\} \text{ for some } x \in \mathbb{Z}^d\}$$

the radius of A as the radius of a covering cube (and put $R = \infty$ if $|A| = \infty$). For $\alpha \geq 0$, $A, F \subset \mathbb{Z}^d$ and $F \neq \emptyset$, we further let

$$\nu_\alpha(A, F) := \min\left\{\sum_{i=1}^{m}\left(\frac{\mathcal{R}(B_i)}{\mathcal{R}(F)}\right)^\alpha : B_1, \ldots, B_m \text{ are finite cubes and } A \cap F \subset \bigcup_{i=1}^{m} B_i\right\}.$$

Let

$$m_\alpha(A) = \sum_{n=1}^{\infty} \nu_\alpha\big(A, Q(2^n) \setminus Q(2^{n-1})\big). \tag{1.9}$$

Mind that $\alpha \mapsto x^\alpha$ is decreasing for $x \in [0, 1]$, and so is $m_\alpha(A)$. We finally define the *discrete Hausdorff dimension*

$$\dim_H(A) := \inf\big\{\alpha \geq 0 \colon m_\alpha(A) < \infty\big\}. \tag{1.10}$$

The definition of discrete Hausdorff dimension is clearly modelled by its continuous counterpart. Similarly to (and yet different from) the spectral dimension, this notion is closely related to the recurrence and transience of random walks on A:

Proposition 1.2 (Thm. 8.3 in [11]) *A set $A \subset \mathbb{Z}^d$ is recurrent if* $\dim_H(A) > d-2$, *and it is transient if* $\dim_H(A) < d-2$.

Comparison with Lemma 1.1 shows that $\dim_H$ and $\dim_s$ often differ. What is the behaviour if $\dim_H(A) = d-2$? If $m_{d-2}(A) < \infty$, then the set is transient as well, but no conclusion is possible when $m_{d-2}(A) = \infty$, because $\dim_H$ is not sensitive enough to decide the matter.

The second example that we discuss here is the *discrete packing dimension* $\dim_p$. Its continuous analogue is the packing dimension as defined by Taylor and Tricot [50], which is the same as Kolmogorov's *metric dimension* and Hawke's *entropy dimension*. To this end, we let $A, F \subset \mathbb{Z}^d$ as before, and $\varepsilon \in (0,1)$. Then we let

$$\mu_\alpha(A,F,\varepsilon) := \max\left\{\sum_{i=1}^m \left(\frac{\mathcal{R}(B_i)}{\mathcal{R}(F)}\right)^\alpha : \begin{array}{c} B_1,\dots,B_m \text{ are finite } \textit{pairwise disjoint} \text{ cubes} \\ \text{centered in } A\cap F \text{ s.t. } \mathcal{R}(B_i) \le \mathcal{R}(F)^{1-\varepsilon} \end{array}\right\},$$

and define the "packing measure"

$$p_\alpha(A,\varepsilon) = \sum_{n=1}^\infty \mu_\alpha\big(A, Q(2^n)\setminus Q(2^{n-1},\varepsilon)\big). \tag{1.11}$$

Then the *discrete packing dimension* is defined as

$$\dim_p(A) := \inf\big\{\alpha \ge 0\colon\ p_\alpha(A,\varepsilon) < \infty \text{ for all } \varepsilon \in (0,1)\big\}. \tag{1.12}$$

Among the results of Barlow and Taylor [11, Lemma 3.1] is the following order of the dimensions: If $A \subset \mathbb{Z}^d$, then

$$0 \le \dim_H(A) \le \dim_m(A) \le \dim_p(A) \le d. \tag{1.13}$$

We return to these notions at the end of this text.

A different approach to the dimensionality of discrete sets has been proposed recently by Bacelli, Haji-Mirsadeghi, and Khezeli [4].

2 Percolation

2.1 Percolation on $\mathbb{L}^d$

Percolation theory studies the geometry of certain random subgraphs of $\mathbb{L}^d$. Let $p \in [0,1]$ be a parameter of the model, and make edges in $\mathbb{E}^d$ *occupied* with probability p (independently of each other), and otherwise vacant. More formally, we consider the probability space $\Omega = \{0, 1\}^{\mathbb{E}^d}$ equipped with the product topology. For a percolation configuration $\omega \in \{0, 1\}^{\mathbb{E}^d}$, an edge $b \in \mathbb{E}^d$ is occupied whenever $\omega(b) = 1$, and it is vacant whenever $\omega(b) = 0$. We equip this space with a family of product measures $(\mathbb{P}_p)_{p\in[0,1]}$ chosen such that $\mathbb{P}_p(b \text{ occupied}) = p$ for any $b \in \mathbb{E}^d$ and $p \in [0, 1]$.

We say that x is *connected* to y and write $x \leftrightarrow y$ when there exists a (finite) path of occupied edges connecting x and y. Formally, $x \leftrightarrow y$ on a configuration $\omega \in \{0, 1\}^{\mathbb{E}^d}$ if there exist $x = v_0, v_1, \dots, v_{m-1}, v_m = y \in \mathbb{Z}^d$ with the property that $\{v_{i-1}, v_i\} \in \mathbb{E}^d$ and $\omega(\{v_{i-1}, v_i\}) = 1$ for all $i = 1, \dots, m$ ($m \in \mathbb{N}$). We further write $\{x \leftrightarrow y\} = \{\omega \colon x \leftrightarrow y \text{ on the configuration } \omega\}$. We let the *cluster* of x be all the vertices that are connected to x, i.e., $C(x) = \{y \colon x \leftrightarrow y\}$. By convention, $x \in C(x)$.

We define the *percolation function* $p \mapsto \theta(p)$ by

$$\theta(p) = \mathbb{P}_p(|C(x)| = \infty), \tag{2.1}$$

where $x \in \mathbb{Z}^d$ is an arbitrary vertex and $|C(x)|$ denotes the number of vertices in $C(x)$. By translation invariance, the above probability does not depend on the choice of x. We therefore often investigate $C = C(\mathbf{0})$ where $\mathbf{0} \in \mathbb{Z}^d$ denotes the origin.

When $\theta(p) = 0$, then the probability that the origin is inside an infinite connected component is 0, so that there is almost surely no infinite connected component. On the other hand, when $\theta(p) > 0$, then (by ergodicity) the proportion of vertices in infinite connected components equals $\theta(p) > 0$, and we say that the system *percolates*.

We define the *percolation critical value* by

$$p_c = \inf\{p \colon \theta(p) > 0\}. \tag{2.2}$$

It is well-known that $p_c = 1$ on the one-dimensional lattice $\mathbb{L}^1$ and $p_c \in (0, 1)$ on $\mathbb{L}^d$ for all $d \geq 2$. For this and other basic properties we refer to the textbooks by Grimmett [26], Bollobas and Riordan [16] and Werner [51]. See Fig. 1 for a simulation of percolation with different values of p.

For every percolation realization $\omega \in \Omega$, we can define a random walk on the cluster C as in (1.4); we denote the corresponding measure by P^ω. Random walk on percolation clusters is a benchmark model of *random walk in (non-elliptic) random environment*.

2.2 Dimension of Percolation Clusters

We now address the question: What is the dimension of the percolation cluster $C = C(\mathbf{0})$? The answer depends on the value of p.

Fig. 1 Three realisations of percolation on $\mathbb{L}^2$

Indeed, if $p < p_c$, then $|C| < \infty$ for $\mathbb{P}_p$-almost all ω, and hence

$$\dim_f(C) = \dim_m(C) = \dim_s(C) = 0 \qquad \mathbb{P}_p - a.s. \tag{2.3}$$

We get a different picture when $p > p_c$, and thus $\theta(p) = \mathbb{P}_p(|C(x)| = \infty) > 0$. We condition on the event that the origin lies in an infinite cluster, and denote the conditional probability by $\mathbb{P}^*_p(\,\cdot\,) = \mathbb{P}_p(\,\cdot \mid 0\leftrightarrow\infty)$. It may be seen by applying the ergodic theorem that

$$\lim_{n\to\infty} \frac{|C \cap Q(n)|}{|Q(n)|} \to \theta(p) \qquad \mathbb{P}^*_p - \text{a.s.}, \tag{2.4}$$

and hence $\dim_m(C) = d$. Furthermore, we get that $\dim_f(C) = d$ (almost surely w.r.t. the measure $\mathbb{P}^*_p$) by exploiting the large deviation bounds on the graphical distance by Antal and Pisztora [3]. Concerning the spectral dimension, Barlow [5] proved the heat kernel bounds

$$c_1 n^{-d/2} \le P^\omega(S_{2n} = \mathbf{0}) \le c_2 n^{-d/2} \tag{2.5}$$

for $\mathbb{P}^*_p$-almost all ω (where the constants $c_1, c_2 > 0$ depend on the value of p), and hence $\dim_s(C) = d$ as well. Barlow's result was strengthened further to get a *quenched invariance principle* [15, 45].

Finally, the critical case $p = p_c$. There is a rather general lower bound on the volume growth of critical clusters. Recall that we denote by $B(n)$ the ball w.r.t. the intrinsic (graph) metric d_C on the cluster $C = C(\mathbf{0})$, and $\partial B(n) = B(n) \setminus B(n-1)$.

Theorem 2.1 *For percolation on* $\mathbb{L}^d$, $d \geq 1$,

$$\mathbb{E}_{p_c}|B(n)| \geq n, \qquad n \geq 1. \tag{2.6}$$

We provide a proof at the end of this chapter. Mind that (2.6) implies that $\liminf_{n\to\infty} \log \mathbb{E}_{p_c}|B(n)|/\log n \geq 1$, and it might be tempting to conjecture that even $\dim_f \geq 1$. This, however, is not true. Even stronger, it is strongly believed that critical infinite clusters do not exist:

Conjecture 2.2 For percolation on $\mathbb{Z}^d$, $d \geq 2$, we have that $\theta(p_c) = 0$, and thus $|C(x)| < \infty$ for all $x \in \mathbb{Z}^d$ $\mathbb{P}_{p_c}$-a.s.

The conjecture is known to be true for $d = 2$ by Kesten [36] as well as in high dimensions by Hara and Slade [29], where the meaning of *high dimensions* is that there exists $d_{\min} > 6$ such that the claim is true for $d \geq d_{\min}$. Fitzner and van der Hofstad [23] optimized the strategy of Hara and Slade and verified that $d_{\min} = 11$ suffices. Proving this conjecture in dimensions $3 \leq d \leq 10$ is a major open problem in percolation theory; see also [26] and [32, Open Problem 1.1].

In view of the presumed result that $\theta(p_c) = 0$, we thus get that all clusters are almost surely finite and hence all dimensions equal 0, precisely as for $p < p_c$. Yet an interesting structure will emerge if we look at the interesting geometry of critical clusters from a different angle. We now investigate this further for the two regimes that we do understand rigorously, namely $d = 2$ and *high dimensions*.

2.3 *The Incipient Infinite Cluster*

When $\theta(p_c) = 0$, this leaves us with a most remarkable situation: At the critical point p_c there are clusters at all length scales, which are, however, all finite. As we then make a density $\varepsilon > 0$ of closed edges open, the large clusters connect up to form a (unique) infinite cluster, no matter how small ε is. At criticality, the critical cluster is therefore *at the verge of appearing*. This observation motivated the introduction of an *incipient infinite cluster* (IIC) as a critical cluster that is *conditioned* to be infinite.

Somewhat simplified, the *incipient infinite cluster* (IIC) is defined as the cluster of the origin under the critical measure $\mathbb{P}_{p_c}$ conditioned on $\{|C(0)| = \infty\}$. Since this would condition on an event of zero probability, a rigorous construction of the IIC requires a limiting argument. The first mathematical construction has been carried out by Kesten [37] in two dimensions, who considered two limiting schemes:

- ▷ under $\mathbb{P}_{p_c}$, condition on the event $\{C(0) \cap \partial\Lambda_n \neq \varnothing\}$, and then let $n \to \infty$;
- ▷ under $\mathbb{P}_p$ $(p > p_c)$, condition on the event $\{|C(0)| = \infty\}$ and let $p \searrow p_c$.

Kesten proved that both limits exist in dimension $d = 2$, and lead to the *same* limiting measure, which he calls the incipient infinite cluster. He was motivated by observations in the physics literature, which indicated anomalous diffusion for random walk on large critical percolation clusters. Kesten [38] confirmed this, and proved that the exit time exponent β satisfies $\beta > 2$ on incipient infinite cluster in two dimensions. It is an open problem to improve this bound.

For percolation on a regular tree, the cluster distribution is precisely that of a Galton-Watson tree with binomial offspring distribution. Hence, the incipient infinite cluster for percolation on a tree is a special case of critical Galton-Watson tree conditioned on non-extinction. It was again Kesten [38] who studied the latter, and proved that it can be constructed in two steps: a single infinite line of descent, casually phrased as "the immortal particle" and more formally as "cluster backbone", and critical trees hanging off this backbone. He further investigated the escape time exponent for this incipient infinite cluster on trees, and proved that $\beta = 3$.

We now come to the case of high-dimensional percolation, where the IIC was constructed by van der Hofstad and Járai:

Theorem 2.3 (IIC Construction [35]) *There is a dimension $d_{\min} > 6$ such that for $d \geq d_{\min}$ and any event E that depends on the status of finitely many edges, the limit*

$$\mathbb{P}_{\mathrm{IIC}}(E) := \lim_{|x|\to\infty} \mathbb{P}_{p_c}\big(E \mid 0 \leftrightarrow x\big) \tag{2.7}$$

exists.

The limitation to events that depend on the status of only finitely many edges is a technical one. In fact, such events form an algebra on Ω which is stable under intersections, and we may thus extend $\mathbb{P}_{\mathrm{IIC}}$ to a measure on the σ-fields generated by the product topology. We denote this measure $\mathbb{P}_{\mathrm{IIC}}$ the *incipient infinite cluster measure*.

It is straightforward to see that indeed $\mathbb{P}_{\mathrm{IIC}}(|C(\mathbf{0})| = \infty) = 1$, as desired. Since $\theta(p_c) = 0$, the IIC is also *one-ended* in the sense that the removal of any finite region of the IIC leaves one infinite part. It can be seen that the infinite path is *essentially unique* in the sense that any pair of infinite self-avoiding paths in the IIC share infinitely many edges.

Van der Hofstad and Járai derive also another construction of the IIC-measure in high dimensions, namely

$$\mathbb{P}_{\mathrm{IIC}}(E) = \lim_{p \nearrow p_c} \frac{\sum_{x\in\mathbb{Z}^d} \mathbb{P}_p(E \cap \{0 \leftrightarrow x\})}{\sum_{x\in\mathbb{Z}^d} \mathbb{P}_p(0 \leftrightarrow x)}. \tag{2.8}$$

A third construction (same as Kesten's first construction in two dimension) was derived with van der Hofstad and Hulshof [33].

Mind that the measure $\mathbb{P}_{\mathrm{IIC}}$ has lost the translation invariance of the percolation measures $\mathbb{P}_p$. Indeed, the origin $\mathbf{0}$ plays a special role, since we have enforced that the cluster $C(\mathbf{0})$ is infinite.

2.4 Lower Bound for the Expected Size of Critical Balls

We now prove Theorem 2.1. One ingredient is an alternative characterization of p_c, namely

$$p_c = \sup\{p \in [0,1]\colon \mathbb{E}_p|C| < \infty\}, \tag{2.9}$$

which is standard in percolation theory [1, 46]; we also refer to the short proof by Duminil-Copin and Tassion [20]. In our proof of Theorem 2.1, we adapt ideas of [20] but use the intrinsic (graph) metric rather than the extrinsic one. It appears that the proof is valid in much wider context, namely all transitive connected graphs whose percolation threshold is strictly between 0 and 1 and for which (2.9) is true.

***Proof** (Proof of Theorem 2.1)* We define the value $\bar{p}_c = \sup M$ with

$$M = \{p \in (0,1)\colon \exists n \in \mathbb{N} \text{ such that } \mathbb{E}_p|\partial B(n)| < 1\}.$$

Fix an arbitrary $p < \bar{p}_c$. Then exists $\varepsilon > 0$ and $n \in \mathbb{N}$ such that $\mathbb{E}_p|\partial B(n)| < 1-\varepsilon$. Fix such ε and n. We now claim that

$$\mathbb{P}_p\big(\partial B(kn) \neq 0\big) \leq (1-\varepsilon)^k, \qquad k \in \mathbb{N}. \tag{2.10}$$

The proof of (2.10) is via induction in k. The initialization of the induction is our assumption. For the inductive step, we assume that (2.10) is true for some k, and aim to prove it for $k+1$. We first condition on the ball $B(n)$:

$$\mathbb{P}_p\big(\partial B((k+1)n) \neq 0\big) = \sum_{A \subset \mathbb{Z}^d} \mathbb{P}_p\big(B(n) = A, \partial B((k+1)n) \neq 0\big). \tag{2.11}$$

We treat the set A as a subgraph of $\mathbb{L}^d$, and denote by ∂A the vertices with maximal graphical distance from $\mathbf{0}$, this allows us to bound

$$\begin{aligned}\mathbb{P}_p\big(\partial B((k+1)n) \neq 0\big) &= \sum_{A \subset \mathbb{Z}^d} \mathbb{P}_p\big(B(n) = A, \bigcup_{y \in \partial A} \partial B_y(kn) \neq 0 \text{ in } (\mathbb{Z}^d \setminus A) \cup \{y\}\big) \\ &\leq \sum_{A \subset \mathbb{Z}^d} \sum_{y \in \partial A} \mathbb{P}_p\big(B(n) = A, \partial B_y(kn) \neq 0 \text{ in } (\mathbb{Z}^d \setminus A) \cup \{y\}\big).\end{aligned} \tag{2.12}$$

The event $\{B(n) = A\}$ depends on the status of all the edges with at least one endpoint in $A \setminus \partial A$. On the other hand, $\{\partial B_y(kn) \neq 0 \text{ in } \mathbb{Z}^d \setminus A \cup \{y\}\}$ depends on the status of the edges not touching $A \setminus \partial A$. Hence, the two events are independent, and we bound further

$$
\begin{aligned}
&\mathbb{P}_p\big(\partial B((k+1)n) \neq 0\big) \\
&\quad \leq \sum_{A\subset\mathbb{Z}^d} \sum_{y\in\partial A} \mathbb{P}_p\big(B(n) = A\big)\,\mathbb{P}_p\big(\partial B_y(kn) \neq 0 \text{ in } (\mathbb{Z}^d \setminus A) \cup \{y\}\big) \\
&\quad \leq \sum_{A\subset\mathbb{Z}^d} \sum_{y\in\partial A} \mathbb{P}_p\big(B(n) = A\big)\,\mathbb{P}_p\big(\partial B_y(kn) \neq 0\big)
\end{aligned}
\tag{2.13}
$$

Transitivity of the underlying lattice gives $\mathbb{P}_p\big(\partial B_y(kn) \neq 0\big) = \mathbb{P}_p\big(\partial B(kn) \neq 0\big)$. Since

$$
\sum_{A\subset\mathbb{Z}^d} \sum_{y\in\partial A} \mathbb{P}_p\big(B(n) = A\big) = \sum_{A\subset\mathbb{Z}^d} |\partial A|\,\mathbb{P}_p\big(B(n) = A\big) = \mathbb{E}_p|\partial B(n)| \leq 1 - \varepsilon,
$$

we can use the induction hypotheses to obtain $\mathbb{P}_p\big(\partial B((k+1)n) \neq 0\big) \leq (1-\varepsilon)^{k+1}$, thus proving (2.10). Consequently,

$$
\mathbb{P}_p(|C| = \infty) \leq \lim_{k\to\infty} \mathbb{P}_p\big(\partial B(kn) \neq 0\big) = 0
$$

and thus $p \leq p_c$. Since $p < \bar{p}_c$ was arbitrary, we conclude $\bar{p}_c \leq p_c$.

We further observe that M is an open subset of $[0, 1]$, and therefore $\bar{p}_c \notin M$. This implies $\mathbb{E}_p|\partial B(n)| \geq 1$ for all $n \in \mathbb{N}$, and thus

$$
\mathbb{E}_{\bar{p}_c}|C| = \sum_{n\in\mathbb{N}_0} \mathbb{E}_{\bar{p}_c}|\partial B(n)| \geq \sum_{n\in\mathbb{N}_0} 1 = \infty,
$$

and via (2.9) we thus get that $\bar{p}_c \geq p_c$. Together with the foregoing, we established $\bar{p}_c = p_c$.

The finishing touch is provided by the partial summation

$$
\mathbb{E}_{p_c}|B(n)| = \mathbb{E}_{\bar{p}_c}|B(n)| = \sum_{k=0}^{n} \mathbb{E}_{\bar{p}_c}|\partial B(k)| \geq \sum_{k=1}^{n} 1 = n. \qquad \square
$$

3 Dimension of the Incipient Infinite Cluster

In this section we come to the main endeavour of this text, which is characterising the various dimensions of incipient infinite cluster.

Let us deal with the planar case first. Kesten [37] proved for various two-dimensional lattices that

$$\lim_{\lambda\to\infty} \mathbb{P}_{\mathrm{IIC}}\left(\lambda^{-1} \le \frac{|C\cap Q(n)|}{n^2\,\mathbb{P}_{p_c}(\mathbf{0}\leftrightarrow\partial Q(n))} \le \lambda\right) = 1 \tag{3.1}$$

uniformly in n. For the case of site percolation on the triangular lattice, it is known that $P_{p_c}(\mathbf{0}\leftrightarrow\partial Q(n)) = n^{-5/48+o(1)}$, cf. [43]. This suggests that the mass dimension equals $\dim_m = 91/96$. However, in view of Lemma 3.5 below, the control of the error terms is not strong enough to conclude that $\dim_m = 91/96$ in an ($\mathbb{P}_{\mathrm{IIC}}$-)almost sure sense. Concerning the fractal dimension $\dim_f$, it is a challenging open problem to derive sharp bounds on the *intrinsic* (graph) distance of critical two-dimensional clusters. For a recent survey of bounds on the intrinsic distance in the planar case, we refer to Damron [19].

We now come to the case of high dimensions, where the results are most complete. For a general survey of results in high-dimensional percolation, we refer to our recent textbook [32].

For the incipient infinite cluster in high dimensions, the results are summarized in the following theorem:

Theorem 3.1 ([18, 40]) *For the incipient infinite cluster in high dimensions, we have that*

$$\dim_s(C) = 4/3, \quad \dim_f(C) = 2, \quad \dim_m(C) = 4 \qquad \mathbb{P}_{\mathrm{IIC}} - a.s.$$

Interestingly, all three dimensions of C are independent of the dimension d of the embedding space, an indication that the geometry of the embedding space is less visible, and the model appears similar as their non-spatial analogues (as predicted by mean-field theory).

In the sequel, we demonstrate the proof for the fractal dimension based on a number of standard results for high-dimensional percolation. Finally, we discuss the necessary adaptations for the other dimensions $\dim_s$ and $\dim_m$.

The analysis of percolation in high dimension is rooted in a technique called the *lace expansion*. For percolation, this was pioneered in a seminal 1990 paper by Hara and Slade [29], who were inspired by earlier work of Brydges and Spencer [17] for self-avoiding walk. For our purpose we need the following estimate on the percolation connectivity: there exist $C, c > 0$ such that for all $x, y \in \mathbb{Z}^d$, $x \neq y$,

$$c|x-y|^{d-2} \le \mathbb{P}_{p_c}(x\leftrightarrow y) \le C|x-y|^{d-2}. \tag{3.2}$$

The estimate (3.2) was first derived for a spread-out version of percolation [30], and adapted by Hara [28] to our setting. Fitzner and van der Hofstad [23] verified that it is valid in dimension $d > 10$.

The upper bound in (3.2) readily implies the famous *triangle conditon*, which in turn implies that various *critical exponents* take on their mean-field values. We need only two implications here, and refer to a general discussion of critical exponents to [32, Section 1.2]: There are constants $c_1, C_1, c_2, C_2 > 0$ such that

$$\frac{c_1}{p_c - p} \le \mathbb{E}_p|C| \le \frac{C_1}{p_c - p}, \qquad p \in (0, p_c), \tag{3.3}$$

and

$$\frac{c_2}{\sqrt{k}} \le \mathbb{P}_{p_c}\big(|C| \ge k\big) \le \frac{C_2}{\sqrt{k}}, \qquad k \in \mathbb{N}. \tag{3.4}$$

The bound (3.3) is due to Aizenman and Newman [2], the bound (3.4) due to Barsky and Aizenman [13].

Our final ingredient is the famous BK-inequality. To this end, we define the *disjoint occurrence* $E \circ F$ of two events E and F as

$$E \circ F = \big\{\omega \colon \exists K \subset \mathbb{E}^d \text{ such that } \omega_K \in E, \omega_{\mathbb{E}^d \setminus K} \in F\big\}, \tag{3.5}$$

where $\omega_K := \{\omega' \colon \omega(e) = \omega'(e) \text{ for all } e \in K\}$ is the "K-cylinder of ω". Then the BK-inequality [14, 48] establishes that

$$\mathbb{P}_p(E \circ F) \le \mathbb{P}_p(E)\, \mathbb{P}_p(F) \tag{3.6}$$

for any $p \in [0, 1]$ and all events E, F that *depend on finitely many edges*. This last confinement can be lifted in many cases, and indeed (3.6) is true for all events that we are considering in the present text (see also Section 2.3 in [26]).

3.1 The Fractal Dimension

We now prove that $\dim_f(C) = 2$ whenever (3.2) is valid. We start by showing that the lower bound (2.6) has a matching upper bound in high dimensions (which is supposedly false in dimension $d < 6$).

Lemma 3.2 (Ball Growth) *Consider percolation in dimension $d > 10$. There exists a constant $C_3 > 0$ such that for all $n \in \mathbb{N}$,*

$$n \le \mathbb{E}_{p_c}|B(n)| \le C_3 n.$$

Proof The lower bound was already contained in (2.6). We follow Sapozhnikov [49] for a proof of the upper bound. Let $p < p_c$. We consider the following coupling of percolation with parameters p and p_c: Starting with a critical percolation configuration (edges are occupied with probability p_c), make every occupied edge vacant with probability $1 - (p/p_c)$. This construction implies that for any $x \in \mathbb{Z}^d$,

$p < p_c$, and $n \in \mathbb{N}$,

$$\mathbb{P}_p\big(d_C(\mathbf{0}, x) \leq n\big) \geq \left(\frac{p}{p_c}\right)^n \mathbb{P}_{p_c}(d_C(\mathbf{0}, x) \leq n).$$

Summing over x and using the inequality $\mathbb{P}_p(d_C(0, x) \leq n) \leq \mathbb{P}_p(0 \leftrightarrow x)$, we obtain

$$\mathbb{E}_{p_c}|B(n)| \leq \left(\frac{p_c}{p}\right)^n \mathbb{E}_p|C| \leq C_1 \left(\frac{p_c}{p}\right)^n (p_c - p)^{-1},$$

where the last bound comes from (3.3). Choosing $p = p_c(1 - \frac{1}{2n})$ proves the claim. □

Lemma 3.3 (Arm Exponents [40]) *Consider percolation in dimension* $d > 10$. *There exists constants* $C, c > 0$ *such that for all* $n \in \mathbb{N}$,

$$\frac{c}{n} \leq \mathbb{P}_{p_c}\big(\partial B(n) \neq \varnothing\big) \leq \frac{C}{n}.$$

Proof We start with the proof of the lower bound, and use the well-known second-moment method. The basic inequality is

$$\mathbb{P}(Z > 0) \geq (\mathbb{E}Z)^2/\mathbb{E}Z^2, \tag{3.7}$$

which is valid for any non-negative random variable Z. We aim to apply this to $Z = \big|B(\lambda n) \setminus B(n)\big|$ with $\lambda = 2C_3$. Now Lemma 3.2 yields

$$\mathbb{E}_{p_c}|B(\lambda n) \setminus B(n)| \geq \lambda n - C_3 n = C_3 n.$$

We now estimate the second moment of $B(\lambda n)$. Indeed, if both x and y are connected with distance $\leq \lambda n$ from $\mathbf{0}$, then there must exist a "branch point" $z \in \mathbb{Z}^d$ such that there are (edge-)disjoint paths from $\mathbf{0}$ to z, from z to x and from z to y. We may use the symbol $\circ$ (recall (3.5)) to write this as

$$\begin{aligned}&\{d_C(\mathbf{0}, x) \leq \lambda n\} \cap \{d_C(\mathbf{0}, y) \leq \lambda n\}\\ &\quad\subseteq \bigcup_z \{d_C(\mathbf{0}, z) \leq \lambda n\} \circ \{d_C(z, x) \leq \lambda n\} \circ \{d_C(z, y) \leq \lambda n\}.\end{aligned}$$

Consequently, the BK-inequality (3.6) and Lemma 3.2 yields

$$\begin{aligned}\mathbb{E}_{p_c}|B(\lambda n)|^2 &= \sum_{x,y} \mathbb{P}_{p_c}(d_C(\mathbf{0}, x) \leq \lambda n, d_C(\mathbf{0}, y) \leq \lambda n)\\ &\leq \sum_{x,y,z} \mathbb{P}_{p_c}(d_C(\mathbf{0}, z) \leq \lambda n)\, \mathbb{P}_{p_c}(d_C(z, x) \leq \lambda n)\, \mathbb{P}_{p_c}(d_C(z, y) \leq \lambda n)\\ &= \Big[\sum_{z \in \mathbb{Z}^d} \mathbb{P}_{p_c}(d_C(\mathbf{0}, z) \leq \lambda n)\Big]^3 = B(\lambda n)^3 \leq C' n^3,\end{aligned} \tag{3.8}$$

for some constant $C' > 0$. Consequently, the bound in (3.7) yields

$$\mathbb{P}_{p_c}\big(\exists x \in \mathbb{Z}^d : d_C(0,x) \geq \lambda n\big) \geq \mathbb{P}_{p_c}\big(|B(\lambda n) \setminus B(n)| > 0\big) \geq \frac{C_3^2 n^2}{C' n^3} = \frac{C_3^2}{C' n},$$

which proves the statement with $c = \frac{C_3^2}{\lambda C'} = \frac{C_3}{2C'}$.

The upper bound uses a clever induction argument. For subgraphs G of the infinite lattice $\mathbb{L}^d$, we denote by $C_G = C_G(\mathbf{0})$ the (restricted) percolation cluster of $\mathbf{0}$ *in the subgraph* G, and denote by $B_{C_G}(n) = \{y \in \mathbb{Z}^d : d_{C_G}(\mathbf{0}, y) \leq n\}$ the corresponding ball w.r.t. the graph metric on the restricted cluster C_G. We further define

$$H(n; G) := \{\partial B_{C_G}(n) \neq \varnothing\}$$

for the "one-arm event" on the graph G, and

$$\Gamma(n) = \sup\{\mathbb{P}_{p_c}(H(n; G)) : G \text{ is subgraph of } \mathbb{E}^d\}.$$

It turns out that working with $\Gamma(n)$ rather than $\mathbb{P}_{p_c}(H(n; \mathbb{L}^d))$ enables us to apply a regeneration argument, which would not work for $\mathbb{P}_{p_c}(H(n; \mathbb{L}^d))$, since it is not monotone.

For C_2 as in (3.4), we choose $C_* \geq 1$ large enough so that

$$3^3 {C_*}^{2/3} + C_2 {C_*}^{2/3} \leq C_*, \tag{3.9}$$

We claim that, for any integer $k \geq 0$,

$$\Gamma(3^k) \leq \frac{C_*}{3^k}. \tag{3.10}$$

This readily implies the upper bound of the lemma, since for any n we choose k such that $3^{k-1} \leq n < 3^k$ and then

$$\mathbb{P}_{p_c}(H(n; \mathbb{L}^d)) \leq \Gamma(n) \leq \Gamma(3^{k-1}) \leq \frac{C_*}{3^{k-1}} \leq \frac{3C_*}{n}.$$

The proof of (3.10) is via induction in k. The claim is trivial for $k = 0$ since $C_* \geq 1$. For the inductive step we assume (3.10) for $k-1$ and prove it for k. Depending on the size $|C_G|$ of the restricted cluster C_G for arbitrary subgraphs G, we estimate

$$\mathbb{P}_{p_c}(H(3^k; G)) \leq \mathbb{P}_{p_c}\big(H(3^k; G), |C_G| \leq {C_*}^{-4/3} 9^k\big) + \mathbb{P}_{p_c}\big(|C_G| > {C_*}^{-4/3} 9^k\big). \tag{3.11}$$

For the second summand, we use (3.4) to obtain

$$\mathbb{P}_{p_c}\big(|C_G| > C_*^{-4/3}9^k\big) \le \mathbb{P}_{p_c}\big(|C_{\mathbb{L}^d}(0)| > C_*^{-4/3}9^k\big) \le C_2 C_*^{2/3} 3^{-k}. \qquad (3.12)$$

For the former, on the other hand, we claim that

$$\mathbb{P}_{p_c}\big(H(3^k; G), |C_G| \le C_*^{-4/3}9^k\big) \le C_*^{-4/3}3^{k+1}\big(\Gamma(3^{k-1})\big)^2. \qquad (3.13)$$

Indeed, if $|C_G| \le C_*^{-4/3}9^k$, then there exists $j \in [\frac{1}{3}3^k, \frac{2}{3}3^k]$ such that $|\partial B_{C_G}(j)| \le C_*^{-4/3}3^{k+1}$. Denote the first such level by j. Then, on the right hand side, we get a factor $\Gamma(j)$ (which is bounded by $\Gamma(3^{k-1})$) from the probability of a connection from the origin to level j, and $C_*^{-4/3}3^{k+1}$ times the probability to go from level j to level 3^k (each of these probabilities is again bounded above by $\Gamma(3^{k-1})$), which shows (3.13).

We combine (3.11), (3.12), and (3.13) with the induction hypothesis, and finally (3.9), to obtain

$$\Gamma(3^k) \le C_*^{-4/3}3^{k+1}\left(\frac{C_*}{3^{k-1}}\right)^2 + \frac{C_2 C_*^{2/3}}{3^k} = \frac{3^3 C_*^{2/3} + C_2 C_*^{2/3}}{3^k} \le \frac{C_*}{3^k},$$

thus proving (3.10). □

While the previous estimates all concern critical percolation, we now turn towards the IIC-measure; and our tool to transfer the results is the construction (2.7).

Lemma 3.4 ([40]) *Consider percolation in dimension $d > 10$. There exist $C > 0$ such that for all $n \in \mathbb{N}$, $\lambda > 1$,*

$$\mathbb{P}_{\text{IIC}}\Big(\frac{1}{\lambda}n^2 \le |B(n)| \le \lambda n^2\Big) \ge 1 - \frac{C}{\lambda}.$$

Proof (Upper Bound) We aim to show that $\mathbb{P}_{\text{IIC}}\big(|B(n)| > \lambda n^2\big) \le C\lambda^{-1}$ for all $\lambda > 0$, $n \in \mathbb{N}$. If $d_C(\mathbf{0}, z) \le n$ and $\mathbf{0}\leftrightarrow x$ (for $x, z \in \mathbb{Z}^d$), then there exists a vertex $y \in \mathbb{Z}^d$ such that

$$\{d_C(\mathbf{0}, y) \le n\} \circ \{d_C(y, z) \le n\} \circ \{y\leftrightarrow x\}.$$

By the BK-inequality (3.6), we can bound this from above as follows:

$$\begin{aligned}
\mathbb{E}_{p_c}[|B(n)|\,\mathbb{1}_{\{\mathbf{0}\leftrightarrow x\}}] &= \sum_z \mathbb{P}_{p_c}(d_C(\mathbf{0}, z) \le n, \mathbf{0}\leftrightarrow x) \\
&\le \sum_{y,z} \mathbb{P}_{p_c}\big(\{d_C(\mathbf{0}, y) \le n\} \circ \{d_C(y, z) \le n\} \circ \{y\leftrightarrow x\}\big) \\
&\le \sum_{y,z} \mathbb{P}_{p_c}(d_C(\mathbf{0}, y) \le n)\, \mathbb{P}_{p_c}(d_C(y, z) \le n)\, \mathbb{P}_{p_c}(y\leftrightarrow x).
\end{aligned} \qquad (3.14)$$

Therefore, we get a bound for the conditional probability

$$\mathbb{E}_{p_c}[|B(n)| \mid \mathbf{0}\leftrightarrow x] \leq \sum_{x,z} \mathbb{P}_{p_c}(d_C(\mathbf{0}, y) \leq n)\, \mathbb{P}_{p_c}(d_C(y,z) \leq n)\, \frac{\mathbb{P}_{p_c}(y\leftrightarrow x)}{\mathbb{P}_{p_c}(0\leftrightarrow x)}. \tag{3.15}$$

The asymptotics (3.2) implies that there is a constant C' such that for all x with $|x-y| \leq 2|x|$, the ratio $\mathbb{P}_{p_c}(y\leftrightarrow x)/\mathbb{P}_{p_c}(0\leftrightarrow x) \leq C'$, thus

$$\mathbb{E}_{p_c}[|B(n)| \mid \mathbf{0}\leftrightarrow x] \leq C \sum_{x,z} \mathbb{P}_{p_c}(d_C(\mathbf{0}, y) \leq n)\, \mathbb{P}_{p_c}(d_C(y,z) \leq n). \tag{3.16}$$

Finally, we use the upper bound in Lemma 3.2 twice to get

$$\mathbb{E}_{p_c}[|B(n)| \mid \mathbf{0}\leftrightarrow x] \leq C'(C_3 n)^2. \tag{3.17}$$

The finishing touch is provided by Markov's inequality:

$$\mathbb{P}_{p_c}(|B(n)| \geq \lambda n^2 \mid \mathbf{0}\leftrightarrow x) \leq \frac{C' C_3^2 n^2}{\lambda n^2} = C' C_3^2 \lambda^{-1}. \tag{3.18}$$

Letting $|x| \to \infty$ yields the claim (as $\{|B(n)| \geq \lambda n^2\}$ is a cylinder event). □

Proof (Lower Bound) For the lower bound, we prove that $\mathbb{P}_{\mathrm{IIC}}\big(|B(n)| < \varepsilon n^2\big) \leq C\varepsilon$ for all $\varepsilon = \lambda^{-1} > 0$, $n \in \mathbb{N}$.

If $|B(n)| < \varepsilon n^2$, then there exists some radius $j \in \{\lceil n/2 \rceil, \dots, n\}$ such that $|\partial B(0, j)| \leq 2\varepsilon n$, and we fix the smallest such j. Then we condition on $\{B(j) = A\}$ for any "j-admissible" subgraph A, which is any finite subgraph A of $\mathbb{L}^d$ containing $\mathbf{0}$ s. t.

- $\mathbb{P}_{p_c}(B(j) = A) > 0$,
- $|\partial A| \leq 2\varepsilon n$, where $|\partial A|$ denote the number of vertices at maximal graphical distance from $\mathbf{0}$
- $|\{y : d_A(\mathbf{0}, y) = k\}| > 2\varepsilon n$ for $k = \lceil n/2 \rceil, \dots, j-1$ (to make sure that j is the "first" level satisfying the above property).

This yields

$$\begin{aligned} \mathbb{P}_{p_c}\big(|B(n)| < \varepsilon n^2, 0\leftrightarrow x\big) &\leq \sum_{j=n/2}^{n} \sum_{A} \mathbb{P}_{p_c}\big(B(j) = A, 0\leftrightarrow x\big) \\ &= \sum_{j=n/2}^{n} \sum_{A} \mathbb{P}_{p_c}\big(0\leftrightarrow x \mid B(j) = A\big)\, \mathbb{P}_{p_c}\big(B(j) = A\big), \end{aligned} \tag{3.19}$$

where the sum is over all j-admissable A. For any such A, we get

$$\mathbb{P}_{p_c}(0\leftrightarrow x \mid B(j)=A) \le \sum_{y\in\partial A} \mathbb{P}_{p_c}(y\leftrightarrow x \text{ with a path avoiding } A\backslash\partial A \mid B(j)=A).$$

However, since $\{y\leftrightarrow x$ with a path avoiding $A\setminus\partial A\}$ only depends on the edges with both endpoints outside $A\setminus\partial A$ and $\{B(j)=A\}$ only depends on the edges with both endpoints in A, the two events are independent, and

$$\begin{aligned}\mathbb{P}_{p_c}(0\leftrightarrow x \mid B(j)=A) &\le \sum_{y\in\partial A} \mathbb{P}_{p_c}(y\leftrightarrow x \text{ with a path avoiding } A\setminus\partial A)\\ &\le \sum_{y\in\partial A} \mathbb{P}_{p_c}(y\leftrightarrow x) \le \sum_{y\in\partial A} C|y-x|^{d-2},\end{aligned}$$

where the last bound uses (3.2). Assuming that x is far away from the origin (again $|x-y|\le 2|x|$ suffices), then there is a constant $C'>0$ such that

$$\mathbb{P}_{p_c}(0\leftrightarrow x \mid B(j)=A) \le C'\,|\partial A|\,|x|^{2-d} \le C'\varepsilon n\,|x|^{2-d}.$$

Furthermore, we have that

$$\sum_{j=n/2}^{n}\sum_{A}\mathbb{P}_{p_c}\big(B(j)=A\big) \le P_{p_c}\big(\partial B(n/2)\neq\varnothing\big).$$

Plugging the previous two bounds in (3.19), we get

$$\begin{aligned}\mathbb{P}_{p_c}\big(|B(n)|<\varepsilon n^2, 0\leftrightarrow x\big) &\le C'\varepsilon n|x|^{2-d}\sum_{j=n/2}^{n}\sum_{A}\mathbb{P}_{p_c}(B(j)=A)\\ &\le C'\varepsilon n|x|^{2-d}\mathbb{P}_{p_c}\big(\partial B(n/2)\neq\varnothing\big),\end{aligned}$$

and now we use the upper bound in Lemma 3.3 to further bound

$$\mathbb{P}_{p_c}\big(|B(n)|<\varepsilon n^2, 0\leftrightarrow x\big) \le C''\varepsilon|x|^{2-d}$$

for a constant $C''>0$. Finally, letting $|x|\to\infty$ and using (2.7) along with the two-point function estimate (3.2) yields the desired result. □

In order to prove that $\dim_f(C)=2$ for the incipient infinite cluster, we combine the previous lemma with the following general criterion:

Lemma 3.5 (Lemma 3.2 in [18]) *Let* $(Z_n)_{n\in\mathbb{N}}$ *be a sequence of positive random variables such that* $Z_1\le Z_2\le\dots$ *. Suppose there are constants* $\alpha,\mu,C>0$ *such* *that for all* $\lambda > 0$ *and* $n \in \mathbb{N}$, *we have*

$$\mathbb{P}(\lambda^{-1} n^{\alpha} \le Z_n \le \lambda n^{\alpha}) \ge 1 - C(\log \lambda)^{-1-\mu}. \tag{3.20}$$

Then

$$\mathbb{P}\left(\lim_{n\to\infty} \frac{\log Z_n}{\log n} = \alpha \right) = 1.$$

Proof We abbreviate $Y_n := \log Z_n / \log n$, and claim that it is sufficient to prove

$$\lim_{k\to\infty} Y_{2^k} = \alpha \qquad \mathbb{P} - a.s. \tag{3.21}$$

Indeed, for $n \in \mathbb{N}$, we choose $k = k(n) \in \mathbb{N}$ such that $2^{k-1} \le n \le 2^k$, and use the monotonicity of the sequence $(Z_n)_n \in \mathbb{N}$ to bound

$$Y_{2^{k-1}} \frac{k-1}{k} = \frac{\log Z_{2^{k-1}}}{\log 2^k} \le \frac{\log Z_n}{\log n} \le \frac{\log Z_{2^k}}{\log 2^{k-1}} = Y_{2^k} \frac{k}{k-1},$$

and then use (3.21) to conclude the claim.

In order to prove (3.21), we define

$$\varepsilon_k := k^{\frac{1+\mu/2}{1+\mu}-1}, \qquad \lambda_k := 2^{k\varepsilon_k},$$

and note that $\varepsilon_k > 0$, $\lambda_k > 1$ for all $k \ge 1$, and $\lim_{k\to\infty} \varepsilon_k = 0$. Then, using (3.20),

$$\begin{aligned}
\sum_{k=1}^{\infty} \mathbb{P}\big(|Y_{2^k} - \alpha| > \varepsilon_k\big) &= \sum_{k=1}^{\infty} \mathbb{P}\big(|\log Z_{2^k} - \log(2^{k\alpha})| > \log \lambda_k\big) \\
&= \sum_{k=1}^{\infty} \mathbb{P}\big(Z_{2^k} < \lambda_k^{-1} 2^{k\alpha}\big) + \mathbb{P}\big(Z_{2^k} > \lambda_k 2^{k\alpha}\big) \\
&\le C \sum_{k=1}^{\infty} \frac{1}{(\log \lambda_k)^{1+\mu}} = \frac{C}{(\log 2)^{1+\mu}} \sum_{k=1}^{\infty} \frac{1}{k^{1+\mu}} < \infty.
\end{aligned}$$

Hence, the Borel-Cantelli lemma implies that

$$\mathbb{P}\big(|Y_{2^k} - \alpha| > \varepsilon_k \text{ for infinitely many } k\big) = 0,$$

which proves (3.21). □

Proof ***(Proof of*** $\dim_f(C) = 2$***)*** We apply Lemma 3.5 with $\mathbb{P}$ being the IIC-measure $\mathbb{P}_{\mathrm{IIC}}$, $\alpha = 2$, $Z_n = B(n)$ and apply Lemma 3.4 to get the desired result. □

3.2 The Spectral Dimension

Control of the return probability of random walk needs two ingredients. The first one is control of the the volume growth, which is achieved in Lemma 3.4. The second ingredient is control of the *effective resistance*. The connection between these two ingredients and random walk behaviour is in the folklore of studying random walks, see in particular Kumagai and Misumi [42] for results in our context. Kozma and Nachmias [40] prove a quantitative estimate on the lower bound on the effective resistance between $\mathbf{0}$ and $\partial B(n)$, and then apply a readily tailored theorem of Barlow at al. [12] to deduce that $\dim_s = 4/3$. Another consequence of this theorem is that the escape time exponent equals $\beta = 3$ $\mathbb{P}_{\mathrm{IIC}}$−almost surely (precisely as for the IIC on trees).

3.3 The Mass Dimension

Already van der Hofstad and Járai [35] showed that

$$\mathbb{E}_{\mathrm{IIC}}|C \cap Q(n)| \approx n^4.$$

From this, we can prove that $\dim_f(C) \leq 4$ rather straightforwardly via Markov's inequality. The challenge is to prove a complementing lower bound, which was achieved by Cames van Batenburg [18] using quantitative bounds on the extrinsic one-arm exponent [41].

Mind that the escape time exponent as defined in (1.6) determines the rate at which a random walk leaves a ball of intrinsic distance n. Unlike on $\mathbb{Z}^d$, the extrinsic and intrinsic distances are not equivalent on the IIC-cluster, and we therefore consider a modified critical exponent β' as

$$E\big[\inf\{n \in \mathbb{N}\colon S_n \in \partial Q(n)\}\big] = n^{\beta' + o(1)}. \tag{3.22}$$

With van der Hofstad and Hulshof [33] we proved that $\beta' = 6$ for $\mathbb{P}_{\mathrm{IIC}}$−almost all realizations ω. This should be contrasted against $\beta = 3$ explained before. This means that the random walk needs order n^3 steps to leave the intrinsic ball $B(n)$, but it needs n^6 steps in order to leave $Q(n)$. The factor 2 between these two exponents is not a coincidence: in high dimensions, the spatial dependency between different parts of a critical cluster is rather weak; in fact so weak that geodesic paths (w.r.t. graph distance) are embedded into $\mathbb{Z}^d$ similar to a random walk path, and thus the graph distance between $\mathbf{0}$ and $\partial Q(n)$ is of the order n^2.

4 Discussion and Outlook

A number of pressing challenges were mentioned en passant, most notably the identification of dimensions of critical percolation clusters in lower dimension. However, in the following we want to bring forward two lines of further research that might be within reach with current techniques.

1. Identify discrete Hausdorff dimension and packing dimension of critical percolation clusters in high dimension. Also for other "natural" random subsets of $\mathbb{Z}^d$. So far the only results are by Barlow and Taylor [11] and Georgiou et al. [25] for the range of (generalised) random walks.
2. Known cases of discrete dimension all deal with subsets of $\mathbb{Z}^d$, and also the focus of the present account is on subsets of the hypercubic lattice. However, there is no obvious need to stick to the lattice setup here—fractal and spectral dimension are meaningful for any locally-finite connected graph, and the others require an embedding of the vertices in some metric space, and $\mathbb{Z}^d$ might appear as an unnecessary limitation.

   From a geometric point of view, it might be more natural to focus on discrete subsets of $\mathbb{R}^d$. Instead of lattice percolation, one might investigate the geometric properties of (critical) continuum percolation clusters. A suitable candidate is the *random connection model*, where vertices are given as a Poisson point process in $\mathbb{R}^d$, and two vertices are linked by an edge with probability depending on the Euclidean distance between the vertices. The critical behaviour of the random connection model in high dimensions has recently been identified [34], paving the way to an investigation of the continuum incipient infinite cluster and its dimension(s).

Acknowledgments The author thanks Martin Barlow and Steffen Winter for providing references and for comments on an earlier version of the manuscript.

References

1. Aizenman, M., Barsky, D.J.: Sharpness of the phase transition in percolation models. Comm. Math. Phys. **108**(3), 489–526 (1987)
2. Aizenman, M., Newman, C.M.: Tree graph inequalities and critical behavior in percolation models. J. Statist. Phys. **36**(1–2), 107–143 (1984)
3. Antal, P., Pisztora, A.: On the chemical distance for supercritical Bernoulli percolation. Ann. Probab. **24**(2), 1036–1048 (1996)
4. Baccelli, F., Haji-Mirsadeghi, M.O., Khezeli, A.: Unimodular Hausdorff and Minkowski dimensions. https://arxiv.org/abs/1807.02980
5. Barlow, M.T.: Random walks on supercritical percolation clusters. Ann. Probab. **32**(4), 3024–3084 (2004). http://dx.doi.org/10.1214/009117904000000748
6. Barlow, M.T.: Which values of the volume growth and escape time exponent are possible for a graph? Rev. Mat. Iberoamericana **20**(1), 1–31 (2004). https://doi.org/10.4171/RMI/378

7. Barlow, M.T.: Random walks and heat kernels on graphs. In: London Mathematical Society Lecture Note Series, vol. 438. Cambridge University Press, Cambridge (2017). https://doi.org/10.1017/9781107415690
8. Barlow, M.T., Bass, R.F.: Random walks on graphical Sierpinski carpets. In: Random Walks and Discrete Potential Theory (Cortona, 1997). Symposium of Mathematical, vol. XXXIX, pp. 26–55. Cambridge University, Cambridge (1999)
9. Barlow, M.T., Perkins, E.A.: Brownian motion on the Sierpiński gasket. Probab. Theory Related Fields **79**(4), 543–623 (1988). https://doi.org/10.1007/BF00318785
10. Barlow, M.T., Taylor, S.J.: Fractional dimension of sets in discrete spaces. J. Phys. A, Math. Gen. **22**(13), 2621–2626 (1989)
11. Barlow, M.T., Taylor, S.J.: Defining fractal subsets of $\mathbb{Z}^d$. Proc. Lond. Math. Soc. (3) **64**(1), 125–152 (1992)
12. Barlow, M.T., Járai, A.A., Kumagai, T., Slade, G.: Random walk on the incipient infinite cluster for oriented percolation in high dimensions. Comm. Math. Phys. **278**(2), 385–431 (2008)
13. Barsky, D.J., Aizenman, M.: Percolation critical exponents under the triangle condition. Ann. Probab. **19**(4), 1520–1536 (1991)
14. Berg, J.v.d., Kesten, H.: Inequalities with applications to percolation and reliability. J. Appl. Probab. **22**(3), 556–569 (1985)
15. Berger, N., Biskup, M.: Quenched invariance principle for simple random walk on percolation clusters. Probab. Theory Related Fields **137**(1–2), 83–120 (2007). http://dx.doi.org/10.1007/s00440-006-0498-z
16. Bollobás, B., Riordan, O.: Percolation. Cambridge University, New York (2006)
17. Brydges, D.C., Spencer, T.: Self-avoiding walk in 5 or more dimensions. Comm. Math. Phys. **97**(1–2), 125–148 (1985)
18. Cames van Batenburg, W.P.S.: The dimension of the incipient infinite cluster. Electron. Commun. Probab. **20**(33), 10 (2015). https://doi.org/10.1214/ECP.v20-3570
19. Damron, M.: Recent work on chemical distance in critical percolation (2016). Preprint arXiv:1602.00775 [math.PR]
20. Duminil-Copin, H., Tassion, V.: A new proof of the sharpness of the phase transition for Bernoulli percolation on $\mathbb{Z}^d$. Enseign. Math. (2) **62**(1–2), 199–206 (2016)
21. Durrett, R.: Probability. Theory and Examples, vol. 49, 5th edn. Cambridge University, Cambridge (2019)
22. Falconer, K.: Fractal geometry. In: Mathematical Foundations and Applications, 3rd edn. Wiley, Hoboken (2014)
23. Fitzner, R., van der Hofstad, R.: Mean-field behavior for nearest-neighbor percolation in $d > 10$. Electron. J. Probab. **22**, 65 (2017). Id/No 43
24. Fraser, J.M.: Interpolating between dimensions. In: Freiberg, U., Hambly, B., Hinz, M., Winter, S. (eds.) Fractal Geometry and Stochastics VI. Progress in Probability, vol. 76, pp. 3–24. Springer, Cham (2020). https://doi.org/10.1007/978-3-030-59649-1_1
25. Georgiou, N., Khoshnevisan, D., Kim, K., Ramos, A.D.: The dimension of the range of a transient random walk. Electron. J. Probab. **23**, 31 (2018). https://doi.org/10.1214/18-EJP201
26. Grimmett, G.: Percolation. In: Grundlehren der Mathematischen Wissenschaften [Fundamental Principles of Mathematical Sciences], vol. 321, 2nd edn. Springer, Berlin (1999)
27. Gromov, M.: Groups of polynomial growth and expanding maps. Inst. Hautes Études Sci. Publ. Math. **53**, 53–73 (1981)
28. Hara, T.: Decay of correlations in nearest-neighbour self-avoiding walk, percolation, lattice trees and animals. Ann. Probab. **36**(2), 530–593 (2008)
29. Hara, T., Slade, G.: Mean-field critical behaviour for percolation in high dimensions. Comm. Math. Phys. **128**(2), 333–391 (1990)
30. Hara, T., Hofstad, R.v.d., Slade, G.: Critical two-point functions and the lace expansion for spread-out high-dimensional percolation and related models. Ann. Probab. **31**(1), 349–408 (2003)
31. Hebisch, W., Saloff-Coste, L.: Gaussian estimates for Markov chains and random walks on groups. Ann. Probab. **21**(2), 673–709 (1993)

32. Heydenreich, M., van der Hofstad, R.: Progress in high-dimensional percolation and random graphs. In: CRM Short Courses. Springer, Cham; Centre de Recherches Mathématiques, Montreal, QC (2017)
33. Heydenreich, M., Hofstad, R.v.d., Hulshof, W.J.T.: Random walk on the high-dimensional IIC. Commun. Math. Phys. **329**(1), 57–115 (2014). https://doi.org/10.1007/s00220-014-1931-2
34. Heydenreich, M., Hofstad, R.v.d., Last, G., Matzke, K.: Lace expansion and mean-field behavior for the random connection model (2019). Preprint arXiv:1908.11356 [math.PR]
35. Hofstad, R.v.d., Járai, A.A.: The incipient infinite cluster for high-dimensional unoriented percolation. J. Statist. Phys. **114**(3–4), 625–663 (2004)
36. Kesten, H.: The critical probability of bond percolation on the square lattice equals 1/2. Comm. Math. Phys. **74**(1), 41–59 (1980)
37. Kesten, H.: The incipient infinite cluster in two-dimensional percolation. Probab. Theory Related Fields **73**(3), 369–394 (1986)
38. Kesten, H.: Subdiffusive behavior of random walk on a random cluster. Ann. Inst. H. Poincaré Probab. Statist. **22**(4), 425–487 (1986)
39. Kigami, J., Lapidus, M.L.: Weyl's problem for the spectral distribution of laplacians on p.c.f. self-similar fractals. Comm. Math. Phys. **158**(1), 93–125 (1993). https://projecteuclid.org:443/euclid.cmp/1104254132
40. Kozma, G., Nachmias, A.: The Alexander-Orbach conjecture holds in high dimensions. Invent. Math. **178**(3), 635–654 (2009)
41. Kozma, G., Nachmias, A.: Arm exponents in high dimensional percolation. J. Am. Math. Soc. **24**(2), 375–409 (2011). http://dx.doi.org/10.1090/S0894-0347-2010-00684-4
42. Kumagai, T., Misumi, J.: Heat kernel estimates for strongly recurrent random walk on random media. J. Theoret. Probab. **21**(4), 910–935 (2008). https://doi.org/10.1007/s10959-008-0183-5
43. Lawler, G.F., Schramm, O., Werner, W.: One-arm exponent for critical 2D percolation. Electron. J. Probab. **7**(2), 13 (2002). https://doi.org/10.1214/EJP.v7-101
44. Lehrbäck, J.: Assouad type dimensions in geometric analysis. In: Freiberg, U., Hambly, B., Hinz, M., Winter, S. (eds.) Fractal Geometry and Stochastics VI. Progress in Probability, vol. 76, pp. 25–46. Springer, Cham (2020). https://doi.org/10.1007/978-3-030-59649-1_2
45. Mathieu, P., Piatnitski, A.: Quenched invariance principles for random walks on percolation clusters. Proc. R. Soc. Lond. Ser. A Math. Phys. Eng. Sci. **463**(2085), 2287–2307 (2007). http://dx.doi.org/10.1098/rspa.2007.1876
46. Menshikov, M.V.: Coincidence of critical points in percolation problems. Dokl. Akad. Nauk SSSR **288**(6), 1308–1311 (1986)
47. Rammal, R., Toulouse, G.: Random walks on fractal structures and percolation clusters. J. de Physique Lettres **44**(1), 13–22 (1983). http://dx.doi.org/10.1051/jphyslet:0198300440101300. https://hal.archives-ouvertes.fr/jpa-00232136
48. Reimer, D.: Proof of the van den Berg-Kesten conjecture. Combin. Probab. Comput. **9**(1), 27–32 (2000). https://doi.org/10.1017/S0963548399004113
49. Sapozhnikov, A.: Upper bound on the expected size of the intrinsic ball. Electron. Comm. Probab. **15**, 297–298 (2010). https://doi.org/10.1214/ECP.v15-1553
50. Taylor, S.J., Tricot, C.: Packing measure, and its evaluation for a Brownian path. Trans. Am. Math. Soc. **288**, 679–699 (1985)
51. Wendelin, W.: Percolation et modèle d'Ising., vol. 16. Société Mathématique de France, Paris (2009)

Asymptotics of Integrals of Betti Numbers for Random Simplicial Complex Processes

Masanori Hino

Abstract We discuss a higher-dimensional analogue of Frieze's $\zeta(3)$-limit theorem for the Erdős–Rényi graph process applied to a family of increasing random simplicial complexes. In particular, we consider the time integrals of Betti numbers, which are interpreted as lifetime sums in the context of persistent homologies. We survey some recent results regarding their asymptotic behavior that answer some questions posed in an earlier study by Hiraoka and Shirai.

Keywords Random simplicial complex · Betti number · Persistent homology · Lifetime sum

Mathematics Subject Classifications (2010) Primary: 60D05; Secondary: 05C80, 55U10, 05E45, 60C05

1 Introduction

Extensive studies on limit behavior of random graphs have their origins in the work of Erdős and Rényi [4, 5]. Graph characteristics such as the threshold probability of connectivity and the limit behavior around the critical probability provide good descriptions of such complicated random discrete objects. In recent studies, the scaling limits of random graphs themselves have attracted attention in pursuit of a more comprehensive understanding; typical limit objects are continuum random trees, which have fractal structures (e.g., see [1, 21] and the references therein). The importance of fractal analysis in the study of random graphs will be emphasized more in future work.

Meanwhile, the homological structures of random simplicial complexes have also attracted interest recently as higher-dimensional counterparts of random

M. Hino (✉)
Department of Mathematics, Kyoto University, Kyoto, Japan
e-mail: hino@math.kyoto-u.ac.jp

U. Freiberg et al. (eds.), *Fractal Geometry and Stochastics VI*,
Progress in Probability 76, https://doi.org/10.1007/978-3-030-59649-1_6

graphs; see Kahle [16] for a survey of recent studies. In this connection, Hiraoka and Shirai [11] studied the asymptotic behavior of persistent homologies of random simplicial complex processes, and Hino and Kanazawa [10] advanced their research by solving some of the problems that they had posed. A natural question to consider next is to characterize suitable scaling limits of random simplicial complexes, which are certain to have fractal structures. However, unlike the case with random graphs, there are as yet no concrete results about this question because the theory and techniques are yet to be developed fully.

In this article, we follow [10, 11] and survey some recent results and new ideas in the study of the homologies of random simplicial complexes. We hope that this survey will serve as a preliminary to studying such objects from the perspective of fractal analysis.

The rest of this article is organized as follows. In Sect. 2, we introduce various concepts regarding random graphs, random graph processes, and their higher-dimensional analogues, and we state some results regarding their asymptotic behavior. In Sect. 3, we provide basic ideas for proving the main theorems. In Sect. 4, we enumerate several problems for future research.

2 Frameworks and Theorems

A typical random graph model is the Erdős–Rényi model $G(n, p)$ [4, 5, 9], which is defined as the distribution of a random graph consisting of n vertices with the edges between each pair of vertices included with probability p independently.[1] In one of the earliest results of random graph theory, Erdős and Rényi proved the following.

Theorem 2.1 ([5]) *Let $\varepsilon > 0$ and $p = p(n)$ depend on n.*

- *If $p < (1 - \varepsilon)(\log n)/n$ for sufficiently large n, then*

$$\mathbb{P}(\textit{the graph is disconnected}) \to 1 \quad \textit{as } n \to \infty.$$

- *If $p > (1 + \varepsilon)(\log n)/n$ for sufficiently large n, then*

$$\mathbb{P}(\textit{the graph is connected}) \to 1 \quad \textit{as } n \to \infty.$$

This theorem shows that the connectivity changes drastically around $p = (\log n)/n$. Since then, there have been many studies of the behavior around the threshold probability, which is one of the central topics of random graph theory.

Meanwhile, there have been other types of studies on the limit behavior of the Erdős–Rényi model. To explain one such type, we introduce a canonical realization

[1] This definition is due to Gilbert [9]. The model that Erdős and Rényi introduced in [4, 5] is slightly different, but the two models behave similarly as the number of vertices tends to infinity.

of the family of Erdős–Rényi models $\{G(n, p)\}_{p\in[0,1]}$ for fixed n. Let $K_n = V_n \sqcup E_n$ be the complete graph with n vertices, where V_n and E_n are the vertex set and the edge set, respectively. We assign independent and identically distributed random variables $\{u_e\}_{e\in E_n}$ that are uniformly distributed on $[0, 1]$. We construct a family of random graphs $\mathcal{X}_n = \{X_n(t)\}_{t\in[0,1]}$ so that each u_e is the birth time of the edge $e \in E_n$. More precisely, for each $t \in [0, 1]$, the random graph $X_n(t)$ is defined as

$$X_n(t) = V_n \sqcup \{e \in E_n \mid u_e \leq t\}.$$

By construction, $X_n(t)$ is nondecreasing with respect to t almost surely, and the law of $X_n(t)$ is equal to $G(n, t)$ for every $t \in [0, 1]$.

Let $L_0(\mathcal{X}_n)$ be the minimal weight of spanning trees[2] of K_n, that is,

$$L_0(\mathcal{X}_n) = \inf\left\{\sum_{e\in T} u_e \;\middle|\; T\text{: a spanning tree of } K_n\right\}.$$

This quantity has several interpretations: By Kruskal's algorithm [18], the identities

$$L_0(\mathcal{X}_n) = \int_0^1 \beta_0(X_n(t))\,dt = \sum_{i=1}^{n-1} t_i \tag{2.1}$$

hold, where $\beta_0(G)$ denotes the number of connected components of the graph G minus one, and t_i denotes the ith random time when the number of connected components of $X_n(t)$ decreases. Frieze [7] proved the asymptotic behavior of $L_0(\mathcal{X}_n)$ as follows.

Theorem 2.2 ([7]) *It holds that*

$$\lim_{n\to\infty} \mathbb{E}[L_0(\mathcal{X}_n)] = \zeta(3)\left(= \sum_{k=1}^{\infty} k^{-3}\right).$$

Moreover, for any $\varepsilon > 0$,

$$\lim_{n\to\infty} \mathbb{P}(|L_0(\mathcal{X}_n) - \zeta(3)| > \varepsilon) = 0.$$

Recently, Hiraoka and Shirai [11] studied a higher-dimensional analogue of (2.1) and Theorem 2.2, with random graphs and the number of connected components replaced by *random simplicial complexes* and the *(reduced) Betti number*, respectively. Let us briefly review the concepts of simplicial complexes and their homologies.

---

[2] A spanning tree of a graph G is, by definition, a tree that includes all the vertices of G.

Let V be a nonempty finite set. A collection X of nonempty subsets of V is called an (abstract) *simplicial complex* over V if the following conditions are satisfied.

- For every $v \in V$, $\{v\}$ belongs to X.
- For any $\sigma \in X$, every nonempty subset of σ belongs to X.

For $\sigma \in X$, $k := \#\sigma - 1$ is called the dimension of σ and is denoted by $\dim \sigma$. We call σ a k-dimensional simplex or, equivalently, a k-simplex, and we call the maximum of $\dim \sigma$ the dimension of X. Any finite graph can be regarded as either a zero- or one-dimensional simplicial complex. If two simplices σ and τ satisfy $\sigma \subset \tau$, then σ is called a face of τ.

For $k \geq 0$, $\sigma = (v_0, v_1, \ldots, v_k) \in V^{k+1}$ is called an ordered k-simplex of X if $\{v_0, v_1, \ldots, v_k\}$ is a k-simplex of X. Two ordered simplices are called equivalent if one is an even permutation of the other. The equivalence class of an ordered k-simplex σ is denoted by $\langle \sigma \rangle$ or $\langle v_0, v_1, \ldots, v_k \rangle$ and is called an oriented k-simplex of X. The space $C_k(X)$ of k-chains on X is defined as the real vector space consisting of all linear combinations of oriented k-simplices under the relation that $\langle v_0, v_1, \ldots, v_k \rangle = -\langle v_1, v_0, \ldots, v_k \rangle$ for any oriented k-simplices.

For $k \geq 1$, the kth boundary operator $\partial_k \colon C_k(X) \to C_{k-1}(X)$ is defined as a linear map such that for any $\langle \sigma \rangle = \langle v_0, v_1, \ldots, v_k \rangle \in C_k(X)$,

$$\partial_k \langle \sigma \rangle = \sum_{i=0}^{k} (-1)^i \langle v_0, \ldots, v_{i-1}, v_{i+1}, \ldots, v_k \rangle.$$

By convention, we define $C_{-1}(X) = \mathbb{R}$, and $\partial_0 \colon C_0(X) \to C_{-1}(X)$ is defined as a linear map such that $\partial_0 \langle v \rangle = 1$ for $v \in V$. Then, it holds that $\partial_k \circ \partial_{k+1} = 0$ for all $k \geq 0$. The *kth homology group* of X over $\mathbb{R}$ and the *kth (reduced) Betti number* are defined as $H_k(X) := \ker \partial_k / \operatorname{Im} \partial_{k+1}$ and $\beta_k(X) := \dim H_k(X)$, respectively.[3] Intuitively, $\beta_k(X)$ is interpreted as the number of k-dimensional holes in X. In particular, $\beta_0(X)$ is equal to the number of connected components of X minus one. In the standard definition, we note that ∂_0 would be defined as the zero operator, which makes our zeroth Betti number defined above equal to the conventional zeroth Betti number minus one.

Research interest has been growing in the higher-dimensional analogue of Theorem 2.1 and related topics; see the survey by Kahle [16] for recent studies. In general, the homological structures of large simplicial complexes are expected to be very complicated. Indeed, as the number of simplices increases, so the effect of creating holes competes against that of filling holes, thereby making the situation more problematic than simply analyzing graphs. A distant goal is to extract nice fractal structures from these simplicial complexes, but initially it would be

[3] In general, we can define the spaces $C_k(X, R)$ and $H_k(X, R)$ as R-modules for a commutative ring R. In this paper, we consider only the case $R = \mathbb{R}$.

meaningful to develop effective tools with which to study the limit behavior as the number of vertices tends to infinity.

We now consider a family $\mathcal{X} = \{X(t)\}_{t\geq 0}$ of subcomplexes of X, and we call it a *right-continuous filtration* of X if $X(s) \subset X(t)$ for $0 \leq s \leq t$ and $X(t) = \bigcap_{t'>t} X(t')$ for $t \geq 0$. Here, $X(t)$ can be an empty set, which is regarded as a (-1)-dimensional simplicial complex. Let $\mathbb{R}[\mathbb{R}_{\geq 0}]$ be a real vector space of formal linear combinations of finite elements of $\mathbb{R}_{\geq 0}$. We describe each element of $\mathbb{R}_{\geq 0}$ as z^t ($t \in \mathbb{R}_{\geq 0}$), where z is indeterminate. The product of two elements of $\mathbb{R}[\mathbb{R}_{\geq 0}]$ is defined so as to be consistent with $az^s \cdot bz^t = abz^{s+t}$ ($a, b \in \mathbb{R}$ and $s, t \in \mathbb{R}_{\geq 0}$). This operation equips $\mathbb{R}[\mathbb{R}_{\geq 0}]$ with a ring structure. For $k \geq 0$, the kth persistent homology $\mathrm{PH}_k(\mathcal{X})$ of $\mathcal{X} = \{X(t)\}_{t\geq 0}$ is defined as

$$\mathrm{PH}_k(\mathcal{X}) = \bigoplus_{t\geq 0} H_k(X(t)),$$

which is regarded as a graded module over $\mathbb{R}[\mathbb{R}_{\geq 0}]$. Here, $H_k(X(s))$ is considered as a subset of $H_k(X(t))$ for $0 \leq s \leq t$ by a natural inclusion from $X(s)$ to $X(t)$. The structure theorem of the persistent homology is stated as follows.

Theorem 2.3 (E.g., See [11, 22]) *For each $k \geq 0$, there exist unique indices $p, q \in \mathbb{Z}_{\geq 0}$ and $\{b_i\}_{i=1}^{p+q}$, $\{d_i\}_{i=1}^{p} \subset \mathbb{R}_{\geq 0}$ such that $b_i < d_i$ for all $i = 1, \ldots, p$, and the following graded module isomorphism holds:*

$$\mathrm{PH}_k(\mathcal{X}) \simeq \bigoplus_{i=1}^{p} \left((z^{b_i})/(z^{d_i})\right) \oplus \bigoplus_{i=p+1}^{p+q} (z^{b_i}),$$

where (z^a) denotes the ideal in $\mathbb{R}[\mathbb{R}_{\geq 0}]$ that is generated by the monomial z^a.

In Theorem 2.3, we call b_i and d_i the kth birth and death times, respectively, which indicate the appearance and disappearance of each k-dimensional "hole" in $\mathcal{X}$. The corresponding lifetime is defined as $l_i := d_i - b_i$. We set $d_i = l_i = \infty$ for $i = p+1, \ldots, p+q$, and we define the lifetime sum $L_k(\mathcal{X})$ as

$$L_k(\mathcal{X}) = \sum_{i=1}^{p+q} (d_i - b_i).$$

The following is a generalization of the second identity of (2.1) to filtrations.

Theorem 2.4 (Lifetime Formula [11, Proposition 2.2]) *It holds that*

$$L_k(\mathcal{X}) = \int_0^\infty \beta_k(X(t))\, dt.$$

Analogously, by defining

$$(L_k(\mathcal{X}))_T = \sum_{i=1}^{p+q} \big((d_i \wedge T) - (b_i \wedge T)\big)$$

for $T > 0$, we have

$$(L_k(\mathcal{X}))_T = \int_0^T \beta_k(X(t))\,dt.$$

An analogue of the first identity of (2.1) has also been obtained by introducing the concept of spanning acycles; see [11] for further details.

Now, we are interested in the asymptotic behavior of $L_k(\mathcal{X})$ for *random* filtrations as the number of vertices tends to infinity. The random models are introduced as follows.

For each $i \in \mathbb{Z}_{\geq 0}$, we take a probability distribution function p_i on $[0, +\infty]$. Let $n \in \mathbb{N}$ and let $K(n)$ denote the complete $(n-1)$-dimensional simplicial complex, namely the family of all nonempty subsets of an n-point set. We take a family of independent random variables $\{u_\tau\}_{\tau \in K(n)}$ such that each u_τ obeys the distribution function $p_{\dim \tau}$. We then define a random simplicial complex process $\mathcal{X}_n = \{X_n(t)\}_{t \geq 0}$ over n vertices by

$$X_n(t) := \{\sigma \in K(n) \mid u_\tau \leq t \text{ for every simplex } \tau (\neq \emptyset) \text{ with } \tau \subset \sigma\}. \tag{2.2}$$

We call this process a multi-parameter random complex process. We can also consider $\mathcal{X}_n = \{X_n(t)\}_{t \in [0,T]}$ for fixed $T > 0$ in an obvious manner. In this case, we write $L_k(\mathcal{X}_n)$ for $(L_k(\mathcal{X}_n))_T$.

We have the following typical examples in mind.

Example 2.5 (cf. [19]) Let $d \in \mathbb{N}$ be fixed. For each $i \in \mathbb{Z}_{\geq 0}$, define

$$p_i(t) = \begin{cases} 1 & (i < d) \\ t \wedge 1 & (i = d) \\ 0 & (i > d) \end{cases} \quad \text{for } t \geq 0.$$

In [10], the corresponding process $\mathcal{K}_n^{(d)} = \{K_n^{(d)}(t)\}_{t \in [0,1]}$ for $n > d$ and $T = 1$ is called the d-Linial–Meshulam complex process. By definition, for each $t \in [0, 1]$, the random simplicial complex $K_n^{(d)}(t)$ $(\subset K(n))$ is described as follows:

- $K_n^{(d)}(t)$ includes every simplex of $K(n)$ whose dimension is less than d.
- $K_n^{(d)}(t)$ includes each d-dimensional simplex of $K(n)$ with probability t independently.
- $K_n^{(d)}(t)$ includes no simplex of $K(n)$ whose dimension is greater than d.

The Erdős–Rényi graph process is identified with $\mathcal{K}_n^{(1)}$.

Example 2.6 (cf. [14]) Let $d \in \mathbb{N}$ be fixed. For each $i \in \mathbb{Z}_{\geq 0}$, define

$$p_i(t) = \begin{cases} 1 & (i < d) \\ t \wedge 1 & (i = d) \\ 1 & (i > d) \end{cases} \quad \text{for } t \geq 0.$$

In [10], the corresponding process $C_n^{(d)} = \{C_n^{(d)}(t)\}_{t \in [0,1]}$ for $n > d$ and $T = 1$ is called the d-flag complex process. By definition, for each $t \in [0, 1]$, the random simplicial complex $C_n^{(d)}(t)$ $(\subset K(n))$ is described as follows:

- $C_n^{(d)}(t)$ includes every simplex of $K(n)$ whose dimension is less than d.
- $C_n^{(d)}(t)$ includes each d-dimensional simplex of $K(n)$ with probability t independently.
- $C_n^{(d)}(t)$ includes each simplex σ of $K(n)$ whose dimension is greater than d if and only if every d-dimensional face of σ belongs to $C_n^{(d)}(t)$.

$C_n^{(1)}$ is also called the random clique complex process.

Our main concern is the asymptotic behavior of $\mathbb{E}[L_k(\mathcal{X}_n)]$ as $n \to \infty$. To state the results, we introduce the following functions:

$$\begin{aligned} q_{-1}(t) &:= 1, \quad q_k(t) := \prod_{i=0}^{k} \{p_i(t)\}^{\binom{k+1}{i+1}} \quad (k \geq 0), \\ r_k(t) &:= \frac{q_{k+1}(t)}{q_k(t)} = \prod_{i=0}^{k+1} \{p_i(t)\}^{\binom{k+1}{i}} \quad (k \geq -1). \end{aligned} \tag{2.3}$$

Note that $q_k(t)$ denotes the probability of a fixed k-simplex appearing at time t. For a k-simplex σ and a $(k+1)$-simplex τ with $\sigma \subset \tau$, $r_k(t)$ represents the conditional probability of τ appearing at time t given σ appearing.

Let $\check{r}_k$ denote the generalized inverse function of r_k, namely

$$\check{r}_k(u) = \inf\{t \geq 0 \mid r_k(t) > u\} \quad \text{for } u < 1,$$

and $\check{r}_k(1) = \infty$. We further define

$$Q_k(t) = \int_0^t q_k(s)\, ds \quad \text{for } t \geq 0,$$

$$\Phi_k(u) = Q_k(\check{r}_k(u)) \text{ and } \Psi_k(u) = Q_k(\check{r}_{k-1}(u)) \quad \text{for } u \in [0, 1).$$

In what follows, we use the standard notations big-O and little-o, and

- $f(u) = \Theta(g(u))$ means that $f(u) = O(g(u))$ and $g(u) = O(f(u))$ as $u \to 0$;
- $a_n \asymp b_n$ means that $a_n = O(b_n)$ and $b_n = O(a_n)$ as $n \to \infty$.

Below, k is a fixed number. The following result is a special case of more-general estimates [10, Theorems 4.3 and 4.4].

Theorem 2.7 ([10, Corollary 4.5]) *Suppose that $\Phi_k(u) = \Theta(u^a)$ for some $a \in [0, \infty)$ and $\Psi_k(u) = o(\Phi_k(u))$ as $u \to 0$. Then, for each $T > 0$,*

$$\mathbb{E}[(L_k(\mathcal{X}_n))_T] \asymp n^{k+1-a}. \tag{2.4}$$

Moreover, if $\int_0^\infty t^{1+\delta} dq_{k+1}(t) < \infty$ for some $\delta > 0$, then

$$\mathbb{E}[L_k(\mathcal{X}_n)] \asymp n^{k+1-a}. \tag{2.5}$$

The following is a rather simple case but is not treated in Theorem 2.7.

Theorem 2.8 ([10, Theorem 4.6]) *If $\Phi_k(u) = \Psi_k(u)$ for all $u \in [0, 1)$, then $L_k(\mathcal{X}_n) = 0$ almost surely for all $n \in \mathbb{N}$.*

We apply these results to Examples 2.5 and 2.6.

Example 2.9 ([10, Example 4.8]) We consider the d-flag complex process $C_n^{(d)} = \{C_n^{(d)}(t)\}_{t\in[0,1]}$ as in Example 2.6. From straightforward computation, we obtain

$$(\Phi_k(u), \Psi_k(u)) = \begin{cases} (0, 0) & (k < d-1), \\ (u, 0) & (k = d-1), \\ \left(\Theta\left(u^{\frac{k+1-d}{d+1} + \binom{k+1}{d}^{-1}}\right), \Theta\left(u^{\frac{k+1}{d+1} + \binom{k}{d}^{-1}}\right)\right) & (k \geq d). \end{cases}$$

From Theorems 2.7 and 2.8, we have

$$\mathbb{E}[L_k(C_n^{(d)})] \asymp \begin{cases} 0 & (k < d-1), \\ n^{\frac{(k+2)d}{d+1} - \binom{k+1}{d}^{-1}} & (k \geq d-1). \end{cases}$$

In particular,

$$\mathbb{E}[L_k(C_n^{(1)})] \asymp n^{k/2+1-1/(k+1)}.$$

This estimate improves Theorem 6.10 in [11] and determines the growth order, thereby answering the question posed in [11, Section 7.4].

Example 2.10 ([10, Example 4.7]) We consider the d-Linial–Meshulam complex process $\mathcal{K}_n^{(d)} = \{K_n^{(d)}(t)\}_{t\in[0,1]}$ as in Example 2.5. It is straightforward to see that

$$(\Phi_k(u), \Psi_k(u)) = \begin{cases} (0,0) & (k < d-1), \\ (u,0) & (k = d-1), \\ (1/2, u^2/2) & (k = d), \\ (0,0) & (k > d). \end{cases}$$

From Theorems 2.7 and 2.8, we have

$$\mathbb{E}[L_k(\mathcal{K}_n^{(d)})] \asymp \begin{cases} 0 & (k \neq d-1, d), \\ n^{d-1} & (k = d-1), \\ n^{d+1} & (k = d). \end{cases}$$

The case $k = d-1$ corresponds to [11, Theorem 1.2].

In fact, we have more-precise asymptotics for $L_{d-1}(\mathcal{K}_n^{(d)})$. Following [11, 20], we introduce the limit constant. Let $t_1^* = c_1^* = 1$. For $d \geq 2$, let t_d^* be the unique root in $(0, 1)$ of

$$(d+1)(1-t) + (1+dt)\log t = 0, \tag{2.6}$$

and define $c_d^* = (-\log t_d^*)/(1-t_d^*)^d > 0$. For $c \geq c_d^*$, let t_c denote the smallest positive root of $(-\log t)/(1-t)^d = c$. Define functions g_d and h_d on $[0,\infty)$ as

$$g_d(c) = \begin{cases} 0 & (c < c_d^*), \\ ct_c(1-t_c)^d + \dfrac{c}{d+1}(1-t_c)^{d+1} - (1-t_c) & (c \geq c_d^*), \end{cases}$$

and

$$h_d(c) = 1 - \frac{c}{d+1} + g_d(c).$$

We also define

$$I_{d-1} := \frac{1}{d!}\int_0^\infty h_d(s)\,ds.$$

Then, the limit behavior of $L_{d-1}(\mathcal{K}_n^{(d)})$ is described as follows.

Theorem 2.11 (Part of [10, Theorem 4.11]) *Let $d \geq 1$. The constant I_{d-1} is finite, and for any $r \in [1, \infty)$,*

$$\lim_{n\to\infty} \mathbb{E}\left[\left|\frac{L_{d-1}(\mathcal{K}_n^{(d)})}{n^{d-1}} - I_{d-1}\right|^r\right] = 0;$$

in particular, $\mathbb{E}[L_{d-1}(\mathcal{K}_n^{(d)})]/n^{d-1}$ converges to I_{d-1} as $n \to \infty$.

This claim is a justification of an informal discussion in [11, Section 7.1]. Note that $I_0 = \zeta(3)$, and Theorem 2.11 with $d = 1$ is consistent with Theorem 2.2. See [10, Section 4.4] for explicit expressions for general I_{d-1} and further information. In particular, we have

$$\begin{aligned} I_1 &= \frac{1}{2}\left[\mathrm{Li}_2(t_2^*) + (\log t_2^*)\log(1-t_2^*) + \frac{t_2^*(\log t_2^*)^2}{2(1-t_2^*)} + \frac{(\log t_2^*)\{\log t_2^* + (1-t_2^*)\}}{4(1-t_2^*)^2}\right] \\ &= \frac{1}{2}\left[\mathrm{Li}_2(t_2^*) + (\log t_2^*)\log(1-t_2^*) + \frac{3(1-t_2^*)(1+3t_2^*)}{2(1+2t_2^*)^2}\right] \end{aligned} \tag{2.7}$$

and

$$\begin{aligned} I_2 &= \frac{1}{12}\left[\mathrm{Li}_2(t_3^*) + (\log t_3^* - 1)\log(1-t_3^*) + \frac{t_3^*(\log t_3^*)(\log t_3^* - 2)}{2(1-t_3^*)}\right. \\ &\quad \left. + \frac{t_3^*(\log t_3^*)^2}{2(1-t_3^*)^2} + \frac{(\log t_3^*)\{\log t_3^* + (1-t_3^*)\}}{3(1-t_3^*)^3}\right] \\ &= \frac{1}{12}\left[\mathrm{Li}_2(t_3^*) + (\log t_3^* - 1)\log(1-t_3^*) + \frac{4((t_3^*)^2 + 5t_3^* + 1)}{(1+3t_3^*)^2}\right], \end{aligned} \tag{2.8}$$

where $\mathrm{Li}_2(x)$ denotes the dilogarithm

$$\mathrm{Li}_2(x) = \sum_{k=1}^{\infty} \frac{x^k}{k^2} \quad (-1 \leq x \leq 1).$$

We remark that the second identities of (2.7) and (2.8) follow from the fact that t_d^* is a root of (2.6).

3 Ideas for Proving the Theorems

In this section, we explain some basic ideas for proving the main results (Theorems 2.7 and 2.11) in the previous section, following [10]. Because

$$\mathbb{E}[(L_k(\mathcal{X}))_T] = \int_0^T \mathbb{E}[\beta_k(X(t))]\,dt \quad \text{and} \quad \mathbb{E}[L_k(\mathcal{X})] = \int_0^\infty \mathbb{E}[\beta_k(X(t))]\,dt$$

from Theorem 2.4, it suffices to obtain a sufficiently sharp estimate of $\mathbb{E}[\beta_k(X(t))]$ for each random simplicial complex $X(t)$. In general, it is a difficult problem to obtain a good estimate of a Betti number for a large simplicial complex X. The following is a basic estimate.

Lemma 3.1 *For every $k \geq 0$,*

$$f_k(X) - f_{k+1}(X) - f_{k-1}(X) \leq \beta_k(X) \leq f_k(X), \tag{3.1}$$

where $f_k(X)$ denotes the number of all k-simplices of X, and $f_{-1}(X) = 1$ by convention.

This is a version of the Morse inequality and is proved by simple application of linear algebra. Lemma 3.1 provides good upper and lower estimates of $\mathbb{E}[\beta_k(X(t))]$ if t is sufficiently small. In fact, as crucially noticed in [11], replacing X in the first inequality of (3.1) by $X(t)$ and integrating with respect to t on a small interval $[0, t_0]$ gives a lower estimate in (2.4) with the correct growth order. Thus, the main difficulty in the proof of Theorem 2.7 is the upper estimate in (2.4) and (2.5).

For general t, we require another strategy for estimating $\beta_k(X(t))$. To explain this strategy, we introduce several concepts from graph theory and topology. Let G be a finite undirected graph with a vertex set V, an edge set E, and with no loops or multiple edges. The degree $\deg(v)$ of a vertex $v \in V$ is defined as the number of $w \in V$ such that $\{v, w\} \in E$. The averaging matrix $A[G] = \{a_{vw}\}_{v,w \in V}$ of G is defined as

$$a_{vw} := \begin{cases} 1/\deg(v) & \text{if } \{v, w\} \in E, \\ 1 & \text{if } \deg(v) = 0 \text{ and } v = w, \\ 0 & \text{otherwise.} \end{cases}$$

This is interpreted as the transition probability of a simple random walk on G. The Laplacian $\mathcal{L}[G]$ of G is defined as $\mathcal{L}[G] = I_V - A[G]$, where I_V is the matrix that acts as the identity operator on V. Let $\{\lambda_i\}_{i=1}^{\#V}$ be all the (not necessarily distinct) eigenvalues of $\mathcal{L}[G]$. Note that $\lambda_i \in [0, 2]$ for all i and at least one λ_i is zero. Define

$$\gamma(G; \alpha) := \#\{i \mid \lambda_i \leq \alpha\} - 1 \, (\geq 0)$$

for $\alpha \geq 0$. By convention, $\gamma(\emptyset; \alpha) := 0$.

Given a D-dimensional simplicial complex X and a j-simplex τ in X with $-1 \leq j \leq D$, the *link* $\mathrm{lk}_X(\tau)$ of τ in X is defined as

$$\mathrm{lk}_X(\tau) := \{\sigma \in X \mid \tau \cap \sigma = \emptyset \text{ and } \tau \cup \sigma \in X\}.$$

This is either an empty set or a simplicial complex whose dimension is at most $D - j - 1$. Let $\mathrm{lk}_X(\tau)^{(1)}$ denote the 1-skeleton of $\mathrm{lk}_X(\tau)$, that is, the totality of the simplices of $\mathrm{lk}_X(\tau)$ whose dimensions are at most 1. This is either an empty set or a graph.

A key estimate is described as follows.

Theorem 3.2 ([10, Theorem 2.5]) *Suppose that the dimension D of X is greater than or equal to* 1. *Then*

$$\beta_{D-1}(X) \le \sum_{\tau} \gamma\big(\mathrm{lk}_X(\tau)^{(1)}; 1 - D^{-1}\big), \tag{3.2}$$

where τ in the summation is taken to be all $(D-2)$-simplices of X.

Informally speaking, this claim says that the Betti number is dominated by the sum of the number of small eigenvalues of the Laplacian on the 1-skeleton of each link of X. In particular, if the right-hand side of (3.2) is zero, then $H_{D-1}(X) = \{0\}$. In this sense, Theorem 3.2 is regarded as a quantitative generalization of the *cohomology vanishing theorem*[4] [2, 8]. The proof of Theorem 3.2 is based on a careful modification of that of [2, Theorem 2.1] and some additional arguments to remove extra assumptions.

From Theorem 3.2, under the assumption that (3.2) provides a sufficiently sharp estimate, the upper estimate of the Betti number is reduced to counting small eigenvalues of Laplacians on graphs. If X is a random simplicial complex, then this is closely related to the study of the eigenvalues of random matrices.

We apply this estimate to the following multi-parameter random simplicial complexes that were introduced in [3, 6]. Let $\{p_i\}_{i=0}^{\infty}$ be fixed parameters with $0 \le p_i \le 1$ for all i. We define a sequence of random simplicial complexes $\{X_n\}_{n \in \mathbb{N}}$ as follows. For each $n \in \mathbb{N}$, we start with a set V of n vertices and retain each vertex with independent probability p_0. Each edge with both ends retained is added with probability p_1, independently. Iteratively, for $i = 1, 2, \dots, n-1$, each i-simplex for which all faces were added by the previous procedures is added with probability p_i, independently. The resulting random simplicial complex is X_n. From the definition, $\{X_n(t)\}_{n \in \mathbb{N}}$ defined in (2.2) for fixed t is nothing but $\{X_n\}_{n \in \mathbb{N}}$ with parameters $\{p_i(t)\}_{i=0}^{\infty}$.

Just as in (2.3), we define

$$q_{-1} := 1, \quad q_k := \prod_{i=0}^{k} p_i^{\binom{k+1}{i+1}} \quad (k \ge 0),$$

$$r_k := \frac{q_{k+1}}{q_k} = \prod_{i=0}^{k+1} p_i^{\binom{k+1}{i}} \quad (k \ge -1).$$

[4]The proof is based on the discussion of the cohomology, not the homology. However, they are isomorphic.

Then, a crucial estimate is described as follows.

Theorem 3.3 ([10, Theorem 3.6]) *Let $k \geq 0$ and $l \in \mathbb{N}$. Then, there exists a positive constant C depending only on k and l such that, for all $n \in \mathbb{N}$,*

$$\mathbb{E}[\beta_k(X_n)] \leq n^{k+1} q_k \left\{1 \wedge C(nr_k)^{-l}\right\}. \tag{3.3}$$

We give a brief outline of the proof of Theorem 3.3. Lemma 3.1 immediately implies the inequality

$$\mathbb{E}[\beta_k(X_n)] \leq n^{k+1} q_k. \tag{3.4}$$

Therefore, it suffices to prove the inequality

$$\mathbb{E}[\beta_k(X_n)] \leq C n^{k+1} q_k (nr_k)^{-l} \tag{3.5}$$

for some C. The proof is decomposed into the following three cases. The constants $K_1 \leq K_2$ below should be taken appropriately.

Case 1 If $r_k \geq \dfrac{K_1}{n} \vee \dfrac{(nr_{k-1})^{1/l}}{n}$, then the effect of "filling k-dimensional holes" is strong; (3.5) follows from a variant of the cohomology vanishing theorem of random simplicial complexes (e.g., [15, Theorem 1.1 (1)] and [6, Theorem 1.1]) that is based on insightful results regarding spectral gaps on random graphs by Hoffman, Kahle, and Paquette [12].

Case 2 If $\dfrac{K_2}{n} \leq r_k \leq \dfrac{(nr_{k-1})^{1/l}}{n}$, then we use a general inequality

$$\begin{aligned}
&\#\{\text{eigenvalues of } L \text{ (counting multiplicities) greater than } \alpha\} \\
&= \#\{\text{eigenvalues of } (L/\alpha)^l \text{ (counting multiplicities) greater than unity}\} \\
&\leq \operatorname{tr}((L/\alpha)^l) = \alpha^{-l} \operatorname{tr}(L^l)
\end{aligned}$$

for nonnegative-definite symmetric matrices L and $\alpha > 0$. Applying this by letting $L = \mathcal{L}[\mathrm{lk}_{X_n}(\tau)]$ with $\tau \in X_n$ and $\alpha = 1 - 1/(k+1)$, and using some combinatorial arguments for estimating $\operatorname{tr}(L^l)$, we can prove (3.5) via Theorem 3.2.

Case 3 If $r_k \leq \dfrac{K_2}{n}$, then (3.4) implies (3.5) for a suitable C.

Remark 3.4 As seen from the above explanation, the novel Betti-number estimate is that in the intermediate range (Case 2). We remark that combinatorial arguments that are similar in spirit are also found in the classical proof of Wigner's semicircle law of random matrices, albeit in a slightly different situation.

Now, we obtain

$$\begin{aligned}\mathbb{E}[L_k(\mathcal{X}_n)] &= \int_0^\infty \mathbb{E}[\beta_k(X_n(t))]\,dt \quad \text{(from Theorem 2.4)}\\ &\le \int_0^\infty n^{k+1} q_k(t)\left\{1 \wedge C(nr_k(t))^{-l}\right\} dt \quad \text{(from Theorem 3.3).}\end{aligned}$$

Taking l to be sufficiently large and performing some elementary calculations, we reach an estimate $\mathbb{E}[L_k(\mathcal{X}_n)] = O(n^{k+1-a})$ as $n \to \infty$. The estimate of $\mathbb{E}[(L_k(\mathcal{X}_n))_T]$ is similarly proved, which completes the proof of Theorem 2.7.

In proving Theorem 2.11, the following is the key fact and follows from the results by Linial and Peled [20] that come from the convergence of a sequence of random graphs induced by $\{K_n^{(d)}(s/n)\}_{n\in\mathbb{N}}$ for fixed $s \ge 0$.

Theorem 3.5 *For any $s \ge 0$ and $\varepsilon > 0$,*

$$\lim_{n\to\infty} \mathbb{P}\left(\left|\frac{\beta_d(K_n^{(d)}(s/n))}{\binom{n}{d}} - g_d(s)\right| > \varepsilon\right) = 0.$$

With the help of the Euler–Poincaré formula, we can prove that, for each $s \ge 0$ and $\varepsilon > 0$,

$$\lim_{n\to\infty} \mathbb{P}\left(\left|\frac{\beta_{d-1}(K_n^{(d)}(s/n))}{\binom{n}{d}} - h_d(s)\right| > \varepsilon\right) = 0. \tag{3.6}$$

We note that

$$\begin{aligned}\left\|\frac{L_{d-1}(\mathcal{K}_n^{(d)})}{n^{d-1}} - I_{d-1}\right\|_{L^r} &= \left\|\int_0^\infty \left(\frac{\beta_{d-1}(K_n^{(d)}(s/n))}{n^d} 1_{[0,n]}(s) - \frac{1}{d!}h_d(s)\right) ds\right\|_{L^r}\\ &\le \int_0^\infty U_n(s)\,ds,\end{aligned}$$

where

$$U_n(s) = \left\|\frac{\beta_{d-1}(K_n^{(d)}(s/n))}{n^d} 1_{[0,n]}(s) - \frac{1}{d!}h_d(s)\right\|_{L^r}.$$

Combining Theorem 3.3 and (3.6), we obtain $\lim_{n\to\infty} U_n(s) = 0$ for each $s \ge 0$ and $\sup_{n\in\mathbb{N}} U_n(s)$ is Lebesgue integrable over $[0, \infty)$. The dominated convergence theorem implies that $\int_0^\infty U_n(s)\,ds$ converges to zero as $n \to \infty$, which finishes the proof of Theorem 2.11.

A similar outline was discussed informally in [11, Section 7.1]. However, because we now have the uniform estimate (3.3), we can provide a rigorous proof.

4 Concluding Remarks

Theorems 2.7 and 2.11 remain at the beginning of the homological study of families of random simplicial complexes. We will describe some potential directions for future research.

1. In [11], discrete Morse theory was used for estimating $L_k(\mathcal{X}_n)$. Although the argument therein did not provide the optimal asymptotics, it may be interesting to investigate that approach further.
2. Work is in progress [17] to prove the existence and identify the limit of scaled expectations of $L_k(\mathcal{X}_n)$ (Theorem 2.11) for general models other than d-Linial–Meshulam complex processes.
3. The limit constants [e.g., (2.7) and (2.8)] for d-Linial–Meshulam complex processes are regarded as "higher-dimensional" analogues of $\zeta(3)$, but the question remains as to whether they have simpler expressions.
4. As already mentioned in [11], the next problem to be considered is proving the central limit theorem for $L_k(\mathcal{X}_n)$. In the case of the Erdős–Rényi process, this has been proved by Janson [13].
5. The sum of the αth power ($\alpha > 0$) of lifetimes was studied in [10, Theorem 4.11] for d-Linial–Meshulam complex processes. In any further investigation, it would not be sufficient to study only the homologies of the simplicial complexes $X_n(t)$ for fixed t: we require the homological structure of the filtration $\{X_n(t)\}_{t\geq 0}$ itself.
6. Regarding item 5 in this list, the scaling limit of graphs in the Gromov–Hausdorff–Prokhorov topology has also been studied extensively (see [1, 21] and the references therein for recent studies). The limit objects in that case would have fractal structures and should provide detailed information about random graphs. Studying the counterpart of random simplicial complexes or their filtrations would be required for more-comprehensive understanding.

Acknowledgments This study was supported by JSPS KAKENHI Grant Numbers JP19H00643 and JP19K21833.

References

1. Addario-Berry, L., Broutin, N., Goldschmidt, C., Miermont, G.: The scaling limit of the minimum spanning tree of the complete graph. Ann. Probab. **45**, 3075–3144 (2017)
2. Ballmann, W., Świątkowski, J.: On L^2-cohomology and property (T) for automorphism groups of polyhedral cell complexes. Geom. Funct. Anal. **7**, 615–645 (1997)
3. Costa, A., Farber, M.: Random simplicial complexes. In: Configuration Spaces. Springer INdAM Series, vol. 14, pp. 129–153. Springer, New York (2016)
4. Erdős, P., Rényi, A.: On random graphs. Publ. Math. Debrecen **6**, 290–297 (1959)
5. Erdős, P., Rényi, A.: On the evolution of random graphs. Publ. Math. Inst. Hungarian Acad. Sci. **5A**, 17–61 (1960)
6. Fowler, C.F.: Homology of multi-parameter random simplicial complexes. Discrete Comput. Geom. **62**, 87–127 (2019)

7. Frieze, A.M.: On the value of a random minimum spanning tree problem. Discrete Appl. Math. **10**, 47–56 (1985)
8. Garland, H.: p-adic curvature and the cohomology of discrete subgroups of p-adic groups. Ann. Math. (2) **97**, 375–423 (1973)
9. Gilbert, E.N.: Random graphs. Ann. Math. Stat. **30**, 1141–1144 (1959)
10. Hino, M., Kanazawa, S.: Asymptotic behavior of lifetime sums for random simplicial complex processes. J. Math. Soc. Jpn. **71**, 765–804 (2019)
11. Hiraoka, Y., Shirai, T.: Minimum spanning acycle and lifetime of persistent homology in the Linial–Meshulam process. Rand. Struct. Algor. **51**, 315–340 (2017)
12. Hoffman, C., Kahle, M., Paquette, E.: Spectral gaps of random graphs and applications. Int. Math. Res. Not. (to appear). https://doi.org/10.1093/imrn/rnz077
13. Janson, S.: The minimal spanning tree in a complete graph and a functional limit theorem for trees in a random graph. Rand. Struct. Algor. **7**, 337–355 (1995)
14. Kahle, M.: Topology of random clique complexes. Discrete Math. **309**, 1658–1671 (2009)
15. Kahle, M.: Sharp vanishing thresholds for cohomology of random flag complexes. Ann. Math. (2) **179**, 1085–1107 (2014)
16. Kahle, M.: Topology of random simplicial complexes: a survey. In: Algebraic Topology: Applications and New Directions. Contemporary Mathematics, vol. 620, pp. 201–221. American Mathematical Society, Providence, RI (2014)
17. Kanazawa, S.: Topology of random simplicial complexes and lifetimes of persistent homology, PhD Thesis, Tohoku University, (2020). http://hdl.handle.net/10097/00128434
18. Kruskal Jr., J.B.: On the shortest spanning subtree of a graph and the traveling salesman problem. Proc. Am. Math. Soc. **7**, 48–50 (1956)
19. Linial, N., Meshulam, R.: Homological connectivity of random 2-complexes. Combinatorica **26**, 475–487 (2006)
20. Linial, N., Peled, Y.: On the phase transition in random simplicial complexes. Ann. Math. (2) **184**, 745–773 (2016)
21. Panagiotou, K., Stufler, B., Weller, K.: Scaling limits of random graphs from subcritical classes. Ann. Probab. **44**, 3291–3334 (2016)
22. Zomorodian, A., Carlsson, G.: Computing persistent homology. Discrete Comput. Geom. **33**, 249–274 (2005)

Part III
Trees and Hyperbolicity

The Continuum Self-Similar Tree

Mario Bonk and Huy Tran

Abstract We introduce the continuum self-similar tree (CSST) as the attractor of an iterated function system in the complex plane. We provide a topological characterization of the CSST and use this to relate the CSST to other metric trees such as the continuum random tree (CRT) and Julia sets of postcritically-finite polynomials.

Keywords Metric tree · Iterated function system · Continuum random tree · Julia set

Mathematics Subject Classifications (2010) Primary: 37C70; Secondary: 37B45

1 Introduction

In this expository paper, we study the topological properties of a certain subset $\mathbb{T}$ of the complex plane $\mathbb{C}$. It is defined as the attractor of an iterated function system. As we will see, $\mathbb{T}$ has a self-similar "tree-like" structure with very regular branching behavior. In a sense it is the simplest object of this type. Sets homeomorphic to $\mathbb{T}$ appear in various other contexts. Accordingly, we give the set $\mathbb{T}$ a special name, and call it the *continuum self-similar tree* (CSST).

To give the precise definition of $\mathbb{T}$ we consider the following contracting homeomorphisms on $\mathbb{C}$:

$$f_1(z) = \tfrac{1}{2}z - \tfrac{1}{2}, \quad f_2(z) = \tfrac{1}{2}\bar{z} + \tfrac{1}{2}, \quad f_3(z) = \tfrac{i}{2}\bar{z} + \tfrac{i}{2}. \tag{1.1}$$

M. Bonk (✉)
Department of Mathematics, University of California, Los Angeles, CA, USA
e-mail: mbonk@math.ucla.edu

H. Tran
Institut für Mathematik, Technische Universität Berlin, Berlin, Germany
e-mail: tran@math.tu-berlin.de

U. Freiberg et al. (eds.), *Fractal Geometry and Stochastics VI*,
Progress in Probability 76, https://doi.org/10.1007/978-3-030-59649-1_7

Then the following statement is true.

Proposition 1.1 *There exists a unique non-empty compact set* $\mathbb{T} \subseteq \mathbb{C}$ *satisfying*

$$\mathbb{T} = f_1(\mathbb{T}) \cup f_2(\mathbb{T}) \cup f_3(\mathbb{T}). \tag{1.2}$$

Based on this fact, we make the following definition.

Definition 1.2 The continuum self-similar tree (CSST) is the set $\mathbb{T} \subseteq \mathbb{C}$ as given by Proposition 1.1.

In other words, $\mathbb{T}$ is the attractor of the iterated function system $\{f_1, f_2, f_3\}$ in the plane. Proposition 1.1 is a special case of well-known more general results in the literature (see [19], [16, Theorem 9.1], or [21, Theorem 1.1.4], for example). We will recall the argument that leads to Proposition 1.1 in Sect. 3.

Spaces of a similar topological type as $\mathbb{T}$ have appeared in the literature before (among the more recent examples is the *antenna set* in [3] or *Hata's tree-like set* considered in [21, Example 1.2.9]). For a representation of $\mathbb{T}$ see Fig. 1.

To describe the topological properties of $\mathbb{T}$, we introduce the following concept.

Definition 1.3 A *(metric) tree* is a compact, connected, and locally connected metric space (T, d) containing at least two points such that for all $a, b \in T$ with $a \neq b$ there exists a unique arc $\alpha \subseteq T$ with endpoints a and b.

Fig. 1 The continuum self-similar tree $\mathbb{T}$

In other words, any two distinct points a and b in a metric tree can be joined by a unique arc α in T. It is convenient to allow $a = b$ here in which case $\alpha = \{a\} = \{b\}$ and we consider α as a *degenerate* arc.

In the following, we will usually call a metric space as in Definition 1.3 a *tree* and drop the word "metric" for simplicity. It is easy to see that the concept of a tree is essentially the same as the concept of a *dendrite* that appears in the literature (see, for example, [29, Chapter V], [23, Section §51 VI], [27, Chapter X]). More precisely, a metric space T is a tree if and only if it is a non-degenerate dendrite (the simple proof is recorded in [6, Proposition 2.2]). If one drops the compactness assumption in Definition 1.3, but requires in addition that the space is geodesic (see below for the definition), then one is led to the notion of a *real tree*. They appear in many areas of mathematics (see [2, 25], for example).

The following statement is suggested by Fig. 1.

Proposition 1.4 *The continuum self-similar tree* $\mathbb{T}$ *is a metric tree.*

If T is a tree, then for $x \in T$ we denote by $\nu_T(x) \in \mathbb{N} \cup \{\infty\}$ the number of (connected) components of $T \backslash \{x\}$. This number $\nu_T(x)$ is called the *valence* of x. If $\nu_T(x) = 1$, then x is called a *leaf* of T. If $\nu_T(x) \geq 3$, then x is a *branch point* of T. If $\nu_T(x) = 3$, then we also call x a *triple point.*

The following statement is again suggested by Fig. 1.

Proposition 1.5 *Each branch point of the tree* $\mathbb{T}$ *is a triple point, and these triple points are dense in* $\mathbb{T}$.

The set $\mathbb{T}$ has an interesting geometric property, namely it is a *quasi-convex* subset of $\mathbb{C}$., i.e., any two points in $\mathbb{T}$ can be joined by a path whose length is comparable to the distance of the points.

Proposition 1.6 *There exists a constant* $L > 0$ *with the following property: if* $a, b \in \mathbb{T}$ *and* α *is the unique arc in* $\mathbb{T}$ *joining* a *and* b*, then*

$$\text{length}(\alpha) \leq L|a - b|.$$

Note that a unique (possibly degenerate) arc $\alpha \subseteq \mathbb{T}$ joining a and b exists, because $\mathbb{T}$ is a tree according to by Proposition 1.4.

Proposition 1.6 implies that we can define a new metric ϱ on $\mathbb{T}$ by setting $\varrho(a, b) = \text{length}(\alpha)$ for $a, b \in \mathbb{T}$, where α is the unique arc in $\mathbb{T}$ joining a and b. Then the metric space $(\mathbb{T}, \varrho)$ is *geodesic*, i.e., any two points in $(\mathbb{T}, \varrho)$ can be joined by a path in $\mathbb{T}$ whose length is equal to the distance of the points. It immediately follows from Proposition 1.6 that metric spaces $\mathbb{T}$ (as equipped with the Euclidean metric) and $(\mathbb{T}, \varrho)$ are bi-Lipschitz equivalent by the identity map.

A natural way to construct $(\mathbb{T}, \varrho)$, at least as an abstract metric space, is as follows. We start with a line segment J_0 of length 2. Its midpoint c subdivides J_0 into two line segments of length 1. We glue to c one of the endpoints of another line segment s of the same length. Then we obtain a set J_1 consisting of three line segment of length 1. The set J_1 carries the natural path metric. We now repeat this procedure inductively. At the nth step we obtain a tree J_n consisting of 3^n line segments of length 2^{1-n}. To pass to J_{n+1}, each of these line segments s is subdivided by its midpoint c_s into two line segment of length 2^{-n} and we glue to c_s one endpoint of another line segment of length 2^{-n}.

In this way, we obtain an ascending sequence $J_0 \subseteq J_1 \subseteq \dots$ of trees equipped with a geodesic metric. The union $J = \bigcup_{n\in\mathbb{N}_0} J_n$ carries a natural path metric ϱ that agrees with the metric on J_n for each $n \in \mathbb{N}_0$. As an abstract space one can define $(\mathbb{T}, \varrho)$ as the completion of the metric space (J, ϱ).

If one wants to realize $\mathbb{T}$ as a subset of $\mathbb{C}$ by this construction, one starts with the initial line segment $J_0 = [-1, 1]$, and adds $s = [0, i]$ in the first step to obtain $J_1 = [-1, 0] \cup [0, 1] \cup [0, i]$. Now one wants to choose suitable Euclidean similarities f_1, f_2, f_3 that copy the interval $[-1, 1]$ to $[-1, 0]$, $[0, 1]$, $[0, i]$, respectively. One hopes to realize J_n as a subset of $\mathbb{C}$ using an inductive procedure based on

$$J_{n+1} = f_1(J_n) \cup f_2(J_n) \cup f_3(J_n), \quad n \in \mathbb{N}_0.$$

In order to avoid self-intersections and ensure that each set J_n is indeed a tree, one has to be careful about the orientations of the maps f_1, f_2, f_3. The somewhat non-obvious choice of these maps as in (1.1) leads to the desired result. See Proposition 4.2 and the discussion near the end of Sect. 4 for precise statements how to use the maps in (1.1) to realize the sets J_n as subsets of $\mathbb{C}$, and obtain $\mathbb{T}$ (as in Definition 1.2) as the closure of $\bigcup_{n\in\mathbb{N}_0} J_n$. A representation of J_5 is shown in Fig. 2.

The conditions in Proposition 1.5 actually characterize the CSST topologically.

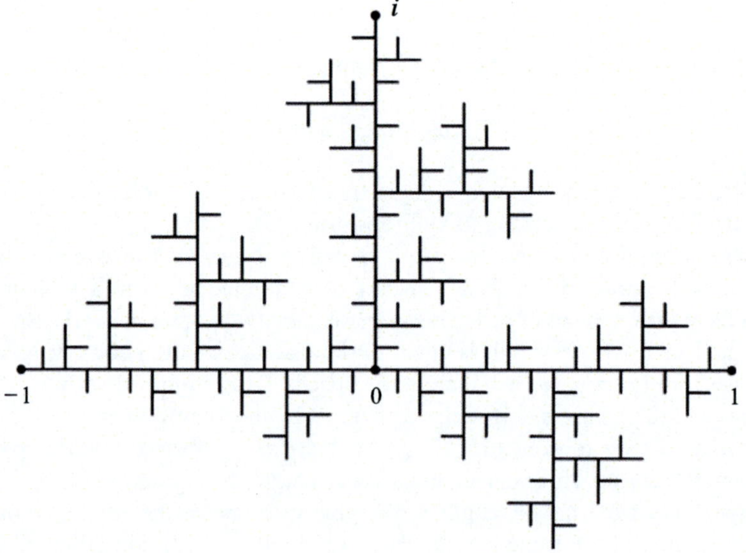

Fig. 2 The set J_5

Theorem 1.7 *A metric tree (T, d) is homeomorphic to the continuum self-similar tree $\mathbb{T}$ if and only if the following conditions are true:*

(i) *For every point $x \in T$ we have $\nu_T(x) \in \{1, 2, 3\}$.*
(ii) *The set of triple points $\{x \in T : \nu_T(x) = 3\}$ is a dense subset of T.*

We will derive Theorem 1.7 from a slightly more general statement. For its formulation let $m \in \mathbb{N}$ with $m \geq 3$. We consider the class $\mathcal{T}_m$ consisting of all metric trees T such that

(i) for every point $x \in T$ we have $\nu_T(x) \in \{1, 2, m\}$, and
(ii) the set of branch points $\{x \in T : \nu_T(x) = m\}$ is a dense subset of T.

Note that by Proposition 1.5 the CSST $\mathbb{T}$ satisfies the conditions in Theorem 1.7 with $m = 3$, and so $\mathbb{T}$ belongs to the class of trees $\mathcal{T}_3$. Now the following statement is true which contains Theorem 1.7 as a special case.

Theorem 1.8 *Let $m \in \mathbb{N}$ with $m \geq 3$. Then all trees in $\mathcal{T}_m$ are homeomorphic to each other.*

Theorems 1.7 and 1.8 are not new. In a previous version of this paper, we considered Theorem 1.7 as a "folklore" statement, but we did not have a reference for a proof. Later, the paper [11] was brought to our attention which contains a more general result which implies Theorem 1.8, and hence also Theorem 1.7 (see [11, Theorem 6.2]; the proof there seems to be incomplete though—the continuity of the map h on the dense subset of X needs more justification). Theorem 1.8 was explicitly stated in [9, (6), p. 490], but it seems that the origins of Theorem 1.8 can be traced back much further to [28] (see also [26, Chapter X], and [10] for more pointers to the relevant older literature about dendrites).

We will give a complete proof of Theorem 1.8. It is based on ideas that are quite different from those in [11], but we consider our method of proof very natural. It is also related to some other recent work, in particular [5, 6]; so one can view the present paper as an introduction to these ideas. We will say more about our motivation below.

Our proof of Theorem 1.8 can be outlined as follows. Fix m as in the statement and consider a tree T in $\mathcal{T}_m$. Then we cut T into m subtrees at a carefully chosen branch point. This process is repeated inductively. One labels the subtrees obtained in this way by finite words consisting of letters in the alphabet $\mathcal{A} = \{1, 2, \ldots, m\}$. The labels are chosen so that if S is another tree in $\mathcal{T}_m$ and one decomposes S in a similar manner, then one has the same combinatorics (i.e., intersection and inclusion pattern) for the subtrees in T and S. The desired homeomorphism between T and S can then be obtained from a general statement that produces a homeomorphism between two spaces, if they admit matching decompositions into pieces satisfying suitable conditions (see Proposition 2.1).

The CSST is related to metric trees appearing in other areas of mathematics. One of these objects is the *(Brownian) continuum random tree* (CRT). This is a random tree introduced by Aldous [1] when he studied the scaling limits of simplicial trees arising from the critical Galton-Watson process. One can describe the CRT as follows. We consider a sample of Brownian excursion $(e_t)_{0 \le t \le 1}$ on the interval $[0, 1]$. For $s, t \in [0, 1]$, we set

$$d_e(s, t) = e(s) + e(t) - 2\inf\{e(r) : \min(s, t) \le r \le \max(s, t)\}.$$

Then d_e is a pseudo-metric on $[0, 1]$. We define an equivalence relation on $[0, 1]$ by setting $s \sim t$ if $d_e(s, t) = 0$. Then d_e descends to a metric on the quotient space $T_e = [0, 1]/\sim$. The metric space (T_e, d_e) is almost surely a metric tree (see [25, Sections 2 and 3]). Curien [13] asked the following question.

Question Is the topology of the CRT almost surely constant, that is, are two independent samples of the CRT almost surely homeomorphic?

This question was the original motivation for the present work and we found a positive answer based on the following statement.

Corollary 1.9 *A sample T of the CRT is almost surely homeomorphic to the CSST $\mathbb{T}$.*

Proof As we discussed, a sample T of the CRT is almost surely a metric tree (see [25, Sections 2 and 3]). Moreover, for such a sample T almost surely for every point $x \in T$, the valence $\nu_T(x)$ is either 1, 2 or 3, and the set $\{x : \nu_T(x) = 3\}$ of triple points is dense in T (see [15, Theorem 4.6] or [25, Proposition 5.2 (i)]). It follows from Proposition 1.5 and Theorem 1.7 that a sample T of the CRT is almost surely homeomorphic to the CSST $\mathbb{T}$. □

Informally, Corollary 1.9 says that the topology of the CRT is (almost surely) constant and given by the topology of a deterministic model space, namely the CSST. In particular, almost surely any two independent samples of the CRT are homeomorphic. This answers Curien's question in the positive. As we found out after we had obtained proofs for Theorem 1.7 and Corollary 1.9, Curien's question had already been answered implicitly in [12]. There the authors used the distributional self-similarity property of the CRT and showed that the CRT is isometric to a metric space with a random metric. This space is constructed similarly to the CSST as the attractor of an iterated function system with maps very similar to (1.1) (they contain an additional parameter though which is unnecessary if one uses the maps in (1.1)).

An important source of trees is given by Julia sets of postcritically-finite polynomials without periodic critical points in $\mathbb{C}$. It follows from [14] (or see [8, Theorem V.4.2]) that the Julia sets of such polynomials are indeed trees. One can show that the Julia set $\mathcal{J}(P)$ of the polynomial $P(z) = z^2 + i$ (see Fig. 3) satisfies the conditions in Theorem 1.7. Accordingly, $\mathcal{J}(P)$ is homeomorphic to the CSST.

There are several directions in which one can pursue these topics further. For example, one can study the topology of more general trees than those in the classes $\mathcal{T}_m$. One may want to replace m with any finite (or even infinite) list of allowed valences for branch points, including branch points of infinite valence. In an earlier

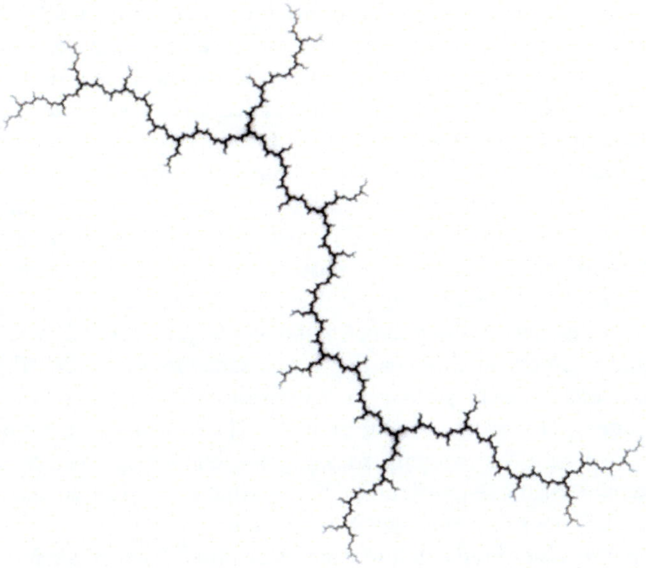

Fig. 3 The Julia set of $P(z) = z^2 + i$

version of our paper, we discussed this in more detail. Since we learned that these results are already contained in [10], we decided to skip this in the present version.

There is one important variant of Theorem 1.8 that we like to mention though. Namely, one can consider the (non-empty) class $\mathcal{T}_\infty$ of trees T such that $\nu_T(x) \in \{1, 2, \infty\}$ for all $x \in T$ and such that the set of branch points of T (i.e., in this case the set $\{x \in T : \nu_T(x) = \infty\}$) is dense in T. Then all trees in $\mathcal{T}_\infty$ are homeomorphic to each other (our method of proof does not directly apply here, but one can use our approach based on a more general version of Proposition 2.1). Moreover, each tree T in $\mathcal{T}_\infty$ is *universal* in the sense that every tree S admits a topological embedding into T. These results are due to Waszewski [28] (see [27, Section 10.4] for a modern exposition of this universality property; see also [9] for a discussion of a universality property of the trees in $\mathcal{T}_m$, $m \in \mathbb{N}$, $m \geq 3$).

Trees in $\mathcal{T}_\infty$ are also interesting, because they naturally arise in probabilistic models. More specifically, the so-called stable trees with index $\alpha \in (1, 2]$ are generalizations of the CRT (see [25, Section 4] for the definition). For fixed $\alpha \in (1, 2)$, a sample T of such a stable tree belongs to $\mathcal{T}_\infty$ almost surely [25, Proposition 5.2 (ii)]. By the previous discussion this implies that two independent samples of stable trees for given $\alpha \in (1, 2)$ are almost surely homeomorphic. Note that the Julia set of a polynomial never belongs to $\mathcal{T}_\infty$. This follows from results due to Kiwi (see [22, Theorem 1.1]).

Another direction for further investigations are questions that are more related to *geometric* properties of metric trees, in contrast to purely *topological* properties. In particular, one can study the *quasiconformal geometry* of the CSST and other trees (for a survey on the general topic of quasiconformal geometry see [4]).

One of the basic notion here is the concept of quasisymmetric equivalence. By definition two metric spaces X and Y are called *quasisymmetrically equivalent* if there exists a quasisymmetry $f: X \to Y$. Roughly speaking, a quasisymmetry is a homeomorphism with good geometric control: it sends metric balls to "roundish" sets with uniformly controlled eccentricity (for the precise definition of a quasisymmetry and other basic concepts of quasiconformal geometry see [17]). Since every quasisymmetry is a homeomorphism, two spaces are homeomorphic if they are quasisymmetrically equivalent. So this gives a stronger type of equivalence for metric spaces that has a more geometric flavor and goes beyond mere topology.

A natural problem in this context is to characterize the CCST $\mathbb{T}$ up to quasisymmetric equivalence, similar to Proposition 1.5 which gives a topological characterization. This problem is solved in [5]. The precise statement is too technical to be included here, but roughly speaking the conditions on a metric tree T to be quasisymmetrically equivalent to $\mathbb{T}$ are similar in sprit to the conditions in Proposition 1.5, but of a more "quantitative" nature.

For example, one of the conditions stipulates that T be *trivalent* (i.e., all branch points of T are triple points), but not only should the branch points of T form a dense subset of T, but T should be *uniformly branching* in the sense that every arc $\alpha \subseteq T$ contains a branch point p of *height* $H_T(p)$ comparable to the diameter of α. Here the height $H_T(p)$ is the diameter of the third largest branch of p (see the discussion around (3.1) for more details).

In our proof of Theorem 1.7 we first realized that this concept of height of a branch point plays a very important role in understanding the geometry and topology of trees. This concept is also used in [5, 6].

The present paper and [5, 6] have another common feature. In all of these works it is important to have good decompositions of the spaces studied, depending on the problem under consideration. This line of thought in the context of quasiconformal geometry can be traced back to [7, Proposition 18.8]. More recently, Kigami [20] has systematically investigated such decompositions in the general framework of partitions of a space given by sets that are labeled by the vertices of a (simplicial) tree. This common philosophy with other recent work is the main motivation why we wanted to present the proof of the known Theorem 1.8 from our perspective.

One can use the characterization of the CSST up to quasisymmetric equivalence established in [5] to prove the following statement (unpublished work by the authors): if the Julia set $\mathcal{J}(P)$ of a postcritically-finite polynomial P with no periodic critical points in $\mathbb{C}$ is homeomorphic to the CSST, then $\mathcal{J}(P)$ is quasisymmetrically equivalent to the CSST.

Finally, we mention in passing that the geometric properties of the continuum random tree (CRT) were considered in the recent paper [24] by Lin and Rohde. Though Lin and Rohde do not study quasisymmetric equivalence, many of their considerations still fit into the general framework of quasiconformal geometry.

The present paper is organized as follows. In Sect. 2 we state and prove a general criterion for two metric spaces to be homeomorphic based on the existence of combinatorially equivalent decompositions of the spaces. In Sect. 3 we collect some general facts about trees that we use later. The CSST is studied in Sect. 4. There we provide proofs of Propositions 1.1, 1.4, 1.5, and 1.6. In Sect. 5 we explain how to decompose trees in $\mathcal{T}_m$ with $m \in \mathbb{N}$, $m \geq 3$. Based on this, we then present a proof Theorem 1.8. Theorem 1.7 is an immediate consequence.

2 Constructing Homeomorphisms Between Spaces

Throughout this paper, we use fairly standard metric space notation. If (X, d) is a metric space, then we denote by $B(a, r) = \{x \in X : d(a, x) < r\}$ the open ball of radius $r > 0$ centered at $a \in X$. If $A, B \subseteq X$, then $\operatorname{diam}(A) = \sup\{d(x, y) : x, y \in A\}$ is the diameter of A and $\operatorname{dist}(A, B) = \inf\{d(x, y) : x \in A,\ y \in B\}$ the (minimal) distance of A and B. Similarly, if $a \in X$, then $\operatorname{dist}(a, B) = \operatorname{dist}(\{a\}, B)$ denotes the distance of the point a to the set B. Finally, if γ is a path in X, then $\operatorname{length}(\gamma)$ stands for its length.

Before we discuss trees in more detail and turn our attention to the CSST, we will establish the following proposition that is the key to showing that two trees are homeomorphic. The statement will also give us some guidance for the desired properties of tree decompositions that we will discuss in the following sections. The proposition is inspired by [7, Proposition 18.8], which provided geometric conditions for the decomposition of a space that can be used to construct quasisymmetric homeomorphisms.

Proposition 2.1 *Let (X, d_X) and (Y, d_Y) be compact metric spaces. Suppose that for each $n \in \mathbb{N}$, the space X admits a decomposition $X = \bigcup_{i=1}^{M_n} X_{n,i}$ as a finite union of non-empty compact subsets $X_{n,i}$, $i = 1, \ldots, M_n \in \mathbb{N}$, with the following properties for all n, i, and j:*

(i) *Each set $X_{n+1,j}$ is the subset of some set $X_{n,i}$.*
(ii) *Each set $X_{n,i}$ is equal to the union of some of the sets $X_{n+1,j}$.*
(iii) $\max_{1 \leq i \leq M_n} \operatorname{diam}(X_{n,i}) \to 0$ *as $n \to \infty$.*

Suppose that for $n \in \mathbb{N}$ the space Y admits a decomposition $Y = \bigcup_{i=1}^{M_n} Y_{n,i}$ as a union of non-empty compact subsets $Y_{n,i}$, $i = 1, \ldots, M_n$, with properties analogous to (i)–(iii) *such that*

$$X_{n+1,j} \subseteq X_{n,i} \text{ if and only if } Y_{n+1,j} \subseteq Y_{n,i} \tag{2.1}$$

and

$$X_{n,i} \cap X_{n,j} \neq \emptyset \text{ if and only if } Y_{n,i} \cap Y_{n,j} \neq \emptyset \tag{2.2}$$

for all n, i, j.

Then there exists a unique homeomorphism $f: X \to Y$ such that $f(X_{n,i}) = Y_{n,i}$ for all n and i.

In particular, under these assumptions the spaces X and Y are homeomorphic.

Proof We define a map $f: X \to Y$ as follows. For each point $x \in X$, by (ii) and (iii) there exists a nested sequence of sets X_{n,i_n}, $n \in \mathbb{N}$, such that $\{x\} = \bigcap_n X_{n,i_n}$. Then the corresponding sets Y_{n,i_n}, $n \in \mathbb{N}$, are also nested by (2.1). Since these sets are non-empty and compact, by condition (iii) for the space Y this implies that there exists a unique point $y \in \bigcap_n Y_{n,i_n}$. We define $f(x) = y$.

Then f is well-defined. To see this, suppose we have another nested sequence X_{n,i'_n}, $n \in \mathbb{N}$, such that $\{x\} = \bigcap_n X_{n,i'_n}$. Then there exists a unique point $y' \in \bigcap_n Y_{n,i'_n}$. Now $x \in X_{n,i_n} \cap X_{n,i'_n}$ and so $Y_{n,i_n} \cap Y_{n,i'_n} \neq \emptyset$ for all $n \in \mathbb{N}$ by (2.2). By condition (iii) for Y, this is only possible if $y = y'$. So $f: X \to Y$ is indeed well-defined.

One can define a map $g: Y \to X$ by a similar procedure. Namely, for each $y \in Y$ we can find a nested sequence Y_{n,i_n}, $n \in \mathbb{N}$, such that $\{y\} = \bigcap_n Y_{n,i_n}$. Then there exists a unique point $x \in \bigcap_n X_{n,i_n}$ and if we set $g(y) = x$, we obtain a well-defined map $g: Y \to X$.

It is obvious from the definitions that the maps f and g are inverse to each other. Hence they define bijections between X and Y.

Conditions (i) and (ii) imply that if $X_{k,i}$ is a set in one of the decompositions of X and $x \in X_{k,i}$, then there exists a nested sequence X_{n,i_n}, $n \in \mathbb{N}$, with $X_{k,i_k} = X_{k,i}$ and $\{x\} = \bigcap_n X_{n,i_n}$. This implies that $f(x) \in Y_{k,i}$ and so $f(X_{k,i}) \subseteq Y_{k,i}$. Similarly, $g(Y_{k,i}) \subseteq X_{k,i}$. Since $g = f^{-1}$, we have $f(X_{k,i}) = Y_{k,i}$ as desired. It is clear that this last condition together with our assumptions determines f uniquely.

It remains to show that f is a homeomorphism. For this it suffices to prove that f and $f^{-1} = g$ are continuous. Since the roles of f and g are completely symmetric, it is enough to establish that f is continuous.

For this, let $\epsilon > 0$ be arbitrary. By (iii) we can choose $n \in \mathbb{N}$ such that

$$\max\{\mathrm{diam}(Y_{n,i}) : 1 \leq i \leq M_n\} < \varepsilon/2.$$

Since the sets $X_{n,i}$ are compact, there exists $\delta > 0$ such that

$$\mathrm{dist}(X_{n,i}, X_{n,j}) > \delta,$$

whenever $i, j \in \{1, \ldots, M_n\}$ and $X_{n,i} \cap X_{n,j} = \emptyset$.

Now suppose that $a, b \in X$ are arbitrary points with $d_X(a, b) < \delta$. We claim that then $d_Y(f(a), f(b)) < \epsilon$. Indeed, we can find $i, j \in \{1, \ldots, M_n\}$ such that $a \in X_{n,i}$ and $b \in X_{n,j}$. Since $d_X(a, b) < \delta$, we then necessarily have $X_{n,i} \cap X_{m,j} \neq \emptyset$ by definition of δ. So $Y_{n,i} \cap Y_{n,j} \neq \emptyset$ by (2.2). Moreover, $f(a) \in f(X_{n,i}) = Y_{n,i}$ and $f(b) \in f(X_{n,j}) = Y_{n,j}$. Hence

$$d_Y(f(a), f(b)) \leq \mathrm{diam}(Y_{n,i}) + \mathrm{diam}(Y_{n,j}) < \epsilon.$$

The continuity of f follows. □

3 Topology of Trees

In this section we fix some terminology and collect some general facts about trees. We do not claim any originality of this material. All of it is standard and well-known, but we did not try to track it down in the literature. Our objective is to make our presentation self-contained, and to have convenient reference points for future work. For general background on trees or dendrites we refer to [29, Chapter V], [23, Section §51 VI], [27, Chapter X]), and the literature mentioned there.

An *arc* α in a metric space is a homeomorphic image of the unit interval $[0, 1] \subseteq \mathbb{R}$. The points corresponding to 0 and 1 are called the *endpoints* of α.

Let T be a tree. Then the last part of Definition 1.3 is equivalent to the requirement that for all points $a, b \in T$ with $a \neq b$, there exists a unique arc in T joining a and b, i.e., it has the endpoints a and b. We use the notation $[a, b]$ for this unique arc. It is convenient to allow $a = b$ here. Then $[a, b]$ denotes the *degenerate arc* consisting only of the point $a = b$. Sometimes we want to remove one or both endpoints from the arc $[a, b]$. Accordingly, we define $(a, b) = [a, b]\backslash\{a, b\}$, $[a, b) = [a, b]\backslash\{b\}$ and $(a, b] = [a, b]\backslash\{a\}$. In Sect. 4 we will not use this notation for arcs in a tree. There $[a, b]$ will always denote the Euclidean line segment joining two points $a, b \in \mathbb{C}$.

A metric space X is called *path-connected* if any two points $a, b \in X$ can be joined by a path in X, i.e., there exists a continuous map $\gamma: [0, 1] \to X$ such that $\gamma(0) = a$ and $\gamma(1) = b$. The space X is *arc-connected* if any two distinct points in X can be joined by an arc in X. The image of a path joining two distinct points in a metric space always contains an arc joining these points (this follows from the fact that every Peano space is arc-connected; see [18, Theorem 3.15, p. 116]). In particular, every path-connected metric space is arc-connected.

Lemma 3.1 *Let (T, d) be a tree. Then for each $\varepsilon > 0$ there exists $\delta > 0$ such that for all $a, b \in T$ with $d(a, b) < \delta$ we have* $\mathrm{diam}([a, b]) < \varepsilon$.

Proof Fix $\varepsilon > 0$. Since T is a compact, connected, and locally connected metric space, it is a *Peano space*. So by the Hahn-Mazurkiewicz theorem there exists a continuous surjective map $\varphi: [0, 1] \to T$ of the unit interval onto T [18, Theorem 3.30, p. 129]. By uniform continuity of φ we can represent $[0, 1]$ as a union $[0, 1] = I_1 \cup \cdots \cup I_n$ of finitely many closed intervals $I_1, \ldots, I_n \subseteq [0, 1]$ with $\mathrm{diam}(X_k) < \varepsilon/2$, where $X_k = \varphi(I_k)$ for $k = 1, \ldots, n$. The sets $X_k = \varphi(I_k)$ are compact. This implies that there exists $\delta > 0$ such that $\mathrm{dist}(X_i, X_j) > \delta$, whenever $i, j \in \{1, \ldots, n\}$ and $X_i \cap X_j = \emptyset$.

Now let $a, b \in T$ with $d(a, b) < \delta$ be arbitrary. We may assume $a \neq b$. Then there exist $i, j \in \{1, \ldots, n\}$ with $a \in X := X_i$ and $b \in Y := X_j$. By choice of δ we must have $X \cap Y \neq \emptyset$. As continuous images of intervals, the sets X and Y are path-connected. Since $X \cap Y \neq \emptyset$, the union $X \cup Y$ that contains the points a and b is also path-connected. This implies that $X \cup Y$ is arc-connected, and so there exists an arc $\alpha \subseteq X \cup Y$ with endpoints a and b. The unique such arc in the tree T is $[a, b]$, and so $[a, b] = \alpha \subseteq X \cup Y$. This implies

$$\operatorname{diam}([a, b]) \leq \operatorname{diam}(X) + \operatorname{diam}(Y) < \varepsilon,$$

as desired. □

Lemma 3.2 *Let (T, d) be a tree and $p \in T$. Then the following statements are true:*

(i) *Each component U of $T \backslash \{p\}$ is an open and arc-connected subset of T.*
(ii) *If U is a component of $T \backslash \{p\}$, then $\overline{U} = U \cup \{p\}$ and $\partial U = \{p\}$.*
(iii) *Two points $a, b \in T \backslash \{p\}$ lie in the same component of $T \backslash \{p\}$ if and only if $p \notin [a, b]$.*

Proof

(i) The set $T \backslash \{p\}$ is open. Since T is locally connected, each component U of $T \backslash \{p\}$ is also open.

For $a, b \in U$ we write $a \sim b$ if a and b can be joined by a path in U. Obviously, this defines an equivalence relation on U. The equivalence classes are open subsets of T. To see this, suppose $a, b \in U$ can be joined by a path β in U. Then for all points x in a sufficiently small neighborhood $V \subseteq U$ of b we have $[b, x] \subseteq U$ as follows from Lemma 3.1. So by concatenating β with (a parametrization of) the arc $[b, x]$, we obtain a path β' in U that joins a and $x \in V$. This shows that every point b in the equivalence class of a has a neighborhood V that also belongs to this equivalence class.

We see that the equivalence classes of $\sim$ partition U into open sets. Since U is connected, there can only be one such set. It follows that U is path-connected and hence also arc-connected.

(ii) Let U be a (non-empty) component of $T \backslash \{p\}$. We choose a point $a \in U$. The set $[a, p)$ is connected, contained in $T \backslash \{p\}$, and meets U in a. Hence $[a, p) \subseteq U$. This implies that $p \in \overline{U}$. On the other hand, the set $U \cup \{p\}$ is closed, because its complement is a union of components of $T \backslash \{p\}$ and hence open by (i). Thus $\overline{U} = U \cup \{p\}$. By (i) no point in U is a boundary point of U, and so $\partial U = \{p\}$.

(iii) If $a, b \in T \backslash \{p\}$ and $p \notin [a, b]$, then $[a, b]$ is a connected subset of $T \backslash \{p\}$. Hence $[a, b]$ lies in a component U of $T \backslash \{p\}$. In particular, $a, b \in [a, b]$ lie in the same component U of $T \backslash \{p\}$.

Conversely, suppose that $a, b \in T \backslash \{p\}$ lie in the same component U of $T \backslash \{p\}$. We know by (i) that U is arc-connected. Hence there exists a (possibly degenerate) arc $\alpha \subseteq U$ with endpoints a and b. But the unique such arc in T is $[a, b]$. Hence $[a, b] = \alpha \subseteq U \subseteq T \backslash \{p\}$, and so $p \notin [a, b]$. □

A subset S of a tree (T, d) is called a *subtree* of T if S equipped with the restriction of the metric d is also a tree as in Definition 1.3. Every subtree S of

T contains two points and hence a non-degenerate arc. In particular, every subtree S of T is an infinite, actually uncountable set.

The following statement characterizes subtrees.

Lemma 3.3 *Let (T, d) be a tree. Then a set $S \subseteq T$ is a subtree of T if and only if S contains at least two points and is closed and connected.*

Proof If S is a subtree of T, then S contains at least two points, and is connected and compact. Hence it is a closed subset of T. Conversely, suppose that S contains at least two points and is closed and connected. Then S is compact, because T is compact.

Suppose that $a, b \in S$, $a \neq b$, are two distinct points in S. We consider the arc $[a, b] \subseteq T$. Suppose there exists a point $p \in [a, b]$ with $p \notin S$. Then $p \neq a, b$, and so by Lemma 3.2 (iii), the points a and b lie in different components of $T\backslash\{p\}$. This is impossible, because the connected set $S \subseteq T\backslash\{p\}$ must be contained in exactly one component of $T\backslash\{p\}$. This shows that $[a, b] \subseteq S$ and so the points a and b can be joined by an arc in S. This arc in S is unique, because it is unique in T.

It remains to show that S is locally connected, i.e., every point in S has arbitrarily small connected relative neighborhoods. To see this, let $a \in S$ and $\varepsilon > 0$ be arbitrary. Then by Lemma 3.1 we can find $\delta > 0$ such that $[a, x] \subseteq B(a, \varepsilon)$ whenever $x \in B(a, \delta)$. Now let M be the union of all arcs $[a, x]$ with $x \in S \cap B(a, \delta)$. These arcs lie in S and so M is a connected set contained in $S \cap B(a, \varepsilon)$. Moreover, $S \cap B(a, \delta) \subseteq M$ and so M is a connected relative (not necessarily open) neighborhood of a in S. This shows that S is locally connected. We conclude that S is indeed a subtree of T. □

Lemma 3.4 *Let (T, d) be a tree, $p \in T$, and U a component of $T\backslash\{p\}$. Then $B = U \cup \{p\}$ is a subtree of T and p is a leaf of B.*

Proof It follows from Lemma 3.2 (i) and (ii) that the set U is connected and that $B = U \cup \{p\} = \overline{U}$. This implies that B is closed and connected. Since $U \neq \emptyset$ and $p \notin U$, the set B contains at least two points. Hence B is a subtree of T by Lemma 3.3. Since $B\backslash\{p\} = U$ is connected, p is a leaf of B. □

If the subtree $B = U \cup \{p\}$ is as in the previous lemma, then we call B a *branch* of p in T (or just a branch of p if T is understood).

Lemma 3.5 *Let (T, d) be a tree, $S \subseteq T$ be a subtree of T, and $p \in S$. Then every branch B' of p in S is contained in a unique branch B of p in T. The assignment $B' \mapsto B$ is an injective map between the sets of branches of p in S and in T. If p is an interior point of S, then this map is a bijection.*

In particular, if under the given assumptions $\nu_T(p)$ is the valence of p in T and $\nu_S(p)$ the valence of p in S, then $\nu_S(p) \leq \nu_T(p)$. Here we have equality if p is an interior point of S.

If $p \in S$ is a leaf of T, then T has only one branch B at p, namely $B = T$. Hence $1 \leq \nu_S(p) \leq \nu_T(p) \leq 1$, and so $\nu_S(p) = 1$. This means that p is also a leaf of S.

More informally, we can say that the property of a point being a leaf in T is passed to subtrees that contain the point.

Proof If B' is a branch of p in S, then $B' = U' \cup \{p\}$, where U' is a component of $S\backslash\{p\}$. Then U' is a connected subset of $T\backslash\{p\}$ and so contained in a unique component U of $T\backslash\{p\}$. Then $B = U \cup \{p\}$ is a branch of p in T with $B' \subseteq B$ and it is clear that B is the unique such branch.

To show injectivity of the map $B' \mapsto B$, let B'_1 and B'_2 be two distinct branches of p in S. Pick points $a \in B'_1\backslash\{p\}$ and $b \in B'_2\backslash\{p\}$. Then a and b lie in different components of $S\backslash\{p\}$ and so $p \in [a, b]$ by Lemma 3.2 (iii) applied to the tree S. Hence a and b lie in different components of $T\backslash\{p\}$, and so in different branches of p in T. This implies that B'_1 and B'_2 must be contained in different branches of p in T. This shows that the map $B' \mapsto B$ is indeed injective.

Now assume in addition that p is an interior point of S. To show surjectivity of the map $B' \mapsto B$, we consider a branch B of p in T. Pick a point $a \in B\backslash\{p\}$. Then $[a, p) \subseteq B\backslash\{p\}$, because B is a subtree of T. Since p is an interior point of S, there exists a point $x \in [a, p)$ close enough to p such that $x \in S\backslash\{p\}$. If B' is the unique branch of p in S that contains x, then we have $x \in B' \cap B$. This implies $B' \subseteq B$. Hence the map $B' \mapsto B$ is also surjective, and so a bijection. □

Lemma 3.6 *Let (T, d) be a tree, $p, a_1, a_2, a_3 \in T$ with $p \neq a_1, a_2, a_3$ and suppose that the sets $[a_1, p)$, $[a_2, p)$, $[a_3, p)$ are pairwise disjoint. Then the points a_1, a_2, a_3 lie in different components of $T\backslash\{p\}$ and p is a branch point of T.*

Proof The arcs $[a_1, p]$ and $[a_2, p] = [p, a_2]$ have only the point p in common. So their union $[a_1, p]\cup[p, a_2]$ is an arc and this arc must be equal to $[a_1, a_2]$. Hence $p \in [a_1, a_2]$ which by Lemma 3.2 (iii) implies that a_1 and a_2 lie in different components of $T\backslash\{p\}$. A similar argument shows that a_3 must be contained in a component of $T\backslash\{p\}$ different from the components containing a_1 and a_2. In particular, $T\backslash\{p\}$ has at least three components and so p is a branch point of T. The statement follows. □

Lemma 3.7 *Let (T, d) be a tree such that the branch points of T are dense in T. If $a, b \in T$ with $a \neq b$, then there exists a branch point $c \in (a, b)$.*

Proof We pick a point $x_0 \in (a, b) \neq \emptyset$. Then x_0 has positive distance to both a and b. This and Lemma 3.1 imply that we can find $\delta > 0$ such that for all $x \in B(x_0, \delta)$ the arc $[x, x_0]$ has uniformly small diameter and so does not contain a or b.

Since branch points are dense in T, we can find a branch point $p \in B(x_0, \delta)$. Then $a, b \notin [p, x_0]$. If $p \in (a, b)$, we are done.

In the other case, we have $p \notin (a, b)$. If we travel from p to $x_0 \in (a, b)$ along $[p, x_0]$, we meet $[a, b]$ in a first point $c \in (a, b)$. Then $a, b, p \neq c$. Moreover, the sets $[a, c)$, $[b, c)$, $[p, c)$ are pairwise disjoint. Hence $c \in (a, b)$ is a branch point of T as follows from Lemma 3.6. □

Lemma 3.8 *Let (X, d) be a compact, connected, and locally connected metric space, J an index set, $p_i \in X$, and U_i a component of $X \backslash \{p_i\}$ for each $i \in J$. Suppose that*

$$U_i \cap U_j = \emptyset$$

for all $i, j \in J$, $i \neq j$. Then J is a countable set. If there exists $\delta > 0$ such that $\operatorname{diam}(U_i) > \delta$ *for each $i \in J$, then J is finite.*

Informally, the space X cannot contain a “comb” with too many long teeth.

Proof We prove the last statement first. We argue by contradiction and assume that $\operatorname{diam}(U_i) > \delta > 0$ for each $i \in J$, where J is an infinite index set. Then we can choose a point $x_i \in U_i$ such that $d(x_i, p_i) \geq \delta/2$. The set $A = \{x_i : i \in J\}$ is infinite and so it must have a limit point $q \in X$, because X is compact. Since X is locally connected, there exists a connected neighborhood N of q such that $N \subseteq B(q, \delta/8)$. Since q is a limit point of A, the set N contains infinitely many points in A. In particular, we can find $i, j \in J$ with $x_i, x_j \in N$ and $i \neq j$. Then

$$\operatorname{dist}(p_i, N) \geq d(p_i, x_i) - \operatorname{diam}(N) \geq \delta/2 - \delta/4 > 0,$$

and so $N \subseteq X \backslash \{p_i\}$. Since the connected set N meets U_i in the point x_i, this implies that $N \subseteq U_i$. Similarly, $N \subseteq U_j$. This is impossible, because we have $i \neq j$ and so $U_i \cap U_j = \emptyset$, while $\emptyset \neq N \subseteq U_i \cap U_j$.

To prove the first statement, note that $\operatorname{diam}(U_i) > 0$ for each $i \in J$. Indeed, otherwise $\operatorname{diam}(U_i) = 0$ for some $i \in J$. Then U_i consists of only one point a. Since X is locally connected, the component U_i of $X \backslash \{p_i\}$ is an open set. So a is an isolated point of X. This is impossible, because the metric space X is connected and so it does not have isolated points.

Now we write $J = \bigcup_{n \in \mathbb{N}} J_n$, where J_n consists of all $i \in J$ such that $\operatorname{diam}(U_i) > 1/n$. Then each set J_n is finite by the first part of the proof. This implies that J is countable. □

We can apply the previous lemma to a tree T and choose for each p_i a fixed branch point p of T. Then it follows that p can have at most countably many distinct complementary components U_i and hence there are only countably many distinct branches $B_i = U_i \cup \{p\}$ of p. Moreover, since $\operatorname{diam}(B_i) = \operatorname{diam}(\overline{U}_i) = \operatorname{diam}(U_i)$, there can only be finitely many of these branches whose diameter exceeds a given positive number $\delta > 0$. In particular, we can label the branches of p by numbers $n = 1, 2, 3, \ldots$ so that

$$\operatorname{diam}(B_1) \geq \operatorname{diam}(B_2) \geq \operatorname{diam}(B_3) \geq \ldots .$$

We now set

$$H_T(p) = \operatorname{diam}(B_3) \tag{3.1}$$

and call $H_T(p)$ the *height* of the branch point p in T. So the height of a branch point p is the diameter of the third largest branch of p.

Lemma 3.9 *Let (T, d) be a tree and $\delta > 0$. Then there are at most finitely many branch points $p \in T$ with height $H_T(p) > \delta$.*

Proof We argue by contradiction and assume that this is not true. Then the set E of branch points p in T with $H_T(p) > \delta$ has infinitely many elements. Since T is compact, the set E has a limit point $q \in T$.

Claim. There exists a branch Q of q such that the set $E \cap Q$ is infinite and has q as a limit point.

Otherwise, q has infinitely many distinct branches $Q_n, n \in \mathbb{N}$, that contain a point $a_n \in E \cap (Q_n \backslash \{q\})$. Then a_n is a branch point with $H_T(a_n) > \delta$ which implies that a_n has at least three branches whose diameters exceed δ. At least one of them does not contain q. If we denote such a branch of a_n by V_n, then V_n is a connected subset of $T \backslash \{q\}$. It meets $Q_n \backslash \{q\}$, because $a_n \in (Q_n \backslash \{q\}) \cap V_n$. It follows that $V_n \subseteq Q_n$ and so $\operatorname{diam}(Q_n) \geq \operatorname{diam}(V_n) > \delta$. Since the branches Q_n of q are all distinct for $n \in \mathbb{N}$, this contradicts Lemma 3.8 (see the discussion after the proof of this lemma). The Claim follows.

We fix a branch Q of q as in the Claim. For each $n \in \mathbb{N}$ we will now inductively construct branch points $p_n \in E \cap (Q \backslash \{q\})$ together with a branch B_n of p_n and an auxiliary compact set $K_n \subseteq T$. They will satisfy the following conditions for each $n \in \mathbb{N}$:

(i) $\operatorname{diam}(B_n) > \delta$,
(ii) the sets $B_1, \ldots, B_n$ are disjoint,
(iii) the set K_n is compact and connected, and

$$B_1 \cup \cdots \cup B_n \subseteq K_n \subseteq Q \backslash \{q\}.$$

We pick an arbitrary branch point $p_1 \in E \cap (Q \backslash \{q\})$ to start. Then we can choose a branch B_1 of p_1 that does not contain q and satisfies $\operatorname{diam}(B_1) > \delta$. We set $K_1 = B_1$. Then K_1 is a compact and connected set that does not contain q and meets Q, because $p_1 \in K_1 \cap Q$. Hence $K_1 \subseteq Q \backslash \{q\}$.

Suppose for some $n \in \mathbb{N}$, a branch point $p_k \in E \cap Q$, a branch B_k of p_k, and a set K_k with the properties (i)–(iii) have been chosen for all $1 \leq k \leq n$.

Since $q \notin K_n$, we have $\operatorname{dist}(q, K_n) > 0$, and so we can find a branch point $p_{n+1} \in E \cap (Q \backslash \{p\})$ sufficiently close to q such that $p_{n+1} \notin K_n$. This is possible, because q is a limit point of $E \cap (Q \backslash \{q\})$. Since the set $K_n \subseteq T \backslash \{p_{n+1}\}$ is connected, it must be contained in a branch of p_{n+1}. Since there are three branches of $p_{n+1} \neq q$ whose diameters exceed δ, we can pick one of them that contains neither q nor K_n. Let B_{n+1} be such a branch of p_{n+1}. Then $\operatorname{diam}(B_{n+1}) > \delta$ and so (i) is true for $n + 1$. We have $B_{n+1} \cap K_n = \emptyset$; so (iii) shows that B_{n+1} is disjoint from the previously chosen disjoint sets $B_1, \ldots, B_n$. This gives (ii).

Since $p_n, p_{n+1} \in Q \backslash \{q\}$, the arc $[p_n, p_{n+1}]$ does not contain q (see Lemma 3.2 (iii)). We also have $p_n \in B_n \subseteq K_n$ and $p_{n+1} \in B_{n+1}$, which implies that the set $K_{n+1} := K_n \cup [p_n, p_{n+1}] \cup B_{n+1} \subseteq Q \backslash \{q\}$ is compact and connected. We have

$$B_1 \cup \cdots \cup B_n \cup B_{n+1} \subseteq K_n \cup B_{n+1} \subseteq K_{n+1} \subseteq Q \backslash \{q\},$$

and so K_{n+1} has property (iii).

Continuing with this process, we obtain disjoint branches B_n for all $n \in \mathbb{N}$ that satisfy (i). The last part of Lemma 3.8 implies that this is impossible and we get a contradiction. □

4 Basic Properties of the Continuum Self-Similar Tree

We now we study the properties of the continuum self-similar tree (CSST). Unless otherwise specified, all metric notions in this section refer to the Euclidean metric on the complex plane $\mathbb{C}$. In this section, i always denotes the imaginary unit and we do not use this letter for indexing as in the other sections. If $a, b \in \mathbb{C}$ we denote by $[a, b]$ the Euclidean line segment in $\mathbb{C}$ joining a and b. We also use the usual notation for open or half-open line segments. So $[a, b) = [a, b] \backslash \{b\}$, etc.

For the proof of Proposition 1.1 we consider a coding procedure of certain points in the complex plane by words in an alphabet. We first fix some terminology related to this. We consider a non-empty set $\mathcal{A}$. Then we call $\mathcal{A}$ an *alphabet* and refer to the elements in $\mathcal{A}$ as the *letters* in this alphabet. In this paper we will only use alphabets of the form $\mathcal{A} = \{1, 2, \ldots, m\}$ with $m \in \mathbb{N}$, $m \geq 3$. We consider the set $W(\mathcal{A}) := \mathcal{A}^{\mathbb{N}}$ of infinite sequences in $\mathcal{A}$ as the set of *infinite words* in the alphabet $\mathcal{A}$ and write the elements $w \in W(\mathcal{A})$ in the form $w = w_1 w_2 \ldots$, where it is understood that $w_k \in \mathcal{A}$ for $k \in \mathbb{N}$. Similarly, we set $W_n(\mathcal{A}) := \mathcal{A}^n$ and consider $W_n(\mathcal{A})$ as the set of all words in the alphabet $\mathcal{A}$ of length n. We write the elements $w \in W_n(\mathcal{A})$ in the form $w = w_1 \ldots w_n$ with $w_k \in \mathcal{A}$ for $k = 1, \ldots, n$. We use the convention that $W_0(\mathcal{A}) = \{\emptyset\}$ and consider the only element $\emptyset$ in $W_0(\mathcal{A})$ as the *empty word* of length 0. Finally,

$$W_*(\mathcal{A}) := \bigcup_{n \in \mathbb{N}_0} W_n(\mathcal{A})$$

is the set of all words of finite length. If $u = u_1 \ldots u_n$ is a finite word and $v = v_1 v_2 \ldots$ is a finite or infinite word in the alphabet $\mathcal{A}$, then we denote by $uv = u_1 \ldots u_n v_1 v_2 \ldots$ the word obtained by concatenating u and v. We call u an *initial segment* and v a *tail* of the word $w = uv$. If the alphabet $\mathcal{A}$ is understood, then we will simply drop $\mathcal{A}$ from the notation. So W will denote the set of infinite words in $\mathcal{A}$, etc.

For the rest of this section, we use the alphabet $\mathcal{A} = \{1, 2, 3\}$. So when we write W, W_n, W_* it is understood that $\mathcal{A} = \{1, 2, 3\}$ is the underlying alphabet. There exists a unique metric d on $W = \{1, 2, 3\}^{\mathbb{N}}$ with the following property. If we have two words $u = u_1u_2\ldots$ and $v = v_1v_2\ldots$ in W and $u \neq v$, then for some $n \in \mathbb{N}_0$ we have $u_1 = v_1, \ldots, u_n = v_n$, and $u_{n+1} \neq v_{n+1}$. Then $d(u, v) = 1/2^n$. More informally, two elements $u, v \in W$ are close in this metric precisely if they share a large number of initial letters. The metric space (W, d) is compact and homeomorphic to a Cantor set.

If $n \in \mathbb{N}_0$ and $w = w_1w_2\ldots w_n \in W_n$, we define

$$f_w := f_{w_1} \circ f_{w_2} \circ \cdots \circ f_{w_n},$$

where we use the maps in (1.1) in the composition. By convention, $f_\emptyset = \mathrm{id}_\mathbb{C}$ is the identity map on $\mathbb{C}$. Note that f_w is a Euclidean similarity on $\mathbb{C}$ that scales Euclidean distances by the factor 2^{-n}. If $a, b \in \mathbb{C}$, then $f_w([a, b]) = [f_w(a), f_w(b)]$. We will use this repeatedly in the following.

Throughout this section we denote by $H \subseteq \mathbb{C}$ the (closed) convex hull of the four points $1, i, -1$, and $\frac{1}{2} - \frac{i}{2}$ (see Fig. 4). We set $H_k = f_k(H)$ for $k = 1, 2, 3$. Then

$$H_1 \cup H_2 \cup H_3 = f_1(H) \cup f_2(H) \cup f_3(H) \subseteq H.$$

This implies that

$$f_w(H) \subseteq H \tag{4.1}$$

for all $w \in W_*$.

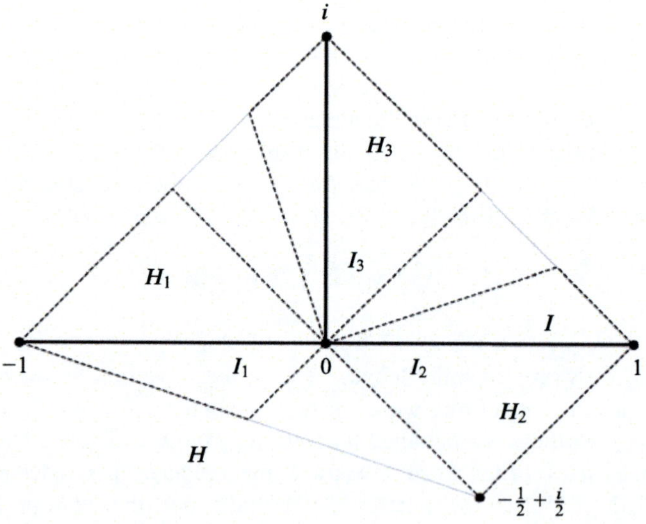

Fig. 4 Illustration of some associated sets

Lemma 4.1 *There exists a well-defined continuous map* $\pi : W \to \mathbb{C}$ *given by*

$$\pi(w) = \lim_{n\to\infty} f_{w_1 w_2 \dots w_n}(z_0)$$

for $w = w_1 w_2 \dots \in W$ *and* $z_0 \in \mathbb{C}$. *Here the limit exists and is independent of the choice of* $z_0 \in \mathbb{C}$.

The existence of such a map π is standard in similar contexts (see, for example, [19, Section 3.1, pp. 426–427]). In the following, $\pi : W \to \mathbb{C}$ will always denote the map provided by this lemma.

Proof Fix $z_0 \in \mathbb{C}$. Then there exists a constant $C \geq 0$ such that

$$|z_0 - f_k(z_0)| \leq C$$

for $k = 1, 2, 3$. If $n \in \mathbb{N}_0$ and $u \in W_n$, then

$$|f_u(a) - f_u(b)| = \frac{1}{2^n}|a - b|$$

for all $a, b \in \mathbb{C}$. This implies that if $w = w_1 w_2 \dots \in W$, $n \in \mathbb{N}$, and $u := w_1 w_2 \dots w_n \in W_n$, then

$$\begin{aligned} |f_{w_1 w_2 \dots w_n}(z_0) - f_{w_1 w_2 \dots w_{n+1}}(z_0)| &= |f_u(z_0) - f_u(f_{w_{n+1}}(z_0))| \\ &= \frac{1}{2^n}|z_0 - f_{w_{n+1}}(z_0)| \leq \frac{C}{2^n}. \end{aligned}$$

It follows that $\{f_{w_1 w_2 \dots w_n}(z_0)\}_{n\in\mathbb{N}}$ is a Cauchy sequence in $\mathbb{C}$. Hence this sequence converges and

$$\pi(w) = \lim_{n\to\infty} f_{w_1 w_2 \dots w_n}(z_0)$$

is well-defined for each $w = w_1 w_2 \dots \in W$.

The limit does not depend on the choice of z_0. Indeed, if $z_0' \in \mathbb{C}$ is another point, then

$$|f_{w_1 w_2 \dots w_n}(z_0) - f_{w_1 w_2 \dots w_n}(z_0')| = \frac{1}{2^n}|z_0 - z_0'|,$$

which implies that

$$\lim_{n\to\infty} f_{w_1 w_2 \dots w_n}(z_0) = \lim_{n\to\infty} f_{w_1 w_2 \dots w_n}(z_0').$$

The definition of π shows that if $w = w_1 w_2 \dots \in W$ and $n \in \mathbb{N}_0$, then

$$\pi(w) = \pi(w_1 w_2 \dots) = f_{w_1 \dots w_n}(\pi(w_{n+1} w_{n+2} \dots)). \tag{4.2}$$

If we pick $z_0 \in H$, then (4.1) and the definition of π imply that $\pi(W) \subseteq H$. If we combine this with (4.2), then we see that if two words $u, v \in W$ start with the same letters $w_1, \dots, w_n$, then

$$|\pi(u) - \pi(v)| \leq \operatorname{diam}(f_{w_1 \dots w_n}(H)) = \frac{1}{2^n} \operatorname{diam}(H).$$

The continuity of the map π follows from this and the definition of the metric d on W. □

We can now establish the result that is the basis of the definition of the CSST. Again arguments along these lines are completely standard.

Proof of Proposition 1.1 Let $\pi: W \to \mathbb{C}$ be the map provided by Lemma 4.1 and define $\mathbb{T} = \pi(W) \subseteq \mathbb{C}$. Since W is compact and π is continuous, the set $\mathbb{T}$ is non-empty and compact. The relation (1.2) immediately follows from (4.2) for $n = 1$. Note that (1.2) implies that

$$f_w(\mathbb{T}) = f_{w1}(\mathbb{T}) \cup f_{w2}(\mathbb{T}) \cup f_{w3}(\mathbb{T}) \tag{4.3}$$

for each $w \in W_n$, $n \in \mathbb{N}_0$. From this in turn we deduce that

$$\bigcup_{w \in W_n} f_w(\mathbb{T}) = \mathbb{T} \tag{4.4}$$

for each $n \in \mathbb{N}_0$.

It remains to show the uniqueness of $\mathbb{T}$. Suppose $\widetilde{\mathbb{T}} \subseteq \mathbb{C}$ is another non-empty compact set satisfying the analog of (1.2). Then the analogs of (4.3) and (4.4) are also valid for $\widetilde{\mathbb{T}}$. This and the definition of π using a point $z_0 \in \widetilde{\mathbb{T}}$ imply that $\mathbb{T} = \pi(W) \subseteq \widetilde{\mathbb{T}}$.

For the converse inclusion, let $a \in \widetilde{\mathbb{T}}$ be arbitrary. Using the relation (4.3) for the set $\widetilde{\mathbb{T}}$, we can inductively construct an infinite word $w_1 w_2 \dots \in W$ such that $a \in f_{w_1 w_2 \dots w_n}(\widetilde{\mathbb{T}})$ for all $n \in \mathbb{N}$. Since

$$\operatorname{diam}(f_{w_1 w_2 \dots w_n}(\widetilde{\mathbb{T}})) = \frac{1}{2^n} \operatorname{diam}(\widetilde{\mathbb{T}}) \to 0 \text{ as } n \to \infty,$$

the definition of π (using a point $z_0 \in \widetilde{\mathbb{T}}$) implies that $a = \pi(w)$. In particular, $a \in \pi(W) = \mathbb{T}$, and so $\widetilde{\mathbb{T}} \subseteq \mathbb{T}$. The uniqueness of $\mathbb{T}$ follows. □

In the proof of Proposition 1.1 we have seen that $\mathbb{T} = \pi(W)$. If $p \in \mathbb{T}$ and $p = \pi(w)$ for some $w \in W$, then we say that the word w *represents* p.

The following statement provides some geometric descriptions of $\mathbb{T}$.

Proposition 4.2 *Let* $I = [-1, 1] \subseteq \mathbb{C}$. *For* $n \in \mathbb{N}_0$ *define*

$$J_n = \bigcup_{w \in W_n} f_w(I) \quad \text{and} \quad K_n = \bigcup_{w \in W_n} f_w(H).$$

Then the sets J_n *and* K_n *are compact and satisfy*

$$J_n \subseteq J_{n+1} \subseteq \mathbb{T} \subseteq K_{n+1} \subseteq K_n \tag{4.5}$$

for $n \in \mathbb{N}_0$. *Moreover, we have*

$$\overline{\bigcup_{n \in \mathbb{N}_0} J_n} = \mathbb{T} = \bigcap_{n \in \mathbb{N}_0} K_n. \tag{4.6}$$

As we will discuss more towards the end of this section, the first identity in (4.6) represents $\mathbb{T}$ as the closure of a union of an ascending sequence of trees as mentioned in the introduction. We will not need the second identity in (4.6) in the following, but included it to show that $\mathbb{T}$ can also be obtained as the intersection of a natural decreasing sequence of compacts sets. This is how many other fractals are constructed.

Proof It is clear that the sets J_n and K_n as defined in the statement are compact for each $n \in \mathbb{N}_0$. Set $I_k = f_k(I)$ for $k = 1, 2, 3$. Then an elementary geometric consideration shows that (see Fig. 4)

$$I \subseteq I_1 \cup I_2 \cup I_3 \subseteq H_1 \cup H_2 \cup H_3 \subseteq H.$$

This in turn implies that

$$\begin{aligned} f_w(I) &\subseteq f_{w1}(I) \cup f_{w2}(I) \cup f_{w3}(I) \\ &\subseteq f_{w1}(H) \cup f_{w2}(H) \cup f_{w3}(H) \subseteq f_w(H) \end{aligned}$$

for each $w \in W_n$, $n \in \mathbb{N}_0$. Taking the union over all $w \in W_n$, we obtain

$$J_n \subseteq J_{n+1} \subseteq K_{n+1} \subseteq K_n \tag{4.7}$$

for all $n \in \mathbb{N}_0$. The set $\widetilde{\mathbb{T}} = \overline{\bigcup_{n \in \mathbb{N}_0} J_n}$ is non-empty, compact, and satisfies

$$\bigcup_{k=1,2,3} f_k(\widetilde{\mathbb{T}}) = \bigcup_{k=1,2,3} f_k\Big(\overline{\bigcup_{n \in \mathbb{N}_0} J_n}\Big) = \overline{\bigcup_{k=1,2,3} f_k\Big(\bigcup_{n \in \mathbb{N}_0} J_n\Big)}$$

$$= \overline{\bigcup_{k=1,2,3} f_k\Big(\bigcup_{n\in\mathbb{N}_0} J_n\Big)} = \overline{\bigcup_{n\in\mathbb{N}_0}\bigcup_{k=1,2,3} f_k(J_n)}$$

$$= \overline{\bigcup_{n\in\mathbb{N}_0} J_{n+1}} = \overline{\bigcup_{n\in\mathbb{N}_0} J_n} = \widetilde{\mathbb{T}}.$$

Hence $\widetilde{\mathbb{T}} = \mathbb{T}$ by the uniqueness statement in Proposition 1.1. So we have the first equation in (4.6).

Since $0 \in H$, we have $f_w(0) \in f_w(H) \subseteq K_n$ for each $w \in W_n$. Since the sets K_n are compact and nested, this implies that for each $w = w_1w_2\ldots \in W$ we have

$$\pi(w) = \lim_{n\to\infty} f_{w_1\ldots w_n}(0) \in \bigcap_{n\in\mathbb{N}_0} K_n.$$

It follows that $\mathbb{T} = \pi(W) \subseteq \bigcap_{n\in\mathbb{N}_0} K_n$.

To show the reverse inclusion, let $a \in \bigcap_{n\in\mathbb{N}_0} K_n$ be arbitrary. Then $a \in K_n$ for each $n \in \mathbb{N}_0$, and so there is a word $u_n \in W_n$ such that $a \in f_{u_n}(H)$. Define $z_n = f_{u_n}(0) \in J_n \subseteq \mathbb{T}$. Since $0 \in H$, we have $z_n \in f_{u_n}(H)$, and so

$$|z_n - a| \le \operatorname{diam}(f_{u_n}(H)) = \frac{1}{2^n}\operatorname{diam}(H).$$

Hence $z_n \to a$ as $n \to \infty$. Since $z_n \in \mathbb{T}$ and $\mathbb{T}$ is compact, it follows that $a \in \mathbb{T}$. We see that $\bigcap_{n\in\mathbb{N}_0} K_n \subseteq \mathbb{T}$. So the second equation in (4.6) is also valid.

The inclusions (4.5) follow from (4.6) and (4.7). □

For a finite word $u \in W_*$ we define

$$\mathbb{T}_u := f_u(\mathbb{T}) \subseteq \mathbb{T}. \tag{4.8}$$

Note that $\mathbb{T}_\emptyset = \mathbb{T}$. Since $\mathbb{T} = \pi(W)$ and $f_u(\pi(v)) = \pi(uv)$ whenever $u \in W_*$ and $v \in W$ (see (4.2)), the set $\mathbb{T}_u$ consists precisely of the points $a \in \mathbb{T}$ that can be represented in the form $a = \pi(w)$ with a word $w \in W$ that has u has an initial segment. This implies that if $v \in W_*$ is a finite word with the initial segment $u \in W_*$, then $\mathbb{T}_v \subseteq \mathbb{T}_u$.

It follows from (4.3) that

$$\mathbb{T}_u = \mathbb{T}_{u1} \cup \mathbb{T}_{u2} \cup \mathbb{T}_{u3}$$

for each $u \in W_*$ and from (4.4) that

$$\mathbb{T} = \bigcup_{u \in W_n} \mathbb{T}_u \tag{4.9}$$

for each $n \in \mathbb{N}_0$.

Since $I = [-1, 1] \subseteq \mathbb{T} \subseteq H$ (as follows from Proposition 4.2) and $\operatorname{diam}(I) = \operatorname{diam}(H) = 2$, we have $\operatorname{diam}(\mathbb{T}) = 2$. If $n \in \mathbb{N}_0$ and $u \in W_n$, then f_u is a similarity map that scales distances by the factor $1/2^n$. Hence

$$\operatorname{diam}(\mathbb{T}_u) = 2^{1-n}. \tag{4.10}$$

We have $0 = f_1(1) = f_2(-1) = f_3(-1)$. This implies

$$0 \in \mathbb{T}_k = f_k(\mathbb{T}) \subseteq f_k(H) = H_k \tag{4.11}$$

for $k = 1, 2, 3$. If $k, \ell \in \{1, 2, 3\}$ and $k \neq \ell$, then (see Fig. 5)

$$H_k \cap H_\ell = \{0\}, \text{ and so } \mathbb{T}_k \cap \mathbb{T}_\ell = \{0\}. \tag{4.12}$$

The next lemma provides a criterion when two infinite words in W represent the same point in $\mathbb{T}$ under the map π. Here we use the notation $\dot{k}$ for the infinite word $kkk\ldots$ for $k \in \{1, 2, 3\}$.

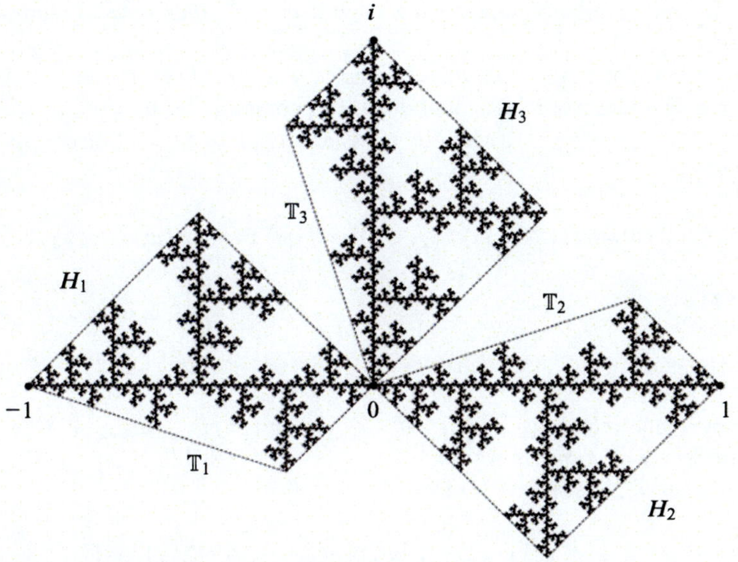

Fig. 5 The CSST $\mathbb{T}$ and its subtrees $\mathbb{T}_1, \mathbb{T}_2, \mathbb{T}_3$

Lemma 4.3 (i) *We have* $\pi^{-1}(0) = \{1\dot{2}, 2\dot{1}, 3\dot{1}\}$.

(ii) *Let* $v, w \in W$ *with* $v \neq w$. *Then* $\pi(v) = \pi(w)$ *if and only if there exists a finite word* $u \in W_*$ *such that* $v, w \in \{u1\dot{2}, u2\dot{1}, u3\dot{1}\}$. *In this case,* $\pi(v) = \pi(w) = f_u(0)$.

Note that if $v \in W$ and $v \in \{u1\dot{2}, u2\dot{1}, u3\dot{1}\}$ for some $u \in W_*$, then u is uniquely determined. This and the lemma imply that each point in $\mathbb{T} = \pi(W)$ has at most three preimages under the map π.

Proof

(i) Note that $1\dot{2} \in \pi^{-1}(0)$ as follows from

$$f_2(1) = 1 \text{ and } f_1(1) = 0.$$

Similarly, $2\dot{1}, 3\dot{1} \in \pi^{-1}(0)$, because

$$f_1(-1) = -1,\ f_2(-1) = 0, \text{ and } f_1(-1) = -1,\ f_3(-1) = 0.$$

Hence $\{1\dot{2}, 2\dot{1}, 3\dot{1}\} \subseteq \pi^{-1}(0)$.

To prove the reverse inclusion, suppose that $\pi(w) = 0$ for some $w = w_1w_2\ldots \in W$. We first consider the case $w_1 = 1$. Then $0 = f_1(a)$, where $a := \pi(w_2w_3\ldots)$, and so $a = 1$. Since $1 \in \mathbb{T}_2\backslash(\mathbb{T}_1 \cup \mathbb{T}_3)$ as follows from (4.11), we must have $w_2 = 2$. Then $1 = f_2(b)$, where $b := w_3w_4\ldots$, and so $b = 1 \in \mathbb{T}_2\backslash(\mathbb{T}_1 \cup \mathbb{T}_3)$. This implies $w_3 = 2$. Repeating the argument, we see that $2 = w_2 = w_3 = \ldots$, and so $w = 1\dot{2}$.

A very similar argument shows that if $w_1 = 2$, then $w = 2\dot{1}$, and if $w_1 = 3$, then $w = 3\dot{1}$.

(ii) Suppose that $\pi(v) = \pi(w)$ for some $v, w \in W_n$, $v \neq w$. Let $u \in W_*$ be the longest initial word that v and w have in common. So $v = uv_{n+1}v_{n+2}\ldots$ and $w = uw_{n+1}w_{n+2}\ldots$, where $n \in \mathbb{N}_0$ and $v_{n+1} \neq w_{n+1}$. Since f_u is bijective and

$$\pi(v) = f_u(\pi(v_{n+1}v_{n+2}\ldots)) = \pi(w) = f_u(\pi(w_{n+1}w_{n+2}\ldots)),$$

we have

$$\pi(v_{n+1}v_{n+2}\ldots) = \pi(w_{n+1}w_{n+2}\ldots).$$

Note that $\pi(v_{n+1}v_{n+2}\ldots) \in \mathbb{T}_{v_{n+1}}$ and $\pi(w_{n+1}w_{n+2}\ldots) \in \mathbb{T}_{w_{n+1}}$. Since $v_{n+1} \neq w_{n+1}$, by (4.12) this is only possible if $\pi(v_{n+1}v_{n+2}\ldots) = \pi(w_{n+1}w_{n+2}\ldots) = 0$. Hence

$$v_{n+1}v_{n+2}\ldots,\ w_{n+1}w_{n+2}\ldots \in \{1\dot{2}, 2\dot{1}, 3\dot{1}\}$$

by (i). The "only if" implication follows. Our considerations also show that $\pi(v) = \pi(w) = f_u(0)$. The reverse implication follows from (i). □

Our next goal is to show that $\mathbb{T}$ is indeed a tree. This requires some preparation.

Lemma 4.4 (i) *For each $p \in \mathbb{T}$ there exists a (possibly degenerate) arc α in $\mathbb{T}$ with endpoints -1 and p.*
(ii) *The sets $\mathbb{T}$, $\mathbb{T}\backslash\{1\}$, and $\mathbb{T}\backslash\{-1\}$ are arc-connected.*

Proof

(i) Let $p \in \mathbb{T}$. Then $p = \pi(w)$ for some $w = w_1 w_2 \ldots \in W$.
Let $v_n = w_1 \ldots w_n$ and define $a_n = f_{v_n}(-1) \in \mathbb{T}$ for $n \in \mathbb{N}_0$. Then $a_0 = f_\emptyset(-1) = -1$. For each $n \in \mathbb{N}_0$ we have

$$[a_n, a_{n+1}) = [f_{v_n}(-1), f_{v_n w_{n+1}}(-1)) = f_{v_n}\big([-1, f_{w_{n+1}}(-1))\big).$$

If $w_{n+1} = 1$, then $f_{w_{n+1}}(-1) = f_1(-1) = -1$; so $a_n = a_{n+1}$ and

$$[a_n, a_{n+1}) = \emptyset.$$

If $w_{n+1} \in \{2, 3\}$, then $f_{w_{n+1}}(-1) = 0$; so

$$[-1, f_{w_{n+1}}(-1)) = [-1, 0) \subseteq \mathbb{T}_1\backslash\{0\},$$

and

$$[a_n, a_{n+1}) = f_{v_n}([-1, 0)) \subseteq f_{v_n}(\mathbb{T}_1\backslash\{0\}) \subseteq \mathbb{T}_{v_n} \subseteq \mathbb{T}.$$

Moreover,

$$\begin{aligned} \text{length}([a_n, a_{n+1})) &= \frac{1}{2^n}\,\text{length}\big([-1, f_{w_{n+1}}(-1))\big) \\ &= \begin{cases} 2^{-n} & \text{if } w_{n+1} = 2, 3, \\ 0 & \text{if } w_{n+1} = 1. \end{cases} \end{aligned} \tag{4.13}$$

Let

$$A_n := \{p\} \cup \bigcup_{k \geq n+1} [a_k, a_{k+1})$$

for $n \in \mathbb{N}_0$. By what we have seen above,

$$[a_k, a_{k+1}) \subseteq \mathbb{T}_{v_k} \subseteq \mathbb{T}_{v_{n+1}}$$

for $k \geq n + 1$. Since $p = \lim_{k\to\infty} a_k$ and $\mathbb{T}_{v_{n+1}}$ is closed, we also have $p \in \mathbb{T}_{v_{n+1}}$, and so

$$A_n \subseteq \mathbb{T}_{v_{n+1}}.$$

This implies that

$$[a_n, a_{n+1}) \cap A_n = \emptyset$$

for each $n \in \mathbb{N}_0$. Indeed, if $w_{n+1} = 1$ this is clear, because then $[a_n, a_{n+1}) = \emptyset$.

If $w_{n+1} = 2$, then

$$A_n \subseteq \mathbb{T}_{v_{n+1}} = f_{v_n}(f_2(\mathbb{T})) = f_{v_n}(\mathbb{T}_2),$$

which implies that

$$[a_n, a_{n+1}) \cap A_n \subseteq f_{v_n}(\mathbb{T}_1 \backslash \{0\}) \cap f_{v_n}(\mathbb{T}_2) = f_{v_n}((\mathbb{T}_1 \backslash \{0\}) \cap \mathbb{T}_2) = \emptyset.$$

If $w_{n+1} = 3$, then $[a_n, a_{n+1}) \cap A_n = \emptyset$ by the same reasoning. This shows that the sets

$$[a_0, a_1), [a_1, a_2), [a_2, a_3), \dots, \{p\}$$

are pairwise disjoint. As $n \to \infty$, we have $a_n \to p$ and also $\operatorname{diam}(A_n) \to 0$ by (4.13). Therefore, the union

$$\alpha = [a_0, a_1) \cup [a_1, a_2) \cup [a_2, a_3) \cup \dots \cup \{p\} \tag{4.14}$$

is an arc in $\mathbb{T}$ joining $a_0 = -1$ and p (if $p = -1$, this arc is degenerate). We have proved (i).

To prepare the proof of (ii), we claim that if $p \neq 1$, then this arc α does not contain 1. Otherwise, we must have $1 \in [a_n, a_{n+1}) \subseteq \mathbb{T}_{v_n}$ for some $n \in \mathbb{N}_0$. This shows that 1 can be written in the form $1 = \pi(u)$, where $u \in W$ is an infinite word starting with the finite word $v := v_n$ (note that this and the statements below are trivially true for $n = 0$). On the other hand, we have $f_2(1) = 1$ which implies that $1 = \pi(\dot{2})$. By Lemma 4.3 (ii) this is only possible if all the letters in v are 2's. Then $f_v(1) = 1$ and it follows that

$$1 = f_v(1) \in [a_n, a_{n+1}) = f_v\big([-1, f_{w_{n+1}}(-1))\big).$$

Since f_v is a bijection, this implies that $1 \in [-1, f_{w_{n+1}}(-1))$. Now $f_{w_{n+1}}(-1) \in \{-1, 0\}$, and we obtain a contradiction. So indeed, $1 \notin \alpha$.

(ii) Let $p, q \in \mathbb{T}$ with $p \neq q$ be arbitrary. In order to show that $\mathbb{T}$ is arc-connected, we have to find an arc γ in $\mathbb{T}$ joining p and q. Now by the construction in (i) we can find arcs α and β in $\mathbb{T}$ joining p and q to -1, respectively. Then the desired arc γ can be found in the union $\alpha \cup \beta$ as follows. Starting from p, we travel long α until we first hit β, say in a point x. Such a point x exists, because $-1 \in \alpha \cap \beta \neq \emptyset$. Let α' be the (possibly degenerate) subarc of α with endpoints p and x, and β' be the subarc of β with endpoints x and q. Then $\gamma = \alpha' \cup \beta'$ is an arc in $\mathbb{T}$ joining p and q.

The arc-connectedness of $\mathbb{T}\backslash\{1\}$ is proved by the same argument. Indeed, if $p, q \in \mathbb{T}\backslash\{1\}$, then by the remark in the last part of the proof of (i), the arcs α and β constructed as in (i) do not contain 1. Then the arc $\gamma \subseteq \alpha \cup \beta$ does not contain 1 either.

Finally, to show that $\mathbb{T}\backslash\{-1\}$ is arc-connected, we assume that $p, q \in \mathbb{T}\backslash\{-1\}$. If x is, as above, the first point on β as we travel along α starting from p, then it suffices to show that $x \neq -1$, because then $-1 \notin \gamma$. This in turn will follow if we can show that α and β have another point in common besides -1.

To find such a point, we revisit the above construction. Pick $w = w_1 w_2 \ldots \in W$ and $u = u_1 u_2 \ldots \in W$ such that $p = \pi(w)$ and $q = \pi(u)$. Let α and β be the arcs for p and q, respectively, as constructed in (i). Then α is as in (4.14) and we can write the other arc β as

$$\beta = [b_0, b_1) \cup [b_1, b_2) \cup [b_2, b_3) \cup \cdots \cup \{q\},$$

where $b_n = f_{u_1 \ldots u_n}(-1)$ for $n \in \mathbb{N}_0$. Since $p \neq q$, we have $w \neq u$, and so there exists a largest $n \in \mathbb{N}_0$ such that $v := w_1 \ldots w_n = u_1 \ldots u_n$ and $w_{n+1} \neq u_{n+1}$. Then $a_n = b_n = f_v(-1) \in \alpha \cap \beta$. If $a_n = b_n \neq -1$, we are done. So we may assume that $a_n = b_n = f_v(-1) = -1$. Then $a_0 = \cdots = a_n = -1$, and so $w_k = u_k = 1$ for $k = 1, \ldots, n$. This shows that all letters in v are equal to 1.

Since the letters w_{n+1} and u_{n+1} are distinct, one of them is different from 1. We may assume $u_{n+1} \neq 1$. Then $f_{u_{n+1}}(-1) = 0$, and so $(b_n, b_{n+1}] = f_v((-1, 0]) \subseteq \beta \setminus \{-1\}$. Here we used that f_v is a homeomorphism with $f_v(-1) = -1$.

Since $p = \pi(w) \neq -1 = \pi(\dot{1})$, we have $w \neq \dot{1}$ and so there exists a smallest $\ell \in \mathbb{N}$ such that $w_{n+\ell} \neq 1$. Then $f_{w_{n+\ell}}(-1) = 0$ and so a simple computation using $w_{n+1} = \cdots = w_{n+\ell-1} = 1$ shows that

$$c := f_{w_{n+1} \ldots w_{n+\ell}}(-1) = f_{w_{n+1} \ldots w_{n+\ell-1}}(0) = 2^{1-\ell} - 1 \in (-1, 0].$$

Hence

$$a_{n+\ell} = f_v(c) \in f_v((-1, 0]) \subseteq \beta \setminus \{-1\}.$$

It follows that $a_{n+\ell} \in \alpha \cap \beta$ and $a_{n+\ell} \neq -1$ as desired. □

The next lemma will help us to identify the branch points of $\mathbb{T}$ once we know that $\mathbb{T}$ is a tree.

Lemma 4.5

(i) *The components of* $\mathbb{T}\backslash\{0\}$ *are given by the non-empty sets* $\mathbb{T}_1\backslash\{0\}$, $\mathbb{T}_2\backslash\{0\}$, $\mathbb{T}_3\backslash\{0\}$.

(ii) *If $u \in W_*$, then $\mathbb{T}\backslash\{f_u(0)\}$ has exactly three components. The sets $\mathbb{T}_{u1}\backslash\{f_u(0)\}$, $\mathbb{T}_{u2}\backslash\{f_u(0)\}$, $\mathbb{T}_{u3}\backslash\{f_u(0)\}$ are each contained in a different component of $\mathbb{T}\backslash\{f_u(0)\}$.*

In the proof we will use the following general facts about components of a subset M of a metric space X. Recall that a set $A \subseteq M$ is relatively closed in M if $A = \overline{A} \cap M$, or equivalently, if each limit point of A that belongs to M also belongs to A. Each component A of M is relatively closed in M, because its relative closure $\overline{A} \cap M$ is a connected subset of M with $A \subseteq \overline{A} \cap M$. Hence $A = \overline{A} \cap M$, because A is a component of M and hence a maximal connected subset of M.

If $A_1, \ldots, A_n \subseteq M$ for some $n \in \mathbb{N}$ are non-empty, pairwise disjoint, relatively closed, and connected sets with $M = A_1 \cup \cdots \cup A_n$, then these sets are the components of M.

Proof

(i) Each of the sets $\mathbb{T}\backslash\{1\}$ and $\mathbb{T}\backslash\{-1\}$ is non-empty, and connected by Lemma 4.4 (ii). Therefore, the sets

$$\mathbb{T}_1\backslash\{0\} = f_1(\mathbb{T}\backslash\{1\}),\ \mathbb{T}_2\backslash\{0\} = f_2(\mathbb{T}\backslash\{-1\}),\ \mathbb{T}_3\backslash\{0\} = f_3(\mathbb{T}\backslash\{-1\})$$

are non-empty and connected. They are also relatively closed in $\mathbb{T}\backslash\{0\}$ and pairwise disjoint by (4.12). Since $\mathbb{T} = \mathbb{T}_1 \cup \mathbb{T}_2 \cup \mathbb{T}_3$ we have

$$\mathbb{T}\backslash\{0\} = (\mathbb{T}_1\backslash\{0\}) \cup (\mathbb{T}_2\backslash\{0\}) \cup (\mathbb{T}_3\backslash\{0\}).$$

This implies that the sets $\mathbb{T}_k\backslash\{0\}$, $k = 1, 2, 3$, are the components of $\mathbb{T}\backslash\{0\}$. The statement follows.

(ii) We prove this by induction on the length $n \in \mathbb{N}_0$ of the word $u \in W_*$. If $n = 0$ and so $u = \emptyset$, this follows from statement (i).

Suppose the statement is true for all words of length $n - 1$, where $n \in \mathbb{N}$. Let $u = u_1 \ldots u_n \in W_n$ be an arbitrary word of length n. We set $\ell := u_1$ and $u' := u_2 \ldots u_n$. Then $u = \ell u'$. To be specific and ease notation, we will assume that $\ell = 1$. The other cases $\ell = 2$ or $\ell = 3$ are completely analogous and we will skip the details.

Note that $f_u(0) \neq 0$. Indeed, if

$$0 = f_u(0) = f_u(\pi(1\dot{2})) = \pi(u1\dot{2}),$$

then $u1\dot{2} \in \{1\dot{2}, 2\dot{1}, 3\dot{1}\}$ by Lemma 4.3 (i). This is only possible $u1 = 1$. This is a contradiction, because u has length $n \geq 1$. Hence $f_u(0) \neq 0$. Since $u_1 = \ell = 1$, we have $f_u(0) \in \mathbb{T}_1\backslash\{0\}$.

By induction hypothesis, $\mathbb{T}\backslash\{f_{u'}(0)\}$ has exactly three connected components V_1, V_2, V_3, and we may assume that $\mathbb{T}_{u'k}\backslash\{f_{u'}(0)\} \subseteq V_k$ for $k = 1, 2, 3$. It follows that

$$f_\ell(\mathbb{T}\backslash\{f_{u'}(0)\}) = f_1(\mathbb{T}\backslash\{f_{u'}(0)\}) = \mathbb{T}_1\backslash\{f_u(0)\}$$

has exactly three connected components $U_k = f_1(V_k) \subseteq \mathbb{T}_1$ with

$$\mathbb{T}_{uk}\backslash\{f_u(0)\} = \mathbb{T}_{1u'k}\backslash\{f_{1u'}(0)\} = f_1(\mathbb{T}_{u'k}\backslash\{f_{u'}(0)\}) \subseteq f_1(V_k) = U_k$$

for $k = 1, 2, 3$.

Let $k \in \{1, 2, 3\}$. Then we have $V_k = \overline{V_k} \cap \mathbb{T}\backslash\{f_{u'}(0)\}$, because V_k is a component of $\mathbb{T}\backslash\{f_{u'}(0)\}$ and hence relatively closed in $\mathbb{T}\backslash\{f_{u'}(0)\}$. This implies that

$$\begin{aligned} U_k = f_1(V_k) &= f_1(\overline{V_k} \cap \mathbb{T}\backslash\{f_{u'}(0)\}) = f_1(\overline{V_k}) \cap \mathbb{T}_1\backslash\{f_u(0)\} \\ &= \overline{f_1(V_k)} \cap \mathbb{T}_1\backslash\{f_u(0)\} = \overline{U_k} \cap \mathbb{T}_1\backslash\{f_u(0)\}. \end{aligned}$$

Since $\mathbb{T}_1 \subseteq \mathbb{T}$ is compact, $U_k \subseteq \mathbb{T}_1$, and so $\overline{U_k} \subseteq \mathbb{T}_1$, this shows that every limit point of U_k distinct from $f_u(0)$ belongs to U_k. Hence U_k is relatively closed in $\mathbb{T}\backslash\{f_u(0)\}$.

Exactly one of the components of $\mathbb{T}_1\backslash\{f_u(0)\}$, say U_1, contains the point $0 \in \mathbb{T}_1\backslash\{f_u(0)\}$. Then $U_1' := U_1 \cup \mathbb{T}_2 \cup \mathbb{T}_3$ is a relatively closed subset of $\mathbb{T}\backslash\{f_u(0)\}$. This set is also connected, because the sets U_1, $\mathbb{T}_2 = f_2(\mathbb{T})$, $\mathbb{T}_3 = f_3(\mathbb{T})$ are connected and have the point 0 in common. Hence the connected sets U_1', U_2, U_3 are pairwise disjoint, relatively closed in $\mathbb{T}\backslash\{f_u(0)\}$, and

$$\mathbb{T}\backslash\{f_u(0)\} = (\mathbb{T}_1\backslash\,\{f_u(0)\}) \cup \mathbb{T}_2 \cup \mathbb{T}_3 = U_1' \cup U_2 \cup U_3.$$

This implies that $\mathbb{T}\backslash\{f_u(0)\}$ has exactly the three connected components U_1', U_2, U_3. Moreover, $\mathbb{T}_{u1}\backslash\{f_u(0)\}$, $\mathbb{T}_{u2}\backslash\{f_u(0)\}$, $\mathbb{T}_{u3}\backslash\{f_u(0)\}$ lie in the different components U_1', U_2, U_3 of $\mathbb{T}\backslash\{f_u(0)\}$, respectively. This provides the inductive step, and the statement follows. □

We can now show that $\mathbb{T}$ is a metric tree.

Proof of Proposition 1.4 We know that $\mathbb{T}$ is compact, contains at least two points, and is arc-connected by Lemma 4.4.

Let $p \in \mathbb{T}$ and $n \in \mathbb{N}$ be arbitrary, and define

$$N = \bigcup\{\mathbb{T}_u : u \in W_n \text{ and } p \in \mathbb{T}_u\}.$$

Since each of the sets $\mathbb{T}_u = f_u(\mathbb{T})$, $u \in W_*$, is a compact and connected subset of $\mathbb{T}$, the set N is connected. Moreover, since each of the finitely many sets $\mathbb{T}_u$, $u \in W_n$, is closed, we can find $\delta > 0$ such that

$$\mathrm{dist}(p, \mathbb{T}_u) \geq \delta$$

whenever $u \in W_n$ and $p \notin \mathbb{T}_u$. Then we have $B(p, \delta) \cap \mathbb{T} \subseteq N$ by (4.9), and so N is a connected relative neighborhood of p in $\mathbb{T}$. It follows from (4.10) that $\mathrm{diam}(N) \leq 2^{2-n}$. This shows that each point in $\mathbb{T}$ has arbitrarily small connected neighborhoods in $\mathbb{T}$. Hence $\mathbb{T}$ is locally connected.

To complete the proof, it remains to show that the arc joining two given distinct points in $\mathbb{T}$ is unique. For this we argue by contradiction and assume that there are two distinct arcs in $\mathbb{T}$ with the same endpoints. By considering suitable subarcs of these arcs, we can reduce to the following situation: there are arcs $\alpha, \beta \subseteq \mathbb{T}$ that have the distinct endpoints $a, b \in \mathbb{T}$ in common, but no other points.

To see that this leads to a contradiction, we represent the points a and b by words in W; so $a = \pi(v)$ and $b = \pi(w)$, where $v = v_1 v_2 \dots$ and $w = w_1 w_2 \dots$ are in W. Since $a \neq b$ and every point in $\mathbb{T}$ has at most three such representations by Lemma 4.3 (ii), we can find a pair v and w representing a and b with the largest common initial word, say $v_1 = w_1, \dots, v_n = w_n$, and $v_{n+1} \neq w_{n+1}$ for some maximal $n \in \mathbb{N}_0$.

Let $u = v_1 \dots v_n = w_1 \dots w_n$ and

$$t = f_u(0) = \pi(u1\dot{2}) = \pi(u2\dot{1}) = \pi(u3\dot{1}).$$

Then $t \neq a, b$. To see this, assume that $t = a$, say. We have $w_{n+1} \in \{1, 2, 3\}$, and so, say $w_{n+1} = 1$. But then $a = t = \pi(u1\dot{2})$ and $b = \pi(u1w_{n+2} \dots)$. So a and b are represented by words with the common initial segment $u1$ that is longer than u. This contradicts the choice of v and w. The cases $w_{n+1} = 2$ or $w_{n+1} = 3$ lead to a contradiction in a similar way.

So indeed $t = f_u(0) \neq a, b$. Moreover $a = \pi(uv_{n+1} \dots) \in \mathbb{T}_{uv_{n+1}} \backslash \{t\}$ and similarly $b \in \mathbb{T}_{uw_{n+1}} \backslash \{t\}$. Since $v_{n+1} \neq w_{n+1}$ the points a and b lie in different components of $\mathbb{T} \backslash \{t\}$ by Lemma 4.5 (ii). So any arc joining a and b must pass through t. Hence $t \in \alpha \cap \beta$, but $t \neq a, b$. This contradicts our assumption that the arcs α and β have no other points than their endpoints a and b in common. □

If $M \subseteq \mathbb{T}$, then we denote by $\partial M \subseteq \mathbb{T}$ the relative boundary of M in $\mathbb{T}$.

Lemma 4.6 *Let $n \in \mathbb{N}$ and $u \in W_n$. Then*

$$\partial \mathbb{T}_u \subseteq \{f_u(-1), f_u(1)\}. \tag{4.15}$$

Moreover, if $p \in \partial \mathbb{T}_u$, then $p = f_w(0)$ for some word $w \in W_*$ of length $\leq n - 1$.

In particular, the set $\partial \mathbb{T}_u$ contains at most two points.

Proof We prove this by induction on n. First consider $n = 1$. So let $u = k \in W_1 = \{1, 2, 3\}$. Then $\mathbb{T}_k \backslash \{0\}$ is a component $\mathbb{T} \backslash \{0\}$ by Lemma 4.5 (i). Hence Proposition 1.4 and Lemma 3.2 (i) imply that $\mathbb{T}_k \backslash \{0\}$ is a relatively open set in

$\mathbb{T}$. So each of its points lies in the relative interior of $\mathbb{T}_k$ and cannot lie in $\partial\mathbb{T}_k$. Therefore, $\partial\mathbb{T}_k \subseteq \{0\}$. Since

$$0 = f_\emptyset(0) = f_1(1) = f_2(-1) = f_3(-1), \tag{4.16}$$

the statement is true for $n = 1$.

Suppose the statement is true for all words in W_n, where $n \in \mathbb{N}$. Let $u \in W_{n+1}$ be arbitrary. Then $u = vk$, where $v \in W_n$ and $k \in \{1, 2, 3\}$. By what we have just seen, the set $\mathbb{T}_k \backslash \{0\}$ is open in $\mathbb{T}$. Hence

$$f_v(\mathbb{T}_k \backslash \{0\}) = f_u(\mathbb{T}) \backslash \{f_v(0)\} = \mathbb{T}_u \backslash \{f_v(0)\}$$

is a relatively open subset of $f_v(\mathbb{T}) = \mathbb{T}_v$. So if $p \in \mathbb{T}_u$ is not an interior point of $\mathbb{T}_u$ in $\mathbb{T}$, then $p = f_v(0)$ or p is not an interior point of $\mathbb{T}_v$ in $\mathbb{T}$ and hence belongs to the boundary of $\mathbb{T}_v$. This and the induction hypothesis imply that

$$\partial\mathbb{T}_u \subseteq \{f_v(0)\} \cup \partial\mathbb{T}_v \subseteq \{f_v(0), f_v(-1), f_v(1)\}.$$

From this we conclude that each point $p \in \partial\mathbb{T}_u \subseteq \{f_v(0)\} \cup \partial\mathbb{T}_v$ can be written in the form $f_w(0)$ for an appropriate word w of length $\leq n$. This is clear if $p = f_v(0)$ and follows for $p \in \partial\mathbb{T}_v$ from the induction hypothesis.

Now $\mathbb{T}_u = f_u(\mathbb{T})$ is compact and so closed in $\mathbb{T}$. Hence $\partial\mathbb{T}_u \subseteq \mathbb{T}_u$. On the other hand, $\mathbb{T}_u$ contains only two of the points $f_v(0), f_v(-1), f_v(1)$. Indeed, if $k = 1$, then $1 \notin \mathbb{T}_1 \subseteq H_1$, and so $f_v(1) \notin f_v(\mathbb{T}_1) = \mathbb{T}_u$. It follows that $\partial\mathbb{T}_u \subseteq \{f_v(-1), f_v(0)\}$. Note that $f_1(-1) = -1$ and $f_1(1) = 0$, and so

$$f_v(-1) = f_v(f_1(-1)) = f_u(-1) \text{ and } f_v(0) = f_v(f_1(1)) = f_u(1).$$

Hence

$$\partial\mathbb{T}_u \subseteq \{f_u(-1), f_u(1)\}.$$

Very similar considerations show that if $k = 2$, then

$$\partial\mathbb{T}_u \subseteq \{f_v(0), f_v(1)\} = \{f_u(-1), f_u(1)\},$$

and if $k = 3$, then

$$\partial\mathbb{T}_u \subseteq \{f_v(0)\} = \{f_u(-1)\}.$$

The statement follows. □

The next lemma shows that all branch points of $\mathbb{T}$ are of the form $f_u(0)$ with $u \in W_*$.

Lemma 4.7 *The branch points of* $\mathbb{T}$ *are exactly the points of the form* $t = f_u(0)$ *for some finite word* $u \in W_*$. *They are triple points of* $\mathbb{T}$.

Proof By Lemma 4.5 (ii) we know that each point $t = f_u(0)$ with $u \in W_*$ is a triple point of the tree $\mathbb{T}$. We have to show that there are no other branch points of $\mathbb{T}$.

So suppose that t is a branch point of $\mathbb{T}$, but $t \neq f_u(0)$ for each $u \in W_*$. Then we can find (at least) three distinct components U_1, U_2, U_3 of $\mathbb{T}\backslash\{t\}$. Pick a point $x_k \in U_k$ and choose $n \in \mathbb{N}$ such that $|x_k - t| > 2^{1-n}$ for $k = 1, 2, 3$. By (4.9) we can find $u \in W_n$ such that $t \in \mathbb{T}_u$. Then t is distinct from the points in the relative boundary $\partial\mathbb{T}_u$, because they have the form $f_w(0)$ for some $w \in W_*$ (see Lemma 4.6). Hence t is contained in the relative interior of $\mathbb{T}_u$ in $\mathbb{T}$. Moreover, $\text{diam}(\mathbb{T}_u) = 2^{1-n}$, and so $x_k \notin \mathbb{T}_u$. For $k = 1, 2, 3$ let α_k be the arc in $\mathbb{T}$ joining x_k and t. As we travel from x_k to t along α_k, there exists a first point $y_k \in \mathbb{T}_u$. Then $y_k \in \partial\mathbb{T}_u$ and so $y_k \neq t$. Let β_k be the subarc of α_k with endpoints x_k and y_k. Then β_k is a connected set in $\mathbb{T}\backslash\{t\}$. Since $x_k \in \beta_k$, it follows that $\beta_k \subseteq U_k$, and so $y_k \in U_k$.

This shows that the points y_1, y_2, y_3 are distinct and contained in the relative boundary $\partial\mathbb{T}_u$. This is impossible, because by Lemma 4.6 the set $\partial\mathbb{T}_u$ consists of at most two points. □

We can now prove Proposition 1.5 which shows that $\mathbb{T}$ satisfies the conditions in Theorem 1.7 and belongs to the class of trees $\mathcal{T}_3$.

Proof of Proposition 1.5 By Lemma 4.7 each branch point of $\mathbb{T}$ is a triple point and each set $\mathbb{T}_u$ for $u \in W_n$ and $n \in \mathbb{N}$ contains the triple point $t = f_u(0)$. The sets $\mathbb{T}_u$, $u \in W_n$, cover $\mathbb{T}$ and have small diameter for n large. It follows that the triple points are dense in $\mathbb{T}$. □

In order to show that $\mathbb{T}$ is a quasi-convex subset of $\mathbb{C}$, we first require a lemma.

Lemma 4.8 *There exists a constant* $K > 0$ *such that if* $p \in \mathbb{T}$ *and* α *is the arc in* $\mathbb{T}$ *joining* 0 *and* p, *then*

$$\text{length}(\alpha) \leq K|p|. \tag{4.17}$$

In particular, the arc α is a rectifiable curve.

Proof Let $p \in \mathbb{T}$ be arbitrary. We may assume that $p \neq 0$. Then $p = \pi(w)$ for some $w = w_1w_2 \ldots \in W$. For simplicity we assume $w_1 = 3$. The other cases, $w_1 = 1$ and $w_1 = 2$, are very similar and we will only present the details for $w_1 = 3$.

Since $p \neq 0 = \pi(3\dot{1})$, we have $w_2w_3 \ldots \neq \dot{1}$. Hence there exists a smallest number $n \in \mathbb{N}$ such that such that $w_{n+1} \neq 1$. Let $v = w_1 \ldots w_n$ be the initial word of w and $w' = w_{n+1}w_{n+2} \ldots$ be the tail of w. The word v has the form $v = 31 \ldots 1$, where the sequence of 1's could possibly be empty. Note that $q := \pi(w') \in \mathbb{T}_{w_{n+1}} \subseteq \mathbb{T}_2 \cup \mathbb{T}_3 \subseteq H_2 \cup H_3$. Since

$$c_0 := \text{dist}(-1, H_2 \cup H_3) > 0$$

(see Fig. 4), for the distance of q and -1 we have $|q+1| \geq c_0$. We also have $f_v(q) = p$, and $f_v(-1) = 0$, because $f_1(-1) = -1$ and $f_3(-1) = 0$. It follows that

$$|p| = |f_v(q) - f_v(-1)| = \frac{1}{2^n}|q+1| \geq \frac{c_0}{2^n}. \tag{4.18}$$

Now define $a_0 = 0 = f_v(-1)$ and $a_k = f_{vw_{n+1}\dots w_{n+k-1}}(0)$ for $k \in \mathbb{N}$ (here $w_{n+1}\dots w_{n+k-1} = \emptyset$ for $k=1$). Note that then

$$a_1 = f_v(0) = f_{w_1\dots w_n}(0) = f_{31\dots 1}(0) = f_3(2^{1-n}-1) = i/2^n,$$

and so

$$[a_0, a_1] = [f_v(-1), f_v(0)] = [0, i/2^n] \subseteq [0, i] \subseteq \mathbb{T}.$$

This also shows that $\text{length}([a_0, a_1]) = 1/2^n$.

For $k \in \mathbb{N}$ we have $f_{w_{n+k}}(0) \in \{-1/2, 1/2, i/2\}$, and $[0, f_{w_{n+k}}(0)] \subseteq \mathbb{T}$. This implies that

$$[a_k, a_{k+1}] = f_{vw_{n+1}\dots w_{n+k-1}}\big([0, f_{w_{n+k}}(0)]\big) \subseteq \mathbb{T}$$

and $\text{length}([a_k, a_{k+1}]) = 1/2^{n+k}$ for $k \in \mathbb{N}$. Since $\lim_{k\to\infty} a_k = \pi(w) = p$, we can concatenate the intervals $[a_k, a_{k+1}] \subseteq \mathbb{T}$ for $k \in \mathbb{N}_0$, add the endpoint p, and obtain a path γ in $\mathbb{T}$ that joins 0 and p with

$$\text{length}(\gamma) = \sum_{k=0}^{\infty} \frac{1}{2^{n+k}} = \frac{1}{2^{n-1}}.$$

The (image of the) path γ will contain the unique arc α in $\mathbb{T}$ joining 0 and p and so $\text{length}(\alpha) \leq 1/2^{n-1}$. If we combine this with (4.18), then inequality (4.8) follows with $K = 2/c_0$. □

We can now show that $\mathbb{T}$ is indeed a quasi-convex subset of $\mathbb{C}$.

Proof of Proposition 1.6 Let $a, b \in \mathbb{T}$ be arbitrary. We may assume that $a \neq b$. Then there are words $u = u_1u_2\dots \in W$ and $v = v_1v_2\dots \in W$ such that $a = \pi(u)$ and $b = \pi(v)$. Since $a \neq b$, we have $u \neq v$ and so there exists a smallest number $n \in \mathbb{N}_0$ such that $u_1 = v_1, \dots, u_n = v_n$ and $u_{n+1} \neq v_{n+1}$. Let $w = u_1\dots u_n = v_1\dots v_n$, $u' = u_{n+1}u_{n+2}\dots \in W$ and $v' = v_{n+1}v_{n+2}\dots \in W$. We define $a' = \pi(u')$ and $b' = \pi(v')$. Set $k = u_{n+1}$ and $\ell = v_{n+1}$. Then $k \neq \ell$, $a' \in \mathbb{T}_k \subseteq H_k$, and $b' \in \mathbb{T}_\ell \subseteq H_\ell$. We now use the following elementary geometric estimate: there exists a constant $c_1 > 0$ such that

$$|x - y| \geq c_1(|x| + |y|),$$

whenever $x \in H_k$, $y \in H_\ell$, $k, \ell \in \{1, 2, 3\}$, $k \neq \ell$. Essentially, this follows from the fact that the sets H_1, H_2, H_3 are contained in closed sectors in $\mathbb{C}$ that are pairwise disjoint except for the common point 0.

In our situation, this means that

$$|a' - b'| \geq c_1(|a'| + |b'|).$$

Let σ and τ be the arcs in $\mathbb{T}$ joining 0 to a' and b', respectively. Then $\sigma \cup \tau$ contains the arc α' in $\mathbb{T}$ joining a' and b'. Then it follows from Lemma 4.8 that

$$\text{length}(\alpha') \leq \text{length}(\sigma) + \text{length}(\tau) \leq K(|a'| + |b'|) \leq L|a' - b'| \tag{4.19}$$

with $L := K/c_1$.

For the similarity f_w we have $f_w(a') = a$ and $f_w(b') = b$. Since $f_w(\mathbb{T}) \subseteq \mathbb{T}$, it follows that $\alpha := f_w(\alpha')$ is the unique arc in $\mathbb{T}$ joining a and b. Since f_w scales distances by a fixed factor (namely $1/2^n$), (4.19) implies the desired inequality $\text{length}(\alpha) \leq L|a - b|$. □

As we already discussed in the introduction, by Proposition 1.6 we can define a new metric ϱ on $\mathbb{T}$ by setting

$$\varrho(a, b) = \text{length}(\alpha) \tag{4.20}$$

for $a, b \in \mathbb{T}$, where α is the unique arc in $\mathbb{T}$ joining a and b. Then the metric space $(\mathbb{T}, \varrho)$ is geodesic, and we have

$$|a - b| \leq \varrho(a, b) \leq L|a - b|$$

for $a, b \in \mathbb{T}$, where L is the constant in Proposition 1.6. This implies that the metric spaces $\mathbb{T}$ (as equipped with the Euclidean metric) and $(\mathbb{T}, \varrho)$ are bi-Lipschitz equivalent by the identity map.

We now want to reconcile Definition 1.2 with the construction of the CSST as an abstract metric space outlined in the introduction. We require an auxiliary statement.

Lemma 4.9 *Let $n \in \mathbb{N}_0$. Then the sets*

$$f_w(\mathbb{T}\backslash\{-1\}), \ w \in W_n, \tag{4.21}$$

are pairwise disjoint and their union is equal to $\mathbb{T}\backslash\{-1\}$.

Proof This is proved by induction on $n \in \mathbb{N}_0$. For $n = 0$ the statement is clear, because then $f_\emptyset(\mathbb{T}\backslash\{-1\}) = \mathbb{T}\backslash\{-1\}$ is the only set in (4.21).

Suppose the statement is true for some $n \in \mathbb{N}$. Then for each $u \in W_n$ the sets

$$\begin{aligned} f_{u1}(\mathbb{T}\backslash\{-1\}) &= f_u(\mathbb{T}_1\backslash\{-1\}), \\ f_{u2}(\mathbb{T}\backslash\{-1\}) &= f_u(\mathbb{T}_2\backslash\{0\}), \\ f_{u3}(\mathbb{T}\backslash\{-1\}) &= f_u(\mathbb{T}_3\backslash\{0\}) \end{aligned}$$

provide a decomposition of $f_u(\mathbb{T}\backslash\{-1\})$ into three pairwise disjoint subsets as follows from (4.9) for $n = 1$, (4.11), and (4.12). This and the induction hypothesis imply that the sets $f_{uk}(\mathbb{T}\backslash\{-1\})$, $u \in W_n$, $k \in \{1, 2, 3\}$, and hence the sets $f_w(\mathbb{T}\backslash\{-1\})$, $w \in W_{n+1}$, are pairwise disjoint, and their union is equal to $\mathbb{T}\backslash\{-1\}$. This is the inductive step, and the statement follows. □

We now consider the sets J_n, $n \in \mathbb{N}_0$, as in Proposition 4.2. Here $J_0 = I = [-1, 1]$ is a line segment of length 2. Since $(-1, 1] \subseteq \mathbb{T}\backslash\{-1\}$, the previous lemma implies that for each $n \in \mathbb{N}_0$, the sets $f_w((-1, 1])$, $w \in W_n$, are pairwise disjoint half-open line segments of length 2^{1-n}. The union of the closures $f_w([-1, 1]) = f_w(I)$, $w \in W_n$, of these line segments is the set J_n. In particular, J_n consists of 3^n line segments of length 2^{1-n} with pairwise disjoint interiors.

Note that for $w \in W_n$ we have

$$\begin{aligned} &f_{w1}((-1, 1]) \cup f_{w2}((-1, 1]) \cup f_{w3}((-1, 1]) \\ &\quad = f_w((-1, 0]) \cup f_w((0, 1]) \cup f_w((0, i]) \\ &\quad = f_w((-1, 1]) \cup f_w([0, i]). \end{aligned}$$

An induction argument based on this shows that for $n \in \mathbb{N}_0$ we have a decomposition

$$J_n\backslash\{-1\} = \bigcup_{w \in W_n} f_w((-1, 1]) \tag{4.22}$$

of $J_n\backslash\{-1\}$ into the pairwise disjoint sets $f_w((-1, 1])$, $w \in W_n$.

In the passage from J_n to J_{n+1} we can think of each line segment $f_w(I) = f_w([-1, 1])$ as being replaced with

$$f_{w1}(I) \cup f_{w2}(I) \cup f_{w3}(I) = f_w([-1, 0]) \cup f_w([0, 1]) \cup f_w([0, i]).$$

So $f_w([-1, 1])$ is split into two intervals $f_w([-1, 0])$ and $f_w([0, 1])$, and at its midpoint $f_w(0)$ a new interval $f_w([0, i])$ is "glued" to $f_w(0)$. This is exactly the procedure described in the introduction. Note that Lemma 4.9 implies that these new intervals $f_w([0, i]) \subseteq f_w(\mathbb{T}\backslash\{-1\})$, $w \in W_n$, are pairwise disjoint. Moreover, each such interval $f_w([0, i])$ meets the set J_n only in the point $f_w(0)$ and in no other point of J_n. Indeed, by (4.22) and Lemma 4.9 we have

$$
\begin{aligned}
f_w((0,i]) \cap J_n &= f_{w3}((-1,1]) \cap J_n = f_{w3}((-1,1]) \cap J_n \backslash \{-1\} \\
&= f_w((0,i]) \cap \bigcup_{u \in W_n} f_u((-1,1]) \\
\subseteq \big(f_w((0,i]) \cap f_w((-1,1])\big) &\cup \bigcup_{u \in W_n,\, u \neq w} f_w(\mathbb{T}\backslash\{-1\}) \cap f_u(\mathbb{T}\backslash\{-1\}) = \emptyset.
\end{aligned}
$$

It is clear that J_n is compact, and one can show by induction based on the replacement procedure just described that J_n is connected. Hence each J_n is a subtree of $\mathbb{T}$ by Lemma 3.3. The metric ϱ in (4.20) restricted to J_n, $n \in \mathbb{N}_0$, and to $J := \bigcup_{n \in N_0} J_n$ is just the natural Euclidean path metric on these sets. In particular, ϱ is a geodesic metric on J. These considerations imply that (J_n, ϱ) for $n \in \mathbb{N}$, and hence (J, ϱ), are isometric to the abstract versions of these spaces defined in the introduction.

By Proposition 4.2 the tree $\mathbb{T}$ is the equal to closure $\overline{J}$ in $\mathbb{C}$. Since on J the Euclidean metric and the metric ϱ are comparable, the set $\mathbb{T} = \overline{J}$ is homeomorphic to the space obtained from the completion of the geodesic metric space (J, ϱ). This is how we described the CSST as an abstract metric space in the introduction.

5 Decomposing Trees in $\mathcal{T}_m$

In the previous section we have seen that for each $n \in \mathbb{N}$ the CSST admits a decomposition

$$
\mathbb{T} = \bigcup_{u \in W_n} \mathbb{T}_u
$$

into subtrees. We will now consider an arbitrary tree in $\mathcal{T}_m$, $m \in \mathbb{N}$, $m \geq 3$, and find similar decompositions into subtrees. Our goal is to have decompositions for each level $n \in \mathbb{N}$ so that the conditions (i)–(iii) in Proposition 2.1 are satisfied.

Note that each tree class $\mathcal{T}_m$ is non-empty. Namely, for each $m \in \mathbb{N}$, $m \geq 3$, a tree in $\mathcal{T}_m$ can be obtained by essentially the same method as for the construction of the CSST as an abstract metric space outlined in the introduction. The only difference is that instead of gluing one line segment of length 2^{-n} to the midpoint c_s of a line segment s of length 2^{1-n} obtained in the nth step, we glue endpoints of $m-2$ such segments to c_s. Since from a purely logical point of view we will not need the fact that $\mathcal{T}_m$ is non-empty for the proof of Theorem 1.8, we will skip further details.

We now fix $m \in \mathbb{N}$, $m \geq 3$, for the rest of this section. We consider the alphabet $\mathcal{A} = \{1, 2, \ldots, m\}$. In the following, words will contain only letters in this fixed alphabet and we use the simplified notation for the sets of words W, W_n, W_* as discussed in Sect. 4.

Let T be an arbitrary tree in the class $\mathcal{T}_m$. We will now define subtrees T_u of T for all levels $n \in \mathbb{N}$ and all $u \in W_n$. The boundary ∂T_u of T_u in T will consist of one or two points that are leaves of T_u and branch points of T. We consider each point in ∂T_u as a *marked leaf* in T_u and will assign to it an appropriate sign $-$ or $+$ so that if there are two marked leaves in T_u, then they carry different signs. Accordingly, we refer to the points in ∂T_u as the *signed marked leaves* of T_u. The same point may carry different signs in different subtrees. We write p^- if a marked leaf p of T_u carries the sign $-$ and p^+ if it carries the sign $+$. To refer to this sign, we also write $\operatorname{sgn}(p, T_u) = -$ in the first and $\operatorname{sgn}(p, T_u) = +$ in the second case. If T_u has exactly one marked leaf, we call T_u a *leaf-tile* and if there are two marked leaves an *arc-tile*.

The reason why we want to use these markings is that it will help us to consistently label the subtrees so that if another tree S in $\mathcal{T}_m$ is decomposed by the same procedure, then we obtain decompositions of our trees T and S into subtrees on all levels n that satisfy the analogs of (2.1) and (2.2) (here $u \in W_n$ will play the role of the index i on each level n). While (2.1) is fairly straightforward to obtain, (2.2) requires a more careful approach and this is where the markings will help us (see Lemma 5.3 (ii) and its proof).

For the construction we will use an inductive procedure on n. As in Sect. 3 (see (3.1) and the discussion before Lemma 3.9), for each branch point $p \in T$, we let $H_T(p)$ be its height, i.e., the diameter of the third largest branch of p in T. If $\delta > 0$, then by Lemma 3.9 there are only finitely many branch points p of T with height $H_T(p) > \delta$, and in particular there is one for which this quantity is maximal.

For the first step $n = 1$, we choose a branch point c of T with maximal height $H_T(c)$. Since T is in the class $\mathcal{T}_m$, this branch point c has $m = \nu_T(c)$ branches in T. So we can enumerate the distinct branches by the letters in our alphabet as T_k, $k \in \mathcal{A}$.

We choose c as the signed marked leaf in each T_k, where we set $\operatorname{sgn}(c, T_1) = +$ and $\operatorname{sgn}(c, T_k) = -$ for $k \neq 1$. So the set of signed marked leaves is $\{c^+\}$ in T_1 and $\{c^-\}$ in T_k, $k \neq 1$. Note that $\partial T_k = \{c\}$ as follows from Lemma 3.2 (ii) and that c is indeed a leaf in T_k by Lemma 3.4 for each $k \in \mathcal{A}$.

Suppose that for some $n \in \mathbb{N}$ and all $u \in W_n$ we have constructed subtrees T_u of T such that ∂T_u consists of one or two signed marked leaves of T_u that are branch points of T. We will now construct the subtrees of the $(n+1)$-th level as follows by subdivision of the trees T_u.

Fix $u \in W_n$. To decompose T_u into subtrees, we will use a suitable branch point c of T in $T_u \backslash \partial T_u$. The choice of c depends on whether ∂T_u contains one or two elements, that is, whether T_u is a leaf-tile or an arc-tile. We will explain this precisely below, but first record some facts that are true in both cases.

Since $c \in T_u \backslash \partial T_u$ is an interior point of T_u, there is a bijective correspondence between the branches of c in T and in T_u (see Lemma 3.5). So $\nu_{T_u}(c) = \nu_T(c) = m$, and we can label the distinct branches of c in T_u by T_{uk}, $k \in \mathcal{A}$. We will choose these labels depending on the signed marked leaves of T_u. Among other things, if

T_u has a marked leaf p^-, then p is passed to T_{u1} with the same sign. Similarly, a marked leaf p^+ of T_u is passed to T_{u2} with the same sign. We will momentarily explain this in more detail (see the *Summary* below).

In any case, we have

$$T_u = \bigcup_{k \in \mathcal{A}} T_{uk}. \tag{5.1}$$

Each set T_{uk} is a subtree of T_u and hence also of T. We call these subtrees the *children* of T_u and T_u the *parent* of its children. Note that two distinct children of T_u have only the point c in common and no other points.

Before we say more about the precise labelings of the children of T_u and their signed leaves, we first want to identify the boundary of each child; namely, we want to show that

$$\partial T_{uk} = \{c\} \cup (\partial T_u \cap T_{uk}) \tag{5.2}$$

for each $k \in \mathcal{A}$.

To see this, first note that T_{uk} is a subtree of T. Hence T_{uk} contains all of its boundary points and so $\partial T_{uk} \subseteq T_{uk}$. We have $c \in \partial T_{uk}$, because $c \in T_{uk}$ and every neighborhood of c contains points in the complement of T_{uk} as follows from Lemma 3.2 (ii) (here it is important that there are at least two branches of c). If $p \in T_{uk} \subseteq T_u$ and $p \notin \{c\} \cup \partial T_u$, then a sufficiently small neighborhood N of p belongs to T_u. Since $T_{uk} \backslash \{c\}$ is relatively open in T_u (this follows from Lemma 3.2 (i)), we can shrink this neighborhood so that $p \in N \subseteq T_{uk}$. So no point p in T_{uk} can be a boundary point of ∂T_{uk} unless it belongs to $\{c\} \cup \partial T_u$. It follows that $\partial T_{uk} \subseteq \{c\} \cup (\partial T_u \cap T_{uk})$.

On the other hand, we know that $c \in \partial T_{uk}$. If $p \in \partial T_u \cap T_{uk}$, then p is a boundary point of T_{uk}, because every neighborhood of p contains elements in the complement of T_u and hence in the complement of $T_{uk} \subseteq T_u$. This gives the other inclusion in (5.2), and (5.2) follows.

The identity (5.2) implies that each point in ∂T_{uk} is a branch point of T, because c is and the points in ∂T_u are also branch points of T by construction on the previous level n. Moreover, each point $p \in \partial T_{uk} \subseteq T_{uk}$ is a leaf of T_{uk}, because if $p = c$, then p is a leaf in T_{uk} by Lemma 3.4. Otherwise, $p \in \partial T_u$. Then p is a leaf of T_u by construction and hence a leaf of T_{uk} by the discussion after Lemma 3.5.

For the choice of the branch point $c \in T_u \backslash \partial T_u$, the precise labeling of the children T_{uk}, and the choice of the signs of the leaves of T_{uk} in ∂T_{uk}, we now consider two cases for the set ∂T_u. See Fig. 6 for an illustration.

Case 1: ∂T_u contains precisely one element, say $\partial T_u = \{a\}$. Note that T_u is a subtree of T and so an infinite set. So $T_u \backslash \partial T_u \neq \emptyset$. All points in $T_u \backslash \partial T_u$ are interior points of T_u. Since branch points in T are dense (here we use that T belongs to $\mathcal{T}_m$), there exist branch points of T in $T_u \backslash \partial T_u$. We choose a branch point $c \in T_u \backslash \partial T_u$ with maximal height $H_T(c)$ among all such branch points. This is possible by Lemma 3.9.

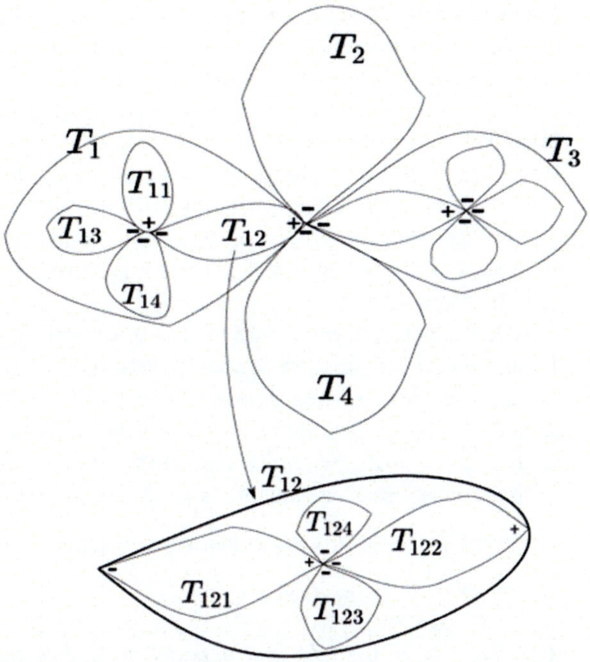

Fig. 6 An illustration for the decomposition of subtrees with one marked leaf (top) or two marked leaves (bottom)

Since $a \in \partial T_u \subseteq T_u \backslash \{c\}$, precisely one of the children of T_u contains a. We now consider two subcases depending on the sign of the marked leaf a.

If $\text{sgn}(a, T_u) = -$, then we choose a labeling of the children so that $a \in T_{u1}$. It then follows from (5.2) that $\partial T_{u1} = \{a, c\}$ and $\partial T_{uk} = \{c\}$ for $k \neq 1$. We choose signs so that the set of signed marked leaves is $\{a^-, c^+\}$ in T_{u1} and $\{c^-\}$ in T_{uk}, $k \neq 1$.

If $\text{sgn}(a, T_u) = +$, then we choose a labeling such that $a \in T_{u2}$. Then again by (5.2) we have $\partial T_{u2} = \{a, c\}$ and $\partial T_{uk} = \{c\}$ for $k \neq 2$. We choose signs so that the set of signed marked leaves is $\{c^+\}$ in T_{u1}, $\{c^-, a^+\}$ in T_{u2}, and $\{c^-\}$ in T_{uk}, $k \neq 1, 2$.

Case 2. ∂T_u contains precisely two elements, say $\partial T_u = \{a^-, b^+\}$. Then we choose a branch point $c \in (a, b)$ of T such that it has the maximal height $H_T(c)$ among all branch points that lie on (a, b). The existence of c is guaranteed by Lemmas 3.7 and 3.9. Note that $(a, b) \subseteq T_u$, because T_u is a subtree of T.

The points a and b lie in different branches of c in T_u as follows from Lemma 3.2 (iii). We choose the labels for the children of T_u so that $a \in T_{u1}$ and $b \in T_{u2}$. Then by (5.2) we have $\partial T_{u1} = \{a, c\}$, $\partial T_{u2} = \{c, b\}$, and $\partial T_{uk} = \{c\}$, $k \neq 1, 2$. We choose signs so that the set of marked leaves is $\{a^-, c^+\}$ in T_{u1}, $\{c^-, b^+\}$ in T_{u2}, and $\{c^-\}$, in T_{uk} for $k \neq 1, 2$.

The most important points of our construction can be summarized as follows.

Summary: T_{uk} is a subtree of T such that ∂T_{uk} consists of one or two points. These points are branch points of T and leaves of T_{uk}. Moreover, the signs of the points in each set ∂T_{uk} are chosen so that these signs differ if ∂T_{uk} contains two points. If c is the branch point used to decompose T_u, then c is a marked leaf in all the children of T_u, namely the marked leaf c^+ in T_{u1} and c^- in T_{uk} for $k \neq 1$.

If T_u has a marked leaf p^-, then p is passed to the child T_{u1} with the same sign. Similarly, a marked leaf p^+ of T_u is passed to T_{u2} with the same sign. So marked leaves are passed to a unique child and they retain their signs.

Since Cases 1 and 2 exhaust all possibilities, this completes the inductive step in the construction of the trees on level $n+1$ and their marked leaves. So we obtain subtrees T_u of T for all $u \in W_*$. Here it is convenient to set $T_\emptyset = T$ with an empty set of marked leaves.

If one applies our procedure to choose signs for the points in $\partial \mathbb{T}_u$ for the subtrees $\mathbb{T}_u$ of the CSST defined in Section 4, then one can recover these signs directly by a simple rule without going through the recursive process. Namely, by Lemma 4.6 we have $\partial \mathbb{T}_u \subseteq \{f_u(-1), f_u(1)\}$. Then it is not hard to see that for $p \in \partial \mathbb{T}_u$, we have $\operatorname{sgn}(\mathbb{T}_u, p) = +$ if $p = f_u(1)$ and $\operatorname{sgn}(\mathbb{T}_u, p) = -$ if $p = f_u(-1)$.

We now summarize some facts about the subtrees T_u of T that we just defined.

Lemma 5.1 *The following statements are true:*

(i) $T = \bigcup_{u \in W_n} T_u$ *for each* $n \in \mathbb{N}$.

(ii) *If* $n \in \mathbb{N}$, $u, v \in W_n$, $u \neq v$, *and* $T_u \cap T_v \neq \emptyset$, *then* $T_u \cap T_v$ *consists of precisely one point* $p \in T$, *which is a marked leaf in both* T_u *and* T_v.

(iii) *For* $n \in \mathbb{N}$, $u \in W_n$, *and* $v \in W_{n+1}$, *we have* $T_v \subseteq T_u$ *if and only if* $u = vk$ *for some* $k \in \mathcal{A}$.

(iv) *For each* $u \in W_*$ *let* c_u *be the branch point chosen in the decomposition of* T_u *into children. Then* $c_u \neq c_v$ *for all* $u, v \in W_*$ *with* $u \neq v$.

Proof

(i) This immediately follows from (5.1) and induction on n.

(ii) We prove this by induction on n. By choice of the subtrees T_k for $k \in \mathcal{A} = W_1$ and their marked leaves this is clear for $n = 1$.

Suppose the statement is true for all distinct words of length $n-1$, where $n \geq 2$. Now consider two words $u, v \in W_n$ of length n with $u \neq v$ and $T_u \cap T_v \neq \emptyset$. Then $u = u'k$ and $v = v'\ell$, where $u', v' \in W_{n-1}$ and $k, \ell \in \mathcal{A}$.

If $u' = v'$, then T_u and T_v are two of the branches obtained from $T_{u'}$ and a suitable branch point $c \in T_{u'}$. In this case, $\{c\} = T_u \cap T_v$ and c is a marked leaf in both T_u and T_v.

In the other case, $u' \neq v'$. Then $T_{u'} \cap T_{v'} \neq \emptyset$, because $T_u \cap T_v \neq \emptyset$, $T_u \subseteq T_{u'}$, and $T_v \subseteq T_{v'}$. By induction hypothesis, $T_{u'} \cap T_{v'}$ consists of precisely one point p, which is a marked leaf in both $T_{u'}$ and $T_{v'}$. Then necessarily $T_u \cap T_v = \{p\}$. Moreover, p is a marked leaf in both T_u and T_v, because marked leaves are passed to children. The statement follows.

(iii) Let $n \in \mathbb{N}$ and $u \in W_n$. Then we have $T_{uk} \subseteq T_u$ for each $k \in \mathcal{A}$ by our construction. Conversely, suppose $T_v \subseteq T_u$, where $v = v'k \in W_{n+1}$ with

$v' \in W_n$ and $k \in \mathcal{A}$. Then $T_{v'} \cap T_u \supseteq T_v$ contains more than one point. By (iii) this implies that $v' = u$. The statement follows.

(iv) If $u \in W_n$, $n \in \mathbb{N}_0$, then by construction $c_u \in T_u$ does not lie in the set ∂T_u of marked leaves of T_u. By (ii) this implies that $c_u \notin T_w$ for each $w \in W_n$, $w \neq u$. It follows that the points c_u, $u \in W_n$, are all distinct, and none of them is contained in the union of sets ∂T_u, $u \in W_n$. By our construction this union is equal to the set of all points c_v, where $v \in W_*$ is a word of length $\leq n-1$. This shows that the branch points c_u, $u \in W_n$, used to define the subtrees of level $n+1$ are all distinct and distinct from any of the previously chosen branch points for levels $\leq n$. The statement follows from this. □

Lemma 5.2 *We have* $\lim_{n\to\infty} \sup\{\operatorname{diam}(T_u) : u \in W_n\} = 0$.

Proof Let $\delta_n := \sup\{\operatorname{diam}(T_u) : u \in W_n\}$ for $n \in \mathbb{N}$. It is clear that the sequence $\{\delta_n\}$ is non-increasing. To show that $\delta_n \to 0$ as $n \to \infty$, we argue by contradiction. Then there exists $\delta > 0$ such that $\delta_n \geq \delta$ for all $n \in \mathbb{N}$. This means that for each $n \in \mathbb{N}$ there exists $u \in W_n$ with

$$\operatorname{diam}(T_u) \geq \delta. \tag{5.3}$$

We now use (5.3) to find an infinite word $w = w_1 w_2 \ldots \in W$ such that

$$\operatorname{diam}(T_{w_1 \ldots w_n}) \geq \delta \tag{5.4}$$

for all $n \in \mathbb{N}$. The word w is constructed inductively as follows. One of the finitely many letters $k \in \mathcal{A}$ must have the property that there are arbitrarily long words u starting with k such that (5.3) is true.

We define $w_1 = k$. Note that then $\operatorname{diam}(T_{w_1}) \geq \delta$. T By choice of w_1, one of the letters $\ell \in \mathcal{A}$ must have the property that there are arbitrarily long words u starting with $w_1 \ell$ such that (5.3) is true. We define $w_2 = \ell$. Then $\operatorname{diam}(T_{w_1 w_2}) \geq \delta$. Continuing in this manner, we can find $w = w_1 w_2 \ldots \in W$ satisfying (5.4).

Obviously,

$$T_{w_1} \supseteq T_{w_1 w_2} \supseteq T_{w_1 w_2 w_3} \supseteq \ldots.$$

So the subtrees $K_n = T_{w_1 \ldots w_n}$, $n \in \mathbb{N}$, of T form a descending family of compact sets with $\operatorname{diam}(K_n) \geq \delta$. This implies that

$$K = \bigcap_{n \in \mathbb{N}} K_n$$

is a non-empty compact subset of T with $\operatorname{diam}(K) \geq \delta$.

In particular, we can choose $p, q \in K$ with $p \neq q$. Then $p, q \in K_n$ for each $n \in \mathbb{N}$. Since K_n is a subtree of T, we then have $[p, q] \subseteq K_n$. Moreover, by Lemma 3.7 there exists a branch point x of T contained in $(p, q) \subseteq K_n$. By Lemma 3.2 (iii) the points p and q lie in different components of $K_n\backslash\{x\}$. In particular, for each $n \in \mathbb{N}$ the point x is not a leaf of K_n and hence distinct from the marked leaves of K_n.

By Lemma 3.9 there are only finitely many branch points $y_1, \dots, y_s$ of T distinct from x with $H_T(y_j) \geq H_T(x) > 0$ for $j = 1, \dots, s$. This implies that at most s of the trees K_n are leaf-tiles, i.e., have only one marked leaf. Indeed, if K_n an leaf-tile, then it is decomposed into branches by use of a branch point $c \in K_n\backslash\partial K_n$ with the largest height $H_T(c)$. The point c is then a marked leaf in each of the children of K_n and in particular in K_{n+1}. Since the branch point $x \in K_n$ is distinct from the marked leaves of K_n and K_{n+1}, we have $x \in K_n\backslash\partial K_n$ and $x \neq c$. So x was not chosen to decompose K_n, and we must have $H_T(c) \geq H_T(x)$. Since the branch points c that appear from leaf-tiles at different levels n are all distinct as follows from Lemma 5.1 (iv), we can have at most s leaf-tiles in the sequence K_n, $n \in \mathbb{N}$. This implies that there exists $N \in \mathbb{N}$ such that K_n for $n \geq N$ is an arc-tile and so has precisely two marked leaves.

Let $a, b \in K_N$ with $a \neq b$ be the marked leaves of K_N. As we travel from x along $[x, a] \subseteq K_N$ towards a, there is a first point x' on $[a, b]$. Then $x' \neq a$. Otherwise, $x' = a$. Then $[x, a]$ and $[a, b]$ have only the point a in common, which implies that $[x, a] \cup [a, b]$ is an arc equal to $[x, b]$. Then $a \in (x, b)$, which by Lemma 3.2 (iii) implies that $x, b \in K_N$ lie in different components of $K_N\backslash\{a\}$. This contradicts the fact that a is a leaf of K_N and so $K_N\backslash\{a\}$ has only one component. Similarly, one can show that $x' \neq b$.

The point x' is a branch point of T. This is clear if $x' = x$. If $x' \neq x$, this follows from Lemma 3.6, because $a, b, x \neq x'$ and the arcs $[a, x')$, $[b, x')$, $[x, x')$ are pairwise disjoint.

The tree K_{N+1} is a branch of K_N obtained from a branch point $c \in (a, b)$ of T with largest height $H_T(c)$ among all branch points on (a, b). We have $x' \neq c$. Otherwise, $x' = c$. Then $x \neq x'$, because $x' = c$ is a marked leaf of K_{N+1} and x is distinct from all the marked leaves in any of the sets K_n. This implies that the points a, b, x lie in different components of $K_N\backslash\{x'\}$ and hence in different branches of x' in K_N. Since a and b are the marked leaves of K_N, the branches containing a and b are arc-tiles and all other branches of $x' = c$ in K_N are leaf-tiles. The unique branch of x' in K_N containing x, which is equal to K_{N+1}, must be a leaf-tile by the way we decomposed T. This is impossible by choice of N and so indeed $x' \neq c$. Note that this implies $H_T(c) \geq H_T(x')$.

Since $x' \neq c$, $c \in (a, b)$, and $[x, x') \cap [a, b] = \emptyset$, we have $[x, x'] \subseteq K_N\backslash\{c\}$. So x' lies in the same branch of c in K_N as x, which is K_{N+1}. Moreover, depending on whether $c \in (a, b)$ lies on the right or left of $x' \in (a, b)$, we have $x' \in (a, c)$ or $x' \in (c, b)$. In the first case, $[a, c] \subseteq K_{N+1}$ and a and c are the marked leaves of K_{N+1}. In the second case, $[c, b] \subseteq K_{N+1}$ and c and b are the marked leaves of K_{N+1}. So in both cases, if a' and b' are the marked leaves of K_{N+1}, then $x' \in (a', b')$, $[x, x'] \subseteq K_{N+1}$, and $[x, x') \cap [a', b'] = \emptyset$.

These facts allow us to repeat the argument for K_{N+1} instead of K_N. Again K_{N+1} is decomposed into branches by choice of a branch point $c' \in (a', b')$. We must have $c' \neq x'$, because otherwise we again obtain a contradiction to the fact that K_{N+2} is not a leaf-tile. This implies that $H_T(c') \geq H_T(x')$. Continuing in this manner, we obtain an infinite sequence of branch points $c, c', \dots$. By construction these branch points are all distinct and have a height $\geq H_T(x')$. This is impossible by Lemma 3.9. We obtain a contradiction that establishes the statement. □

The previous argument shows that each branch point x of T will eventually be chosen as a branch point in the decomposition of T into the subtrees T_u, $u \in W_*$. Indeed, otherwise x is distinct from all the marked leaves of any of the subtrees T_u, $u \in W_*$. This in turn implies that there exists a unique infinite word $w = w_1 w_2 \ldots \in W$ such that $x \in K_n := T_{w_1 \ldots w_n}$ for $n \in \mathbb{N}$. From this one obtains a contradiction as in the last part of the proof of Lemma 5.2.

Lemma 5.3 *Let $m \in \mathbb{N}$, $m \geq 3$, and suppose T and S are trees in $\mathcal{T}_m$. Assume that subtrees T_u of T and S_u of S with signed marked leaves have been defined for $u \in W_*$ by the procedure described above. Then the following statements are true:*

(i) *Let $n \in \mathbb{N}$, $u \in W_n$, and $v \in W_{n+1}$. Then $T_v \subseteq T_u$ if and only if $S_v \subseteq S_u$.*
(ii) *For $n \in \mathbb{N}$ and $u, v \in W_n$ with $u \neq v$ we have $T_u \cap T_v \neq \emptyset$ if and only if $S_u \cap S_v \neq \emptyset$. Moreover, if these intersections are non-empty, then they are singleton sets, say $\{p\} = T_u \cap T_v$ and $\{\widetilde{p}\} = S_u \cap S_v$. The point p is a signed marked leaf in T_u and T_v, the point $\widetilde{p}$ is a signed marked leaf in S_u and S_v, $\operatorname{sgn}(p, T_u) = \operatorname{sgn}(\widetilde{p}, S_u)$, and $\operatorname{sgn}(p, T_v) = \operatorname{sgn}(\widetilde{p}, S_v)$.*

In (ii) we are actually only interested in the statement that $T_u \cap T_v \neq \emptyset$ if and only if $S_u \cap S_v \neq \emptyset$. The additional claim in (ii) will help us to prove this statement by an induction argument.

Proof

(i) This follows from Lemma 5.1 (iii) applied to the decompositions of T and S. Indeed, we have $T_v \subseteq T_u$ if and only if $v = uk$ for some $k \in \mathcal{A}$ if and only if $S_v \subseteq S_u$.
(ii) We prove this by induction on $n \in \mathbb{N}$. The case $n = 1$ is clear by how the decompositions were chosen.

Suppose the claim is true for words of length $n - 1$, where $n \geq 2$. Now consider two words $u, v \in W_n$ of length n with $u \neq v$. Then $u = u'k$ and $v = v'\ell \in W_n$, where $u', v' \in W_{n-1}$ and $k, \ell \in \mathcal{A}$. Since the claim is symmetric in T and S, we may assume that $T_u \cap T_v \neq \emptyset$.

If $u' = v'$, then T_u and T_v are two of the branches obtained from $T_{u'}$ and a branch point $c \in T_{u'}$. In this case, $T_u \cap T_v = \{c\}$ and c is a marked leaf in both T_u and T_v. Similarly, S_u and S_v are two of the branches obtained from $S_{u'}$ and a branch point $\widetilde{c} \in S_{u'}$. We have $S_u \cap S_v = \{\widetilde{c}\}$ and $\widetilde{c}$ is a marked leaf in both S_u and S_v. Moreover, c has the same sign in T_u as $\widetilde{c}$ in S_u. Indeed, by the choice of labeling in the decomposition, this sign is $+$ if $k = 1$ and $-$ otherwise. Similarly, c has the same sign in T_v as $\widetilde{c}$ in S_v. This shows that the statement is true in this case.

In the other case, $u' \neq v'$. Then $T_{u'} \cap T_{v'} \neq \emptyset$, because $T_u \cap T_v \neq \emptyset$, $T_u \subseteq T_{u'}$, and $T_v \subseteq T_{v'}$. Then by induction hypothesis, $T_{u'} \cap T_{v'}$ consists of precisely one point p that is a marked leaf in both $T_{u'}$ and $T_{v'}$. The set $S_{u'} \cap S_{v'}$ consists of one point $\widetilde{p}$

that is a marked leaf in $S_{u'}$ and $S_{v'}$. Moreover, we have $\operatorname{sgn}(p, T_{u'}) = \operatorname{sgn}(\widetilde{p}, S_{u'})$ and $\operatorname{sgn}(p, T_{v'}) = \operatorname{sgn}(\widetilde{p}, S_{v'})$. Since $\emptyset \neq T_u \cap T_v \subseteq T_{u'} \cap T_{v'} = \{p\}$, we then have $T_u \cap T_v = \{p\}$.

If $\operatorname{sgn}(p, T_{u'}) = \operatorname{sgn}(\widetilde{p}, S_{u'}) = -$, then $u = u'1$, because $p \in T_u$. Hence $\widetilde{p} \in S_{u'1} = S_u$, because the marked leaf $\widetilde{p}$ of $S_{u'}$ with $\operatorname{sgn}(\widetilde{p}, S_{u'}) = -$ is passed to the child $S_{u'1}$. If $\operatorname{sgn}(p, T_{u'}) = \operatorname{sgn}(\widetilde{p}, S_{u'}) = +$, then $u = u'2$ and $\widetilde{p} \in S_{u'2} = S_u$.

Similarly, if $\operatorname{sgn}(p, T_{v'}) = \operatorname{sgn}(\widetilde{p}, S_{v'}) = -$, then $v = v'1$ and if $\operatorname{sgn}(p, T_{v'}) = \operatorname{sgn}(p, S_{v'}) = +$, then $v = v'2$, because $p \in T_v$. In both cases, $\widetilde{p} \in S_v$.

In each of these cases, p is a marked leaf in T_u and T_v, and $\widetilde{p}$ is a marked leaf in S_u and S_v. In particular, $\{\widetilde{p}\} \subseteq S_u \cap S_v \subseteq S_{u'} \cap S_{v'} = \{\widetilde{p}\}$ and so $S_u \cap S_v = \{\widetilde{p}\}$. So both $T_u \cap T_v = \{p\}$ and $S_u \cap S_v = \{\widetilde{p}\}$ are singleton sets consisting of marked leaves as claimed. Since signed marked leaves are passed to children with the same sign, we have

$$\operatorname{sgn}(p, T_u) = \operatorname{sgn}(p, T_{u'}) = \operatorname{sgn}(\widetilde{p}, S_{u'}) = \operatorname{sgn}(\widetilde{p}, S_u).$$

Similarly, we conclude that $\operatorname{sgn}(p, T_v) = \operatorname{sgn}(\widetilde{p}, S_v)$. The statement follows. □

We are now ready to prove Theorem 1.8, and Theorem 1.7 as an immediate consequence.

Proof of Theorem 1.8 Let m be as in the statement, and consider arbitrary trees T and S in the class $\mathcal{T}_m$. For each $n \in \mathbb{N}$ we consider the decompositions $T = \bigcup_{u \in W_n} T_u$ and $S = \bigcup_{u \in W_n} S_u$ as defined earlier in this section. Here of course, $W_n = W_n(\mathcal{A})$, where $\mathcal{A} = \{1, 2, \dots, m\}$.

We want to show that decompositions of T and S for different levels $n \in \mathbb{N}$ have the properties in Proposition 2.1. In this proposition the index i for fixed level n corresponds to the words $u \in W_n$.

The spaces T and S are trees and hence compact. The sets T_u and S_u appearing in their decompositions are subtrees and hence non-empty and compact. Conditions (i), (ii), and (iii) in Proposition 2.1 follow from Lemma 5.1 (iii), (5.1), and Lemma 5.2, respectively. Finally, (2.1) and (2.2) follow from Lemma 5.3 (i) and (ii).

Proposition 2.1 implies T and S are homeomorphic as desired. □

Proof of Theorem 1.7 As we have seen in Sect. 4, the CSST $\mathbb{T}$ is a metric tree with the properties (i) and (ii) as in the statement (see Proposition 1.4 and Proposition 1.5). In particular, $\mathbb{T}$ belongs to the class $\mathcal{T}_3$. Since these properties (i) and (ii) are obviously invariant under homeomorphisms, every metric tree T homeomorphic to $\mathbb{T}$ has these properties.

Conversely, suppose that T is a metric tree with properties (i) and (ii). Then T belongs to the class $\mathcal{T}_3$. So Theorem 1.8 for $m = 3$ implies that T and $\mathbb{T}$ are homeomorphic. □

The method of proof for Theorem 1.8 can be used to a establish a slightly stronger result for $m = 3$.

Theorem 5.4 *Let T and S be trees in $\mathcal{T}_3$. Suppose $p_1, p_2, p_3 \in T$ are three distinct leaves of T, and $q_1, q_2, q_3 \in S$ are three distinct leaves of S. Then there exists a homeomorphism $f: T \to S$ such that $f(p_k) = q_k$ for $k = 1, 2, 3$.*

Note that $-1, 1 \in \mathbb{T}$ are leaves of $\mathbb{T}$ as follows from Lemma 4.4 (ii). Moreover, $i \in \mathbb{T}$ is also a leaf of $\mathbb{T}$, because the set

$$\mathbb{T} \setminus \{i\} = \mathbb{T}_1 \cup \mathbb{T}_2 \cup (\mathbb{T}_3 \setminus \{i\}) = \mathbb{T}_1 \cup \mathbb{T}_2 \cup g_3(\mathbb{T} \setminus \{1\})$$

is connected. Hence $\mathbb{T}$, and so by Theorem 1.8 every tree in $\mathcal{T}_3$, has at least three leaves (actually infinitely many). If we apply Theorem 5.4 to $S = \mathbb{T}$, then we see that if T is a tree in $\mathcal{T}_3$ with three distinct leaves p_1, p_2, p_3, then there exists a homeomorphism $f: T \to \mathbb{T}$ such that $f(p_1) = -1$, $f(p_2) = 1$, and $f(p_3) = i$.

Proof of Theorem 5.4 We will employ a slight modification of our decomposition and coding procedure. The underlying alphabet corresponds to the case $m = 3$, and so $\mathcal{A} = \{1, 2, 3\}$. We describe this for the tree T. Essentially, one wants to use the leaves p_1, p_2, p_3 of T as additional marked leaves for any of the inductively defined subtrees T_u for $n \in \mathbb{N}$ and $u \in W_n = W_n(\mathcal{A})$ if it contains any of these leaves. Here p_1 carries the sign $-$, while p_2 and p_3 carry the sign $+$.

Instead of starting the decomposition process with a branch point $c \in T$ of maximal height, one chooses a branch point c so that the leaves p_1, p_2, p_3 lie in distinct branches T_1, T_2, T_3 of c in T, respectively. To find such a branch point, one travels from p_1 along $[p_1, p_2]$ until one first meets $[p_2, p_3]$ in a point c. Then the sets $[p_1, c)$, $[p_2, c)$, $[p_3, c)$ are pairwise disjoint. For $k, \ell \in \mathcal{A}$ with $k \neq \ell$ the set $[p_k, c) \cup \{c\} \cup (c, p_\ell]$ is an arc with endpoints p_k and p_ℓ, and so it must agree with $[p_k, p_\ell]$. In particular, $c \in [p_k, p_\ell]$. Since each point p_k is a leaf, it easily follows from Lemma 3.2 (iii) that $c \neq p_1, p_2, p_3$. Indeed, if $c = p_1$ for example, then $c = p_1 \in [p_2, p_3]$ and so p_2 and p_3 would lie in different components of $T \setminus \{p_1\}$. This is impossible, because p_1 is a leaf of T and so $T \setminus \{p_1\}$ is connected.

We conclude that the connected sets $[p_1, c)$, $[p_2, c)$, $[p_3, c)$ are non-empty and must lie in different branches T_1, T_2, T_3 of c. In particular, c is a branch point of T. We can choose the labels so that $p_k \in T_k$ for $k = 1, 2, 3$. The point c is a marked leaf in each of theses branches with a sign chosen as before. With the additional signs for the distinguished leaves, we then have the set of marked leaves $\{p_1^-, c^+\}$ in T_1, $\{c^-, p_2^+\}$ in T_2, and $\{c^-, p_3^+\}$ in T_3.

We now continue inductively as before. If we have already constructed a subtree T_u for some $n \in \mathbb{N}$ and $u \in W_n$ with one or two signed marked leaves, then we decompose T_u into three branches labeled T_{u1}, T_{u2}, T_{u3} by using a suitable branch point $c \in T_u$. Namely, if T_u is a leaf-tile and has one marked leaf $a \in T_u$, we choose a branch point $c \in T_u \setminus \{a\}$ with maximal height $H_T(c)$. If T_u is an arc-tile with two marked leaves $\{a, b\} \subseteq T_u$ we choose a branch point $c \in T_u$ of maximal height on $(a, b) \subseteq T_u$.

Marked leaves and their signs are assigned to the children T_{u1}, T_{u2}, T_{u3} of T_u as before. In particular, a marked leaf x^- of T_u is passed to T_{u1} with the same sign. Similarly, a marked leaf x^+ of T_u is passed to T_{u2} with the same sign. If we continue in this manner, we obtain subtrees T_u with one or two signed marked leaves for all levels $n \in \mathbb{N}$ and $u \in W_n$.

We apply the same procedure for the tree S and its leaves q_1, q_2, q_3. Then Lemmas 5.1, 5.2, and 5.3 are true (with almost identical proofs) for the decompositions of T and S obtained in this way. The argument in the proof of Theorem 1.8 based on Proposition 2.1 now guarantees the existence of a homeomorphism $f : T \to S$ such that

$$f(T_u) = S_u \text{ for all } n \in \mathbb{N} \text{ and } u \in W_n. \tag{5.5}$$

In our construction $p_1 \in T_1$ carries the sign $-$ and is hence passed to T_{11} with the same sign; so $p_1 \in T_{11}$. Repeating this argument, we see the

$$p_1 \in T_1 \cap T_{11} \cap T_{111} \cap \dots .$$

The latter nested intersection of compact sets cannot contain more than one point, because by Lemma 5.3 the diameters of our subtrees T_u, $u \in W_n$, approach 0 uniformly as $n \to \infty$. Thus, $\{p_1\} = T_1 \cap T_{11} \cap T_{111} \cap \dots$. The same argument shows that $\{q_1\} = S_1 \cap S_{11} \cap S_{111} \cap \dots$, and so (5.5) implies that $f(p_1) = q_1$.

Similarly, the points p_2, p_3, q_2, q_3 carry the sign $+$ in their respective trees. This leads to

$$\{p_2\} = T_2 \cap T_{22} \cap T_{222} \cap \dots, \quad \{q_2\} = S_2 \cap S_{22} \cap S_{222} \cap \dots,$$
$$\{p_3\} = T_3 \cap T_{32} \cap T_{322} \cap \dots, \quad \{q_3\} = S_3 \cap S_{32} \cap S_{322} \cap \dots,$$

which by (5.5) gives $f(p_2) = q_2$ and $f(p_3) = q_3$.

We have shown the existence of a homeomorphism $f : T \to S$ with the desired normalization. □

Acknowledgments The authors would like to thank Daniel Meyer for many valuable comments on this paper. The first author was partially supported by NSF grants DMS-1506099 and DMS-1808856.

References

1. Aldous, D.: The continuum random tree. I. Ann. Probab. **19**, 1–28 (1991)
2. Bestvina, M.: $\mathbb{R}$-trees in topology, geometry and group theory. In: Daverman, R.J., Sher, R.B. (eds.) Handbook of Geometric Topology, pp. 55–91. North-Holland, Amsterdam (2002)
3. Bishop, C.J., Tyson, J.T.: Conformal dimension of the antenna set. Proc. Am. Math. Soc. **129**, 3631–3636 (2001)
4. Bonk, M.: Quasiconformal geometry of fractals. In: Proceedings of the International Congress of Mathematicians (Madrid 2006), vol. II, pp. 1349–1373. European Mathematical Society, Zürich (2006)
5. Bonk, M., Meyer, D.: Uniformly branching trees (2020). Preprint. arXiv:2004.07912

6. Bonk, M., Meyer, D.: Quasiconformal and geodesic trees. Fund. Math. **250**, 253–299 (2020)
7. Bonk, M., Meyer, D.: Expanding Thurston Maps. Mathematical Surveys and Monographs, vol. 225. American Mathematical Society, Providence, RI (2017)
8. Carleson, L., Gamelin, T.W.: Complex Dynamics. Springer, New York (1993)
9. Charatonik, J.J.: Open mappings of universal dendrites. Bull. Acad. Polon. Sci. Ser. Sci. Math. **28**, 489–494 (1980)
10. Charatonik, J., Charatonik, W.J.: Dendrites. XXX National Congress of the Mexican Mathematical Society (Aguascalientes, 1997). Aportaciones Mat. Comun., vol. 22, pp. 227–253. Sociedad Matemática Mexicana, Mexico (1998)
11. Charatonik, W.J., Dilks, A.: On self-homeomorphic spaces. Topol. Appl. **55**, 215–238 (1994)
12. Croydon, D., Hambly, B.: Self-similarity and spectral asymptotics for the continuum random tree. Stoch. Process. Appl. **118**(5), 730–754 (2008)
13. Curien, N.: Dissecting the circle, at random. In: Journées MAS 2012. ESAIM Proceedings, vol. 44, pp. 129–139. EDP Sciences, Les Ulis (2014)
14. Douady, A., Hubbard, J.H.: Étude dynamique des polynômes complexes. Partie I. Publications Mathématiques d'Orsay, vol. 84. Université de Paris-Sud, Département de Mathématiques, Orsay (1984)
15. Duquesne, T., Le Gall, J.-F.: Probabilistic and fractal aspects of Lévy trees. Probab. Theory Relat. Fields **131**, 553–603 (2005)
16. Falconer, K.: Fractal Geometry, 2nd edn. Wiley, Hoboken, NJ (2003)
17. Heinonen, J.: Lectures on Analysis on Metric Spaces. Springer, New York (2001)
18. Hocking, J.G., Young, G.S.: Topology. Addison-Wesley Publishing Co., Inc., Reading, MA/London (1961)
19. Hutchinson, J.E.: Fractals and self-similarity. Indiana Univ. Math. J. **30**, 713–747 (1981)
20. Kigami, J.: Weighted partition of a compact metrizable space, its hyperbolicity and Ahlfors regular conformal dimension (2018). Preprint. arXiv:1806.06558
21. Kigami, J.: Analysis on Fractals. Cambridge Tracts in Mathematics, vol. 143. Cambridge University Press, Cambridge (2001)
22. Kiwi, J.: Wandering orbit portraits. Trans. Am. Math. Soc. **354**, 1473–1485 (2002)
23. Kuratowski, K.: Topology, vol. II. Academic, New York/London; Polish Scientific Publishers, Warsaw (1968)
24. Lin, P., Rohde, S.: Conformal welding of dendrites (2019). Preprint
25. Le Gall, J.-F.: Random real trees. Ann. Fac. Sci. Toulouse Math. (6) **15**, 35–62 (2006)
26. Menger, K.: Kurventheorie. Teubner, Leipzig (1932)
27. Nadler Jr., S.B.: Continuum Theory. Monographs and Textbooks in Pure and Applied Mathematics, vol. 158. Marcel Dekker, New York (1992)
28. Wazewski, T.: Sur les courbes de Jordan ne renfermant aucune courbe simple fermeé de Jordan. Ann. Soc. Polonaise Math. **2**, 49–170 (1923)
29. Whyburn, G.T.: Analytic Topology. Colloquium Publications, vol. 28. American Mathematical Society, Providence, RI (1963)

p-Hyperbolicity of Ends and Families of Paths in Metric Spaces

Nageswari Shanmugalingam

Abstract The purpose of this note is to give an expository survey on the notions of p-parabolicity and p-hyperbolicity of metric measure spaces of locally bounded geometry. These notions are extensions of the notions of recurrence and transience to non-linear operators such as the p-Laplacian (with the standard Laplacian or the 2-Laplacian associated with recurrence and transience behaviors). We discuss characterizations of these notions in terms of potential theory and in terms of moduli of families of paths in the metric space.

Keywords Recurrence · p-hyperbolic · Singular function · Modulus of curve families · Ends

Mathematics Subject Classifications (2010) Primary: 31E05; Secondary: 43A85, 65M80

1 Introduction

It is now a well-known fact that Brownian motion is recurrent in $\mathbb{R}$ and $\mathbb{R}^2$ but is transient in $\mathbb{R}^n$ for $n \geq 3$. In other words, a Brownian motion, starting from a closed ball in $\mathbb{R}^n$, will almost surely return infinitely often to that ball when $n \leq 2$ but almost surely will eventually not return to the ball when $n \geq 3$. This dichotomous behavior of recurrence versus transience can be seen in more general Riemannian manifolds, leading to a classification of manifolds as parabolic (Brownian motion is recurrent, returning infinitely often to a ball) or hyperbolic (where the Brownian motion is transient). The works [11, 29] demonstrated that the recurrence or transience of the Brownian motion is intimately connected with the

N. Shanmugalingam (✉)
University of Cincinnati, Department of Mathematical Sciences, Cincinnati, OH, USA
e-mail: shanmun@uc.edu

U. Freiberg et al. (eds.), *Fractal Geometry and Stochastics VI*,
Progress in Probability 76, https://doi.org/10.1007/978-3-030-59649-1_8

existence of global singular functions, also known as Green's functions. A manifold is transient if and only if it supports a non-negative singular function.

During the past 20 years the notion of first order calculus has been developed for more general non-smooth metric measure spaces where the metric space is complete and the measure is a locally doubling Radon measure supporting a local Poincaré type inequality. For such spaces, it is not clear what the Brownian motion is, but thanks to Kakutani's theorem, we know that Brownian motion on a Riemannian manifold is a probabilistic approach to harmonic functions and the Laplace–Beltrami operator on the manifold. Physics and the theory of Markovian process as described in [10] also back this up, with the link provided through the heat equation. Using this as a motivation, we can study recurrence or transience of a metric measure space in terms of the existence of a singular function associated with the so-called 2-harmonicity.

Indeed, the recurrence and transience properties of the space seem to be associated with a "large scale" dimension of the underlying space. To explore the effect of the geometry of a space on curves in the space, we also move away from the realm of linear operators (Laplace–Beltrami operators) to non-linear p-Laplace type operators. In his dissertation [16], Holopainen gave a definition of p-parabolicity and p-hyperbolicity in Riemannian manifolds and their connections to p-harmonic functions. In this note we will describe some of the connections between the geometry of curves in the setting of metric measure spaces, which should be thought of as a non-linear analog of recurrence versus transience, and p-harmonic functions in the space. Metric measure spaces that correspond to transient spaces for p-harmonic functions are said to be *p-hyperbolic* while those that are not are said to be *p-parabolic*.

2 Background Notions

The context of this note is that of metric measure spaces that need not be smooth (Riemannian). Here (X, d, μ) denotes a metric measure space with the measure μ assumed to be a Radon measure such that balls have positive and finite measure. In this section we will give a brief account of the basic notions used in the study of parabolicity versus hyperbolicity of the space in terms of first order analysis. For details on these notions, we recommend [15] and the references therein.

To understand p-parabolicity (recurrence) and p-hyperbolicity (transience), we need to have a concept of a "size" of families of curves in X. To this end, let Γ be a collection of curves in X, and we set $\mathcal{A}(\Gamma)$ to be the collection of all non-negative Borel measurable functions ρ on X such that for each locally rectifiable path $\gamma \in \Gamma$ we have

$$\int_\gamma \rho\, ds \geq 1.$$

Here a path is locally rectifiable if it maps an interval $I \subset \mathbb{R}$ continuously into X and for each compact subinterval $J \subset I$ we have that $\gamma|_J$ has finite length. An excellent introduction to the notion of path integrals in metric setting can be found in [15, Chapter 5], [1, Chapters 4, 6] and [13, Chapter 7].

Definition 2.1 Given $1 \le p < \infty$, the *p-modulus* of the collection Γ is the number

$$\mathrm{Mod}_p(\Gamma) = \inf_{\rho \in \mathcal{A}(\Gamma)} \int_X \rho^p \, d\mu.$$

Observe that if Γ consists only of paths that are not locally rectifiable, then by definition $\mathrm{Mod}_p(\Gamma) = 0$, whereas if Γ includes a constant curve, then $\mathrm{Mod}_p(\Gamma) = \infty$. It is not too difficult to verify that Mod_p is an outer measure on the collection of all paths, and that if Γ has even one constant path then $\mathrm{Mod}_p(\Gamma) = \infty$. It is a result of Fuglede [9] that the only sets that are Mod_p-measurable are those of zero p-modulus and their complements. In this note we only consider Mod_p to the extent of verifying whether a family Γ satisfies $\mathrm{Mod}_p(\Gamma) > 0$ or not. To this end, the following result of Koskela and MacManus [23] is useful.

Lemma 2.2 *Let Γ be a family of paths in X, and $1 < p < \infty$. Then $Mod_p(\Gamma) = 0$ if and only if there is a nonnegative Borel function $\rho \in L^p(X)$ such that for each $\gamma \in \Gamma$,*

$$\int_\gamma \rho \, ds = \infty.$$

Definition 2.3 Given two sets $E, F \subset X$, by $\Gamma(E, F)$ we mean the collection of all curves in X with one end point in E and the other in F.

The following definition is based on the dissertation [16].

Definition 2.4 We say that X is *p-hyperbolic* if there is a closed ball $B = \overline{B}(x_0, R) \subset X$ and a strictly monotone increasing sequence of real numbers $R_n > R$ with $\lim_n R_n = \infty$ and

$$\lim_n \mathrm{Mod}_p(\Gamma(\overline{B}(x_0, R), X \setminus B(x_0, R_n)) > 0.$$

We say that X is *p-parabolic* if it is not p-hyperbolic.

There are now at least five available notions of Sobolev spaces in the metric setting: Poincaré–Sobolev, Korevaar–Schoen, Hajłasz–Sobolev, Newton–Sobolev, and Dirichlet domain spaces, see for example [12, 14, 15]. In this paper we will focus on the notion of Newton–Sobolev spaces as they are the closest aligned to the study of paths in a metric space, though for the case $p = 2$ one can replace this with Dirichlet forms and the corresponding Dirichlet domains whenever they are available, by considering the corresponding heat equation, see [11] for example.

Given a function $f : X \to \mathbb{R}$, we say that a non-negative Borel measurable function g on X is an *upper gradient* of f if for each non-constant compact rectifiable curve $\gamma : [a, b] \to X$ we have

$$|f(\gamma(b)) - f(\gamma(a))| \le \int_\gamma g\, ds.$$

We say that g is a p-*weak* upper gradient of f if the collection Γ of non-constant compact rectifiable curves for which the above inequality fails satisfies $\mathrm{Mod}_p(\Gamma) = 0$. With $D_p(f)$ denoting the collection of all p-weak upper gradients of f that also belong to $L^p(X)$, we say that $f \in N^{1,p}(X)$ if $f \in L^p(X)$ (that is, the *function* f belongs to an equivalence class in $L^p(X)$) and $D_p(f)$ is nonempty. The set $D_p(f)$ is a convex lattice subset of $L^p(X)$, and by a result in [23], it is also closed in $L^p(X)$. We set

$$\|f\|_{N^{1,p}(X)} := \|f\|_{L^p(X)} + \inf_{g\in D_p(f)} \|g\|_{L^p(X)}.$$

For $1 < p < \infty$, by the uniform convexity of $L^p(X)$ and the lattice property of $D_p(f)$ we know that there is a unique element $g_f \in D_p(f)$ with the property that for each $g \in D_p(f)$, $g_f \le g$ almost everywhere. Thus

$$\|f\|_{N^{1,p}(X)} = \|f\|_{L^p(X)} + \|g_f\|_{L^p(X)}.$$

Equipped with the norm $\|\cdot\|_{N^{1,p}(X)}$, the space $N^{1,p}(X)$ is a Banach space, see for example [26] and [15]. Classically, the measure of the set where two functions disagree determines whether the two functions belong to the same equivalence class in $L^p(X)$. In the setting of $N^{1,p}(X)$ the notion of p-capacity of a set plays this role, and here the value of p determines what sets are of zero p-capacity. Given a set $E \subset X$, we set

$$\mathrm{Cap_p}(E) := \inf_f \int_X [|f|^p + g_f^p] d\mu,$$

where the infimum is over all functions $f \in N^{1,p}(X)$ such that $f \ge 1$ on E. A more pertinent notion related to parabolicity and hyperbolicity is that of *relative* p-*capacity*.

Definition 2.5 Given two closed sets $E, F \subset X$ such that $E \cap F$ is empty,

$$\mathrm{cap}_p(E, F) := \inf_f \int_X g_f^p d\mu,$$

where the infimum is over all functions $f \in N^{1,p}(X)$ such that $f \ge 1$ on E and $f \le 0$ on F.

There is a close connection between $\mathrm{cap_p}(E, F)$ and $\mathrm{Mod}_p(\Gamma(E, F))$. Indeed, if $\rho \in \mathcal{A}(\Gamma(E, F))$, then the function u defined by

$$u(y) = \inf_{\gamma_y} \int_{\gamma_y} \rho ds$$

with infimum taken over all locally rectifiable curves in X with one endpoint y and the other end point in E, is measurable (see for example [20]) and satisfies $u = 0$ on E and $u \geq 1$ on F. If then $X \setminus E$ is bounded, we would have $u \in N^{1,p}(X)$ with $\rho \in D_p(u)$, and thus we would have

$$\mathrm{cap_p}(E, F) \leq \mathrm{Mod}_p(\Gamma(E, F)).$$

Typically in this note E would be $X \setminus B(x_0, R)$ for some $x_0 \in X$ and $R > 0$, and F would be a compact subset of the ball $B(x_0, R)$.

Definition 2.6 We say that the measure μ is *uniformly locally doubling* on X if there is a constant $C_D \geq 1$ and a scale $0 < R_0 \leq \infty$ such that whenever $x \in X$ and $0 < r < R_0$, we have

$$\mu(B(x, 2r)) = \mu(\{y \in X : d(x, y) < 2r\}) \leq C_D \mu(B(x, r)).$$

We say that (X, d, μ) supports a *uniformly local p-Poincaré inequality* if there are constants $C > 0$, $\lambda \geq 1$ and a scale $0 < R_1 \leq \infty$ such that whenever $x \in X$, $0 < r < R_0$, and $f \in N^{1,p}(B(x, 2\lambda r))$, we have

$$\fint_{B(x,r)} |f - f_{B(x,r)}| \, d\mu \leq C r \left(\fint_{B(x,\lambda r)} g_f^p \, d\mu \right)^{1/p}.$$

Here

$$f_B := \fint_B f d\mu := \frac{1}{\mu(B)} \int_B f \, d\mu.$$

It is known that if X is complete, μ is uniformly locally doubling, and (X, d, μ) supports a uniformly local p-Poincaré inequality, then for compact sets $E, F \subset X$,

$$\mathrm{cap_p}(E, F) = \mathrm{Mod}_p(\Gamma(E, F)). \tag{2.1}$$

A proof of this can be obtained by adapting the proof found in [21] where it was assumed that $R_0 = R_1 = \infty$. It follows immediately that $\mathrm{cap_p}(E, F) = \mathrm{cap_p}(F, E)$, even though this was not at all obvious merely from considering the definition of $\mathrm{cap_p}(E, F)$.

Standing Assumptions We will assume in this note that $1 < p < \infty$, X is complete, μ is uniformly locally doubling, and (X, d, μ) supports a uniformly local p-Poincaré inequality.

3 Potential Theoretic Characterization of p-Hyperbolicity via p-Singular Functions

In this section we will discuss a Grigor'yan-type characterization of p-hyperbolicity in terms of existence of global singular functions. A p-singular function is a *non-negative* p-superharmonic function u on X such that there is a point $x_0 \in X$ for which u is p-harmonic in $X \setminus \{x_0\}$, $u \in N^{1,p}(X \setminus B(x_0, r))$ for each $r > 0$, and satisfies $\lim_{y \to x_0} u(y) = \infty$. As described in [11], a manifold X is transient (that is, it is 2-hyperbolic) if and only if X supports a 2-singular function. In the setting of manifolds, the dissertation [16] extends this result to the non-linear setting of all $1 < p < \infty$.

Following [27], for a non-empty open set $\Omega \subset X$ and a function u on Ω, we say that u is *p-harmonic* in Ω if $u \in N^{1,p}_{loc}(\Omega)$ and for each open set $V \subset \Omega$ with $\overline{V} \subset \Omega$ compact and each $v \in N^{1,p}(X)$ with $v = 0$ in $X \setminus V$ we have

$$\int_V g_u^p \, d\mu \le \int_V g_{u+v}^p \, d\mu.$$

We say that u is *p-superharmonic* in Ω if whenever $V \subset \Omega$ with $\overline{V} \subset \Omega$ a compact set and $v \in N^{1,p}(X)$ is p-harmonic in a neighborhood of $\overline{V}$ with $v \le u$ on ∂V, we must have $v \le u$ on V.

Definition 3.1 Let Ω be a nonempty open subset of X with $X \setminus \Omega$ nonempty and $x_0 \in \Omega$. We say that a non-negative function u on X is a p-singular function on Ω with singularity at x_0 if

1. u is p-harmonic in $\Omega \setminus \{x_0\}$,
2. $\lim_{\Omega \setminus \{x_0\} \ni y \to x_0} u(y) = \mathrm{cap}_p(\{x_0\}, X \setminus \Omega)^{1/(1-p)}$,
3. $u \in N^{1,p}(X \setminus B(x_0, r))$ for each $r > 0$, and $u = 0$ in $X \setminus \Omega$,
4. and finally,

$$\left(\frac{p-1}{p}\right)^{2(p-1)} (b-a)^{1-p} \le \mathrm{cap}_p(\{u \ge b\}, \{u > a\}) \le p^2 (b-a)^{1-p}$$

whenever $0 \le a < b$ such that $\{u > a\} \subset B(x_0, R_0/2)$.

In the above definition, Condition 2 is equivalent to enforcing the condition $\lim_{\Omega \setminus \{x_0\} \ni y \to x_0} u(y) = \infty$ if $\mathrm{cap}_p(\{x_0\}, X \setminus \Omega) = 0$ (which is the case for values of p that are not larger than the dimension of the space). Thus the first three properties would be satisfied by positive scalar multiples of a p-singular function. The fourth condition dictates the the condensers $(\{u \geq b\}, \{u > a\})$ for $b > a$, or more specifically the value of $\mathrm{Mod}_p(\Gamma(\{u \geq b\}, \{u \leq a\})$, in terms of $(b-a)^{1-p}$. Hence this condition narrows the candidates for p-singular functions. Indeed, from the arguments in [16], this fourth condition guarantees uniqueness of p-singular functions in the context of Riemannian manifolds and other spaces where there is an Euler–Lagrange equation corresponding to the p-energy minimization property. A combination of the above second and fourth conditions guarantee then that the p-Laplacian type operator, corresponding to the Euler-Lagrange equation, acts on the p-singular function to give the unit atomic measure δ_{x_0} supported at x_0.

From [22] we know that functions that are p-harmonic on an open set satisfy local Hölder continuity and (if they are non-negative) a Harnack inequality. Namely, we know that given a p-harmonic function h on a domain $U \subset X$ and $x \in U$, there are constants $\alpha, C_h > 0$ such that if $r > 0$ with $B(x, 2r) \subset U$ and whenever $z, w \in B(x, r)$ we have $|h(z) - h(w)| \leq C_h d(z, w)^{\alpha_x}$; this is the local Hölder continuity ([22, Theorem 5.2]). Moreover, it is shown in [22, Corollary 7.3] that there is a constant $C > 0$ so that if h is p-harmonic and non-negative on U and $B(x, 2r) \subset U$, then $\sup_{B(x,r)} h \leq C \inf_{B(x,r)} h$. Using this Harnack inequality for non-negative p-harmonic functions in $\Omega \setminus \{x_0\}$, it is shown in [19] that if Ω is a relatively compact subset of X, then for each $x_0 \in \Omega$ we always have a p-singular function on Ω with singularity at x_0. Therefore the non-trivial aspect of the existence of singular functions is when Ω is unbounded.

Definition 3.2 A function u on X is said to be a *p-singular function on* X with singularity at $x_0 \in X$ if

1. u is p-harmonic in $X \setminus \{x_0\}$ with $u > 0$ there,
2. there is a sequence of bounded open sets $\Omega_j \subset X$ with
3. $\overline{\Omega}_j \subset \Omega_{j+1}$ and $X = \bigcup_j \Omega_j$ and $r_0 > 0$ such that for $0 < r < r_0$ and $x \in X$ with $d(x, x_0) = r$, $\lim_{X\setminus\{x_0\}\ni y\to x_0} u(y) \simeq \lim_j \mathrm{cap}_p(\overline{B}(x_0, r), X \setminus \Omega_j)^{1/(1-p)}$,
4. $u \in N^{1,p}_{loc}(X \setminus \{x_0\})$,
5. and finally,

$$\left(\frac{p-1}{p}\right)^{2(p-1)} (b-a)^{1-p} \leq \mathrm{cap}_p(\{u \geq b\}, \{u > a\}) \leq p^2 (b-a)^{1-p}$$

whenever $0 \leq a < b \leq \lim_j \mathrm{cap}_p(\{x_0\}, X \setminus \Omega_j)^{1/(1-p)}$ with b sufficiently large.

Note that the notation adopted in [19] is slightly different from that here; there the relative capacity $\mathrm{cap}_p(E, F)$ is computed with respect to functions $u \in N^{1,p}(X)$ with $u = 0$ in $X \setminus F$ and $u \geq 1$ on E, with $E \subset F$. Hence to interpret the notation of [19] here, we should substitute the second component of $\mathrm{cap}_p(E, \Omega)$, namely Ω there, with $X \setminus \Omega$ in this current paper. In the setting of metric measure spaces, the following theorem was established in [19, Theorem 3.14].

Theorem 3.3 (X, d, μ) *is p-hyperbolic if and only if there is a point* $x_0 \in X$ *and a p-singular function on X with singularity at* x_0*. If* (X, d, μ) *is p-hyperbolic, then for* every $x_0 \in X$ *there is a p-singular function with singularity at* x_0*.*

The idea for the proof is simple, though the details are cumbersome; we refer the interested reader to [19] for the details, and merely give a sketch of the proof now.

Sketch of Proof Suppose first that X is p-hyperbolic; then there is some $x_0 \in X$, $R > 0$, and a strictly monotone increasing sequence of positive real numbers R_n, $n \in \mathbb{N}$, with $R_1 > R$, such that

$$\lim_n \operatorname{Mod}_p(\Gamma(\overline{B}(x_0, R), X \setminus B(x_0, R_n)) > 0.$$

Since each curve in $\Gamma(\overline{B}(x_0, R), X \setminus B(x_0, R_{n+1})$ has a subcurve that belongs to the family $\Gamma(\overline{B}(x_0, R), X \setminus B(x_0, R_n)$, it follows that

$$\operatorname{Mod}_p(\Gamma(\overline{B}(x_0, R), X \setminus B(x_0, R_{n+1})) \le \operatorname{Mod}_p(\Gamma(\overline{B}(x_0, R), X \setminus B(x_0, R_n)),$$

and so the above limit is well-defined. Then by (2.1) we know that

$$0 < \lim_n \operatorname{cap}_\mathrm{p}(X \setminus B(x_0, R_n), \overline{B}(x_0, R)) \le \operatorname{cap}_\mathrm{p}(X \setminus B(x_0, R_1), \overline{B}(x_0, R)) < \infty.$$

For each n let u_n be a p-singular function in $B(x_0, R_n)$ with singularity at x_0; Thanks to the uniformly local version of Harnack's inequality and the definition of p-singular functions, for each $n \in \mathbb{N}$ the sequence u_m, $m \ge n$, is locally uniformly bounded in $B(x_0, R_n) \setminus \{x_0\}$. A stability result for p-harmonic functions (see [28]) then gives us a subsequence of u_m, and a function u_∞, such that u_m converges locally uniformly in $X \setminus \{x_0\}$ to u_∞ with u_∞ a p-harmonic function on $X \setminus \{x_0\}$. A direct argument would show that u_∞ is a p-singular function on X with singularity at x_0.

Now suppose that X supports a p-singular function u with singularity at some $x_0 \in X$. Then for sufficiently small $r > 0$ and a nested sequence of open sets Ω_j with $X = \bigcup_j \Omega_j$ such that

$$u \simeq \lim_j \operatorname{cap}_\mathrm{p}(X \setminus \Omega_j, \overline{B}(x_0, r))^{1/(1-p)}$$

on the sphere $S(x_0, r) = \{y \in X \,:\, d(x_0, y) = r\}$. Thus

$$\lim_j \operatorname{cap}_\mathrm{p}(X \setminus \Omega_j, \overline{B}(x_0, r)) > 0.$$

By passing to a subsequence if necessary, we may assume that $R_j := \operatorname{dist}(x_0, X \setminus \Omega_j)$ is a strictly monotone increasing sequence; as $X = \bigcup_j \Omega_j$, it follows that $\lim_j R_j = \infty$, and so

$$\operatorname{cap}_\mathrm{p}(X \setminus B(x_0, R_j), \overline{B}(x_0, r)) \ge \operatorname{cap}_\mathrm{p}(X \setminus \Omega_j, \overline{B}(x_0, r)).$$

Hence we now have

$$\lim_j \mathrm{cap_p}(X \setminus B(x_0, R_j), \overline{B}(x_0, r)) > 0,$$

that is, X is p-hyperbolic. □

Note that here we require the singular functions to be non-negative. Reverting back to the setting of Euclidean spaces $\mathbb{R}^n$, we know that $\mathbb{R}^n$ supports p-singular functions for $1 < p < n$, but does not support a p-singular function for $p = n$; in the case of $p = n$ we have Green's functions, which are functions u that are p-harmonic in $\mathbb{R}^n \setminus \{x_0\}$, $\lim_{y\to x_0} u(y) = \infty$, and $\Delta_n u = \delta_{x_0}$; however, in this case u is not non-negative, and indeed we have that $\lim_{y\to\infty} u(y) = -\infty$. For more on singular functions and p-parabolicity, see for example [2, 3, 8, 11, 16, 17].

4 p-Hyperbolicity and p-Modulus of a Family of Curves Connecting a Ball to ∞

In the setting of manifolds and with $p = 2$, we know from [11] that a manifold M is 2-parabolic if and only if the (Brownian) probability measure of the collection of all Brownian paths γ in M that eventually never return to a given ball in M is zero; that is, if B is a ball in M and Γ is the collection of all Brownian paths $\gamma : [0, \infty) \to M$ such that $\gamma(t) = x_0$ and $\gamma(t) \notin B$ for all $t \geq t_B \in [0, \infty)$, then $\mathbb{P}(\Gamma) = 0$. In the non-linear setting of $p \neq 2$, and even when $p = 2$ but in the setting of metric measure spaces where the upper gradient structure does not come from an inner product structure on the space, the connection to Brownian motion is more tenuous. However, there is a connection between p-parabolicity and p-modulus of families of curves connecting B to ∞; the focus of this section is to explore this idea further.

From Definition 2.4, a metric measure space X is p-hyperbolic if there is a ball $B = B(x_0, R_0)$ and a positive number $\tau > 0$ such that whenever $R > R_0$, the p-modulus of the collection of all paths connecting B to $X \setminus B(x_0, R)$ is at least τ. Let $\Gamma(R)$ denote this collection of paths. Set $\Gamma := \bigcap_{R>0} \Gamma(R)$. Then Γ consists of all paths that have one end point in B and leave each bounded subset of X. Moreover, for $R_0 < R < T$ we have $\Gamma(T) \subset \Gamma(R)$, and so the family $(\Gamma(R))_{R>R_0}$ is a decreasing sequence of families of paths. However, in general it is not true that if Γ_n, $n \in \mathbb{N}$, is a decreasing sequence of families of curves, then $\lim_n \mathrm{Mod}_p(\Gamma_n) = \mathrm{Mod}_p(\bigcap_n \Gamma_n)$. However, we will see in this section that we can still conclude that $\mathrm{Mod}_p(\Gamma) > 0$. As far as I know, this fact is not proven in currently existing literature on analysis on metric spaces, we provide a complete proof of this here. Note that this result is new even in the Euclidean setting. We first need the following lemma.

Lemma 4.1 *There is a non-negative Borel measurable function $h \in L^p(X)$ such that for each $x_0 \in X$ and $R > 0$,*

$$\inf_{B(x_0,R)} h := \beta_R > 0.$$

Proof We fix $x_0 \in X$ and $R_0 > 0$, and set

$$h := \sum_{k\in\mathbb{N}} \frac{1}{2^k\,\mu(\overline{B}(x_0,(k+2)R_0)\setminus B(x_0,kR_0))^{1/p}} \chi_{B(x_0,(k+2)R_0)\setminus\overline{B}(x_0,kR_0)}.$$

Then h is lower semicontinuous, and satisfies the desired requirements. □

Now we are ready to prove the main result of this section.

Theorem 4.2 *Let B be a ball in X and let Γ be the collection of all paths $\gamma : [0,\infty) \to X$ such that $\gamma(0) \in B$ and for each $R > 0$ there is some $t_{\gamma,R} > 0$ such that $\gamma(t) \notin B(x_0, R)$ whenever $t > t_{\gamma,R}$. Then X is p-hyperbolic if and only if $Mod_p(\Gamma) > 0$.*

Proof Suppose first that X is p-hyperbolic. Then

$$\lim_{R\to\infty} \mathrm{Mod}_p(\Gamma(R)) =: \tau > 0. \tag{4.1}$$

Suppose that $\Gamma = \bigcap_{R>R_0} \Gamma(R)$ satisfies $\mathrm{Mod}_p(\Gamma) = 0$. Then we know from Fuglede's theorem (see the discussion following Definition 2.1) that there is a non-negative Borel function $\rho \in L^p(X)$ such that $\int_\gamma \rho\, ds = \infty$ for each locally rectifiable path $\gamma \in \Gamma$. An application of the Vitali-Carathéodory theorem allows us to assume that ρ is also lower semicontinuous. Moreover, by replacing ρ with $\max\{\rho, h\}$ with h as in Lemma 4.1, we may also assume that for each $R > 0$,

$$\inf_{B(x_0,R)} \rho := \beta_R > 0.$$

Scaling ρ by a positive constant if necessary, we can also assume that

$$\int_X \rho^p\, d\mu \le \tau/2.$$

Then by (4.1) and by the fact that $R \mapsto \mathrm{Mod}_p(\Gamma(R))$ is monotone *decreasing*, we know that $\rho \notin \mathcal{A}(\Gamma(R))$ for each $R > R_0$. Thus, for each positive integer $n \ge 2$, there is a rectifiable curve $\gamma_n \in \Gamma(nR_0)$ such that $\int_{\gamma_n} \rho ds < 1$.

For each positive integer $n \ge 2$ and each positive integer $k \ge n$, we now have that

$$\ell(\gamma_k \cap B(x_0, nR_0)) \le \frac{1}{\beta_{nR_0}} \int_{\gamma_n} \rho\, ds < \frac{1}{\beta_{nR_0}} < \infty.$$

It follows that the sequence γ_n, $n \in \mathbb{N}$, of paths in X (using arc-length parametrization) is locally equicontinuous and locally equibounded. Given that the metric space X is complete and doubling, it follows that X is proper (that is, closed and bounded subsets of X are compact, see for example [13]). Therefore we can invoke the Arzelà–Ascoli theorem and a Cantor diagonalization argument to obtain a subsequence of paths, denoted γ_{n_j}, $j \in \mathbb{N}$, and and a locally rectifiable path γ with one end point in $B = B(x_0, R_0)$, such that $\gamma_{n_j} \to \gamma$ locally uniformly in $[0, \infty)$. Recall that ρ is lower semicontinuous. Hence an adaptation of the argument found in [13, Page 13–14], we have

$$\int_\gamma \rho\, ds \le \liminf_k \int_{\gamma_k} \rho\, ds \le 1.$$

On the other hand, as $\gamma_n \in \Gamma(nR_0)$, it follows that $\gamma \in \Gamma(kR_0)$ for each positive integer $k \ge 2$; whence we have that $\gamma \in \Gamma$. This violates our choice of ρ as a function in $L^p(X)$ such that for each $\widetilde{\gamma} \in \Gamma$ we have $\int_{\widetilde{\gamma}} \rho\, ds = \infty$. We can therefore conclude that we must have $\mathrm{Mod}_p(\Gamma) > 0$ as desired.

Finally, if $\mathrm{Mod}_p(\Gamma) > 0$, then for each $R > R_0$ we must have

$$\mathrm{Mod}_p(\Gamma(R)) \ge \mathrm{Mod}_p(\Gamma) > 0,$$

and therefore X is p-hyperbolic. This concludes the proof of the theorem. □

Note that the outer measure Mod_p, on the family of all paths in X, sees only locally rectifiable paths. Hence p-hyperbolicity of the metric measure space X (or a Riemannian manifold M) tells us that there is a plenitude of locally rectifiable curves γ in X beginning from a given ball B and eventually leaving every bounded subset of X. The key here is that these curves are locally rectifiable. In the event that $p = 2$ and we are in the setting of Riemannian manifolds M, this perspective is dual to the perspective of Brownian paths which are almost surely not even locally rectifiable (though they are almost surely locally Hölder continuous). It would be interesting to know whether there is an object analogous to Brownian motion for the non-linear setting of $p \neq 2$ that sees locally non-rectifiable paths. One possible process associated with the p-Laplacian, called tug-of-war with noise in [25], might shed some light on this, but this direction of study has so far not focused on properties of paths associated with the tug-of-war with noise process. The paper [24] gives a nice introduction to the tug-of-war process, and the regularity theory associated with the tug-of-war with noise is explored in [4].

5 p-Parabolicity and a Liouville-Type Theorem

The classical Liouville theorem states that there is no non-constant bounded complex-analytic function on the entire complex plane. A version of this theorem states that there is no non-constant positive harmonic function on the Euclidean space $\mathbb{R}^n$. In the non-smooth setting, if μ is globally doubling and supports a global p-Poincaré inequality, then by the results in [22] we know that non-negative p-harmonic functions satisfy a Harnack inequality, and hence there are no non-constant positive p-harmonic functions on such metric measure spaces. The situation is different when considering metric measure spaces equipped with a measure that is locally doubling and supports a local p-Poincaré inequality. The hyperbolic spaces $\mathbb{H}^n$ are examples of such spaces, as are infinite trees with bounded degree that are not homeomorphic to $\mathbb{R}$. As we know, $\mathbb{H}^n$ does support a non-constant positive harmonic function. It was shown in [6] that if the measure is globally doubling and supports a global p-Poincaré inequality, and in addition the metric space is *annular quasiconvex*, then there are no global non-constant p-harmonic functions (whether non-negative or not) with finite energy. Here a metric space X is annular quasiconvex if there is a constant $C \geq 1$ such that whenever $x_0 \in X$ and $r > 0$, and whenever $x, y \in B(x_0, r) \setminus B(x_0, r/2)$, there is a rectifiable path γ in $B(x_0, Cr) \setminus B(x_0, r/C)$ with end points x and y, and with length $\ell(\gamma) \leq Cd(x, y)$. This version of Liouville theorem (finite energy Liouville theorem) is not equivalent to the standard Liouville theorem described above. In this section we discuss the effect of p-hyperbolicity on the existence of global non-constant positive/finite energy p-harmonic functions.

Note that when $1 < p < n$, the Euclidean space $\mathbb{R}^n$ is p-hyperbolic, but does not have a non-constant positive p-harmonic function nor a non-constant finite energy p-harmonic function; here we say that a function u on a metric space X has finite energy if it has an upper gradient $g_u \in L^p(X)$. Hence p-hyperbolicity of a space does not guarantee existence of non-constant global p-harmonic functions. The results of [6] indicate that we need the space to fail to be annular quasiconvex, and strongly so. The following notion of ends of a metric space is a direct analog of the theory of ends of Riemannian manifolds as described in [2].

Definition 5.1 A sequence of connected sets $\{E_k\}_k$ is said to be an end (or end at infinity) of X if there is a sequence of balls $B_k \subset X$ with $B_k \subset B_{k+1}$ such that E_k is a component of $X \setminus B_k$ and $E_{k+1} \subset E_k$ for each positive integer k. We say that an end $\{E_k\}$ is a p-hyperbolic end if

$$\liminf_{k\to\infty} \operatorname{Mod}_p(\Gamma(\overline{B}_1, E_k)) > 0.$$

We say that an end is p-parabolic if it is not p-hyperbolic.

It is possible for a metric measure space to be p-hyperbolic but have only p-parabolic ends. Indeed, if X is a K-regular tree (that is, each vertex has exactly K number of edges attached to it) with $K \geq 3$, with the edges of unit length and equipped with the Lebesgue measure $\mathcal{L}^1$, then the measure on X is uniformly locally doubling and supports a uniformly local 1-Poincaré inequality, see for example [5]. Observe that each end of X corresponds to a geodesic ray starting from a vertex in X. Fix such an end, and we list the vertices that make up the corresponding geodesic ray by x_k, $k \in \mathbb{N}$. We fix $B = B(x_1, 1)$. The function ρ_k given by setting $\rho_k = 0$ on all the edges except on the edges $[x_2, x_3], \cdots, [x_{k-1}, x_k]$, where it is set to take on the value of $1/(k-1)$. Then $\rho_k \in \mathcal{A}(\Gamma(B, X_k))$ with X_k the connected component of $X \setminus x_k$ containing x_{k+1}. Therefore

$$\mathrm{Mod}_p(\Gamma(B, X_k)) \leq \int_X \rho_k^p \, d\mu = \frac{1}{(k-1)^p} k,$$

and so

$$\lim_{k\to\infty} \mathrm{Mod}_p(\Gamma(B, X_k)) = 0.$$

Therefore the end is a p-parabolic end of X. However, X itself is p-hyperbolic for each $p > 1$. This is a consequence of the following result from [7, Theorem 1.2] together with the fact that there is a non-constant p-harmonic function on X with finite energy (see [6]). Indeed, fixing a base vertex v_0, we set $u = 0$ at v_0. We will define the value of u at each vertex, with the understanding that a linear interpolation will extend the function to the edges that make up X. For ease of computation, we will focus on $p = 2$ and $K = 3$. Then with $v_{1,1}$, $v_{1,2}$ and $v_{1,3}$ denoting the three vertices that are neighbors of v_0, we set $u(v_{1,1}) = 0$, $u(v_{1,2}) = 1/2$, and $u(v_{1,3}) = -1/2$. On the connected component of $X \setminus \{v_0\}$ containing $v_{1,1}$ we set $u = 0$. We can then extend u to vertices in the connected component of $X \setminus \{v_0\}$ containing $v_{1,2}$ by setting $u(w) = \sum_{j=1}^{k} 2^{-j}$ where w is a vertex in this component that is a distance k from $v_{1,2}$. We set $u(w) = -\sum_{j=1}^{k} 2^{-j}$ where w is a vertex in the component of $X \setminus \{v_0\}$ containing $v_{1,3}$, with k the distance between w and $v_{1,3}$. A direct computation shows that u is 2-harmonic in X with finite energy $\sum_{j=1}^{\infty} 2^{-kp} 2^k$ with $p = 2$.

Theorem 5.2 *Suppose that in addition to being uniformly locally doubling and supporting a uniformly local p-Poincaré inequality, we have that X is unbounded and proper. Then*

- *if X has a non-constant p-harmonic function with finite energy, then X is p-hyperbolic.*
- *If X has at least two p-hyperbolic ends, then it has a non-constant bounded p-harmonic function with finite energy.*

Observe that when $n > 1$, the hyperbolic space $\mathbb{H}^n$ has only one end, and this end is indeed p-hyperbolic when $p < n$; the Euclidean space $\mathbb{R}^n$ also has only one end, and this end is p-hyperbolic when $p < n$. On the other hand, $\mathbb{R}^n$ supports no non-constant positive p-harmonic functions while $\mathbb{H}^n$ does.

Unlike the property of supporting a p-singular function, the property of supporting a non-constant positive or finite energy p-harmonic function does not characterize p-hyperbolic spaces; however, the above discussion shows that there is a connection between the existence of non-constant positive/finite energy p-harmonic functions and p-hyperbolicity. A deeper understanding of the structures of p-hyperbolic ends and p-parabolic ends of X might lead to a characterization of the property of supporting a non-constant positive or finite energy p-harmonic functions, and this field of enquiry is still under development. For other partial characterizations of p-hyperbolicity using volume growth conditions see [17] (Riemannian manifold setting) and [18] (metric setting). It was shown in [17, Proposition 1.7] that if X is a non-compact complete Riemannian manifold and

$$\int_1^\infty \left(\frac{1}{\mu(B(x_0,t))}\right)^{1/(p-1)} dt = \infty,$$

then it is p-parabolic. Moreover, it is shown in [17, Corollary 2.29] that if there is a constant $C > 0$ and a point $x_0 \in X$ such that each sequence $x_k \in X$ with $2 < d(x_k, x_0) \to \infty$ as $k \to \infty$ can be connected to x_0 by geodesics γ_k with the property that

$$\int_1^{d(x_k,x_0)} \left(\frac{1}{\mu(B(\gamma_k(t), t/8))}\right)^{1/(p-1)} dt \le C,$$

then X is p-hyperbolic. Versions of these results in the metric setting can be found in [18], where large-scale dimension conditions are given to guarantee p-parabolicity and p-hyperbolicity of the space.

Acknowledgments The author's research was partially supported by grants from the National Science Foundation (U.S.), DMS# 1500440 and DMS# 1800161. The author thanks the kind referee for helpful suggestions that improved the exposition of this article.

References

1. Ambrosio, L., Tilli, P.: Topics on Analysis in Metric Spaces. Oxford Lecture Series in Mathematics and its Applications, vol. 25. Oxford University Press, Oxford (2004)
2. Ancona, A.: Negatively curved manifolds, elliptic operators, and the Martin boundary. Ann. Math. **125**, 495–536 (1987)
3. Anderson, M.T., Schoen, R.: Positive harmonic functions on complete manifolds of negative curvature. Ann. Math. **121**, 429–461 (1985)
4. Bjorland, C., Caffarelli, L., Figalli, L.: Non-local gradient dependent operators. Adv. Math. **230**(4–6), 1859–1894 (2012)
5. Björn, A., Björn, J., Gill, J.T., Shanmugalingam, N.: Geometric analysis on Cantor sets and trees. J. Reine Angew. Math. **725**, 63–114 (2017)
6. Björn, A., Björn, J., Shanmugalingam, N.: The Liouville theorem for p-harmonic functions and quasiminimizers with finite energy. Math. Z. **297**, 827–854 (2021)
7. Björn, A., Björn, J., Shanmugalingam, N.: Existence of global p-harmonic functions and p-parabolic ends, p-hyperbolic ends (in preparation)
8. Coulhon, T., Holopainen, I., Saloff-Coste, L.: Harnack inequality and hyperbolicity for subelliptic p-Laplacians with applications to Picard type theorems. Geom. Funct. Anal. **11**, 1139–1191 (2001)
9. Fuglede, B.: Extremal length and functional completion. Acta Math. **98**, 171–219 (1957)

10. Fukushima, M., Oshima, Y., Takeda, M.: Dirichlet Forms and Symmetric Markov Processes. De Gruyter Studies in Mathematics, vol. 19, 2nd edn. (2011)
11. Grigoryan, A.: Analytic and geometric background of recurence and non-explosion of the brownian motion on riemannian manifolds. Bull. AMS **36**(2), 135–249 (1999)
12. Hajłasz, P., Koskela, P.: Sobolev Met Poincaré. Memoirs of the American Mathematical Society, vol. 145. American Mathematical Society, Providence (2000)
13. Heinonen, J.: Lectures on Analysis on Metric Spaces. Springer, New York (2001)
14. Heinonen, J., Koskela, P.: Quasiconformal maps in metric spaces with controlled geometry. Acta Math. **181**, 1–61 (1998)
15. Heinonen, J., Koskela, P., Shanmugalingam, N., Tyson, J.T.: Sobolev spaces on metric measure spaces. In: An Approach Based on Upper Gradients. New Mathematical Monographs, vol. 27. Cambridge University Press, Cambridge (2015)
16. Holopainen, I.: Nonlinear potential theory and quasiregular mappings on Riemannian manifolds. Ann. Acad. Sci. Fenn. Ser. AI Math. Dissertationes **74**, 1–45 (1990)
17. Holopainen, I.: Volume growth, Green's functions, and parabolicity of ends. Duke Math. J. **97**, 319–346 (1999)
18. Holopainen, I., Koskela, P.: A note on Lipschitz functions, upper gradients, and the Poincaré inequality. New Zealand J. Math. **28**, 37–42 (1999)
19. Holopainen, I., Shanmugalingam, N.: Singular functions on metric measure spaces. Collect. Math. **53**, 313–332 (2002)
20. Järvenpää, E., Järvenpää, M., Rogovin, K., Rogovin, S., Shanmugalingam, N.: Measurability of equivalence classes and MEC_p-property in metric spaces. Rev. Mat. Iberoamericana **23**(3), 811–830 (2007)
21. Kallunki, S., Shanmugalingam, N.: Modulus and continuous capacity. Ann. Acad. Sci. Fenn. Ser. A1 Math. **26**, 455–464 (2001)
22. Kinnunen, J., Shanmugalingam, N.: Regularity of quasi-minimizers on metric spaces. Manuscripta Math. **105**, 401–423 (2001)
23. Koskela, P., MacManus, P.: Quasiconformal mappings and Sobolev spaces. Stud. Math. **131**, 1–17 (1998)
24. Manfredi, J., Parviainen, M., Rossi, J.: Dynamic programming principle for tug-of-war games with noise. ESAIM Control Optim. Calc. Var. **18**(1), 81–90 (2012)
25. Peres, Y., Sheffield, S.: Tug-of-war with noise: a game-theoretic view of the p-Laplacian. Duke Math. J. **145**(1), 91–120 (2008)
26. Shanmugalingam, N.: Newtonian spaces: an extension of Sobolev spaces to metric measure spaces. Rev. Mat. Iberoam. **16**(2), 243–279 (2000)
27. Shanmugalingam, N.: Harmonic functions on metric spaces. Ill. J. Math. **45**(3), 1021–1050 (2001)
28. Shanmugalingam, N.: Some convergence results for p-harmonic functions on metric measure spaces. Proc. Lond. Math. Soc. **87**, 226–246 (2003)
29. Troyanov, M.: Parabolicity of manifolds. Siberian Adv. Math. **9**, 125–150 (1999)

Part IV
Physical Models and Fractals

Breaking of Continuous Scale Invariance to Discrete Scale Invariance: A Universal Quantum Phase Transition

Omrie Ovdat and Eric Akkermans

Abstract We provide a review on the physics associated with phase transitions in which continuous scale invariance is broken into discrete scale invariance. The rich features of this transition characterized by the abrupt formation of a geometric ladder of eigenstates, low energy universality without fixed points, scale anomalies and Berezinskii–Kosterlitz–Thouless scaling are described. The important role of this transition in various celebrated single and many body quantum systems is discussed along with recent experimental realizations. Particular focus is devoted to a recent realization in graphene.

Keywords Discrete scale invariance · Continuous scale invariance · Universality · Limit cycles · Graphene · Berezinskii–Kosterlitz–Thouless

Mathematics Subject Classifications (2010) Primary: 28A80, Secondary: 28A75, 60G22

1 Introduction

Continuous scale invariance (CSI)—a common property of physical systems—describes the invariance of a physical quantity $f(x)$ (e.g., the mass) when changing a control parameter x (e.g., the length). This property is expressed by a simple scaling relation,

$$f(ax) = b\, f(x), \tag{1.1}$$

satisfied $\forall a > 0$ and corresponding $b(a)$, whose general solution is the power law

$$f(x) = C\, x^{\gamma} \tag{1.2}$$

O. Ovdat · E. Akkermans (✉)
Israel Institute of Technology, Haifa, Israel
e-mail: somrie@campus.technion.ac.il; eric@physics.technion.ac.il

U. Freiberg et al. (eds.), *Fractal Geometry and Stochastics VI*,
Progress in Probability 76, https://doi.org/10.1007/978-3-030-59649-1_9

with $\gamma = \ln b/\ln a$. Other physical systems possess the weaker discrete scale invariance (DSI) expressed by the same scaling relation (1.1) but now satisfied for fixed values a, b and whose solution becomes

$$f(x) = x^{\gamma}\, G\,(\ln x/\ln a)\,, \tag{1.3}$$

where $G(u+1) = G(u)$. Since $G\,(u)$ is a periodic function, one can expand it in Fourier series $G\,(u) = \sum c_n e^{2\pi i n u}$, thus,

$$f\,(x) = \sum_{n=-\infty}^{\infty} c_n x^{\gamma + i\frac{2\pi n}{\ln a}}\,. \tag{1.4}$$

If $f\,(x)$ is required to obey CSI, $G\,(u)$ would be constraint to fulfill the relation $G\,(u) = G\,(u + a_0)\ \forall a_0 \in \mathbb{R}$. In this case, $G\,(u)$ can only be a constant function, that is, $c_n = 0$ for all $n \neq 0$ eliminating all terms with complex exponents in (1.4). Therefore, real exponents are a signature of CSI and complex exponents are a signature of DSI. DSI is a typical property of a class of fractal systems [1, 3, 4, 18, 25, 26, 76, 84] and it underlines the construction of special geometric objects such as the Cantor set, Sierpinski triangle, Koch snowflake and others.

In this article we describe a variety of distinct quantum systems in which a sharp transition initiates the breaking of CSI into DSI. Essential to all these cases is a DSI phase characterized by a sudden appearance of a low energy spectrum arranged in an infinite geometric series. Accordingly, each transition is associated with exponents that change from real to complex valued at the critical point. We describe the universal properties of this transition. Particularly, in the framework of the renormalization group it is shown that universality in this case is not associated with trajectories terminating at a fixed point but with periodic flow known as a limit cycle. Intrinsic to this phenomena is a special type of scale anomaly in which residual discrete scaling symmetry remains at the quantum level.

We discuss the physical realizations of the CSI to DSI transition and present recent experimental observations which provide evidence for the existence of the critical point and for the universal low energy features of the DSI phase. We discuss the basic ingredients that underline these features and the possibility of their occurrence in other, yet to be studied systems.

2 The Schrödinger $1/r^2$ Potential

A well studied example exhibiting the breaking of CSI to DSI is given by the problem of a quantum particle in an attractive inverse square potential [17, 53] described by the Hamiltonian ($\hbar = 1,\ m = 1/2$)

$$H_S = p^2 - \lambda/r^2. \tag{2.1}$$

This system constitutes an effective description of the "Efimov effect" [23, 24] and plays a role in various other systems [16, 19, 47, 54, 67, 79].

2.1 The Spectral Properties of H_S

The Hamiltonian H_S has an interesting yet disturbing property – the power law form of the potential matches the order of the kinetic term. As a result, the Schrödinger equation

$$H_S\psi = E\psi \tag{2.2}$$

depends on the single dimensionless parameter λ which raises the question of the existence of a characteristic energy to express the eigenvalues E_n. This absence of characteristic scale implies the invariance of $H_S\psi = E\psi$ under the scale transformation [43]

$$x^i \to ax^i,\ E \to a^{-2}E \tag{2.3}$$

which indicates that if there is one negative energy bound state E_n then there is an unbounded continuum of bound states which render the Hamiltonian nonphysical and mathematically not self-adjoint [32, 59].

The eigenstates of H_S can be solved in terms of Bessel functions which confirm these assertions in more detail. For $E < 0$ and lowest orbital angular momentum subspace $l = 0$, the most general decaying solution is described by the radial function

$$\psi(r) \approx r^{-\frac{d-2}{2}} \left((kr)^{-\sqrt{\lambda_c-\lambda}} \left(a_1 + O(kr)^2 \right) + (kr)^{\sqrt{\lambda_c-\lambda}} \left(a_2 + O(kr)^2 \right) \right) \tag{2.4}$$

where $k \equiv \sqrt{-E}$, a_1, a_2 are energy independent coefficients, d is the space dimension and $\lambda_c \equiv (d-2)^2/4$.[1] As seen in (2.4), for $\lambda > \lambda_c - 1$, $\psi_0(r)$ is normalizable $\forall \mathrm{Re}(E) < 0$ which constitutes a continuum of complex valued bound states of H_S. Thus, for $\lambda > \lambda_c - 1$, H_S is no longer self-adjoint, a property that originates from the strong singularity of the potential and is characteristic of a general class of potentials with high order of singularity [17].

A simple, physically instructive procedure to deal with the absence of self-adjointness is to remove the singular $r = 0$ point by introducing a short distance

[1] For higher angular momentum channels λ_c is larger and given by $(d-2)^2/4 + l(l+d-2)$.

cutoff L and to apply a boundary condition at $r = L$ [5, 10, 13, 19, 37, 62, 63]. The most general boundary condition is the mixed condition

$$L\frac{\psi'(L)}{\psi(L)} = g, \tag{2.5}$$

$g \in \mathbb{R}$, for which there is an infinite number of choices each describing different short range physics.

Equipped with condition (2.5) the operator H_S is now a well defined self-adjoint operator on the interval $L < r < \infty$. The spectrum of H_S exhibits two distinct features in the low energy $kL \ll 1$ regime. For $\lambda < \lambda_c \equiv (d-2)^2/4$, the expression of $L\psi'(L)/\psi(L)$ as given from (2.4) is independent of k to leading order in kr. As a result, Eq. (2.5) does not hold for a general choice of g. For $\lambda > \lambda_c$, the insertion of (2.4) into (2.5) leads to

$$(kL)^{2i\sqrt{\lambda-\lambda_c}} = e^{i\gamma} \tag{2.6}$$

where $\gamma(g, \lambda)$ is a phase that can be calculated (the explicit expression of γ is not important for the purpose of this section). The solution of (2.6) yields a set of bound states with energies

$$k_n = k_0 e^{-\frac{\pi n}{\sqrt{\lambda-\lambda_c}}} \tag{2.7}$$

where $n \in Z$, such that $k_n L \ll 1$ and $k_0 \equiv \frac{1}{L} e^{\frac{\gamma}{2\sqrt{\lambda-\lambda_c}}}$. Thus, for $\lambda < \lambda_c \equiv (d-2)^2/4$, the spectrum contains no bound states close to $E = 0$, however, as λ goes above λ_c, an infinite series of bound states appears. Moreover, in this "over-critical" regime, the states arrange in a geometric series such that

$$k_{n+1}/k_n = e^{-\frac{\pi}{\sqrt{\lambda-\lambda_c}}}. \tag{2.8}$$

The absence of any states for $\lambda < \lambda_c$ is a signature of CSI while the geometric structure of (2.7) for $\lambda > \lambda_c$ is a signature of DSI since k_n is invariant under $\{k_n\} \to \left\{\exp\left(-\pi/\sqrt{\lambda-\lambda_c}\right) k_n\right\}$. Accordingly, as seen in (2.4), the characteristic behavior of the eigenstates for $kr \ll 1$ manifests an abrupt transition from real to complex valued exponents as λ exceeds λ_c. Thus, H_S exhibits a quantum phase transition (QPT) at λ_c between a CSI phase and a DSI phase. The characteristics of this transition are independent of the values of L, g which enter only into the overall factor k_0 in (2.7). The functional dependence of k_n on $\sqrt{\lambda-\lambda_c}$ is characteristic of Berezinskii–Kosterlitz–Thouless (BKT) transitions as was identified in [44, 45, 47, 50]. Finally, the breaking of CSI to DSI in the $\lambda > \lambda_c$ regime constitutes a special type of scale anomaly since a residual symmetry remains even after regularization (see Table 1).

Table 1 Summary of the properties associated with the transition occurring at $\lambda = \lambda_c$ for the Hamiltonian H_S given in Eq. (2.1) on the interval $L < r < \infty$

	$\lambda < \lambda_c$	$\lambda > \lambda_c - 1$	$\lambda > \lambda_c$	
Formal Hamiltonian	CSI	CSI	CSI	Scale anomaly $\Downarrow$
Self-adjointness	$H = H^{\dagger}$	$H \neq H^{\dagger}$	$H \neq H^{\dagger}$	
Regularization with L	Redundant	Essential	Essential	
Symmetry of eigenspace	CSI	CSI	DSI	
Quantum Phase Transition $\Longrightarrow$				

2.2 *Physical Realizations of H_S*

A well known realization of H_S for $\lambda > \lambda_c$ is the "Efimov effect" [12, 23, 24]. In 1970, Efimov studied the quantum problem of three identical nucleons of mass m interacting through a short range (r_0) potential. He pointed out that when the scattering length a of the two-body interaction becomes very large, $a \gg r_0$, there exists a scale-free regime for the low-energy spectrum, $\hbar^2/ma^2 \ll E \ll \hbar^2/mr_0^2$, where the corresponding bound-states energies follow the geometric series $E_n = -E_0 e^{-2\pi n/s_0}$ where $s_0 \approx 1.00624$ is a dimensionless number and $E_0 > 0$ a problem-dependent energy scale. Efimov deduced these results from an effective Schrödinger equation in $d = 3$ with the radial ($l = 0$) attractive potential $V(r) = -\lambda/r^2$ with $\lambda = s_0 + 1/4 > \lambda_c$ ($\lambda_c = 1/4$ for $d = 3$). Despite being initially controversial, Efimov physics has turned into an active field especially in atomic and molecular physics where the universal spectrum has been studied experimentally [35, 51, 52, 56, 65, 72, 73, 87] and theoretically [12]. The observation of the Efimov geometric spectral ratio $e^{2\pi/s_0} \approx 515.028$ have been recently determined using an ultra-cold gas of caesium atoms [42].

In addition to the Efimov effect, the inverse square potential also describes the interaction of a point like dipole with an electron in three dimensions. In this case, the dipole potential is considered as an inverse square interaction with non-isotropic coupling [16]. The Klein Gordon equation for a scalar field on an Euclidean AdS $d+1$ space time can be written in the form of (2.2). The over-critical regime $\lambda > \lambda_c$ corresponds to the violation of the Breitenlohner–Freedman bound [47].

3 Massless Dirac Coulomb System

The inverse square Hamiltonian (2.1), a simple system exhibiting a rich set of phenomena, inspires studying the ingredients which lead to the aforementioned DSI and QPT and whether they are found in other systems. One such candidate system is described by a massless Dirac fermion in an attractive Coulomb potential [27–29, 36, 60, 71, 80, 81] with the scale invariant Hamiltonian ($\hbar = c = 1$)

$$H_D = \gamma^0\gamma^j p_j - \beta/r \tag{3.1}$$

where β specifies the strength of the electrostatic potential, d is the space dimension and γ^μ are $d+1$ matrices satisfying the anti-commutation relation

$$\{\gamma^\mu, \gamma^\nu\} = 2\eta^{\mu\nu} \tag{3.2}$$

with $\eta^{00} = \eta^{ii} = -1, i = 1, \ldots, d$ and $\eta^{\mu\nu} = 0$ for $\mu \neq \nu$.

Based on the previous example, it may be anticipated that, like H_S, H_D will exhibit a sharp spectral transition at some critical β in which the singularity of the potential will ruin self-adjointness. As detailed below, the analog analysis of the Dirac equation

$$H_D\psi = E\psi \tag{3.3}$$

confirm these assertions and details a remarkable resemblance between the low energy features of the two systems.

3.1 The Spectral Properties of H_D

Utilizing rotational symmetry, the angular part of Eq. (3.3) can be solved and the radial dependence of ψ is given in terms of two functions $\Psi_1(r), \Psi_2(r)$ [21] determined by the following set of equations

$$\begin{aligned} \Psi_2'(r) + \frac{(d-1+2K)}{2r}\Psi_2(r) &= \left(E + \frac{\beta}{r}\right)\Psi_1(r) \\ -\Psi_1'(r) - \frac{(d-1-2K)}{2r}\Psi_1(r) &= \left(E + \frac{\beta}{r}\right)\Psi_2(r) \end{aligned} \tag{3.4}$$

where

$$K \equiv \begin{cases} \pm\left(l + \frac{d-1}{2}\right) & d > 2 \\ m + 1/2 & d = 2 \end{cases}, \tag{3.5}$$

$l = 0, 1, \ldots$ and $m \in \mathbb{Z}$ are orbital angular momentum quantum numbers. In terms of these radial functions, the scalar product of two eigenfunctions $\psi, \tilde{\psi}$ is given by

$$\int dV\, \psi^\dagger\tilde{\psi} = \int dr\, r^{d-1}\left(\Psi_1^*(r)\,\tilde{\Psi}_1(r) + \Psi_2^*(r)\,\tilde{\Psi}_2(r)\right). \tag{3.6}$$

Unlike H_S in Sect. 2, the spectrum of H_D does not contain any bound states, a property that reflects the absence of a mass term. As a result, the spectrum is a continuum of scattering states spanning $-\infty < E < \infty$. In the absence of bound states we explore the possible occurrence of "quasi-bound" states. Quasi bound states are pronounced peaks in the density of states $\rho(E)$, embedded within the continuum spectrum. These resonances describe a scattering process in which an almost monochromatic wave packet is significantly delayed by $V(r)$ as compared to the same wave packet in free propagation [31].

An elegant procedure for calculating the quasi-bound spectrum [31] is to allow the energy parameter to be complex valued $E \rightarrow \epsilon \equiv E_R - i\frac{W}{2}$ and look for solutions of (3.4) with no outgoing e^{-iEr} plane wave component for $r \rightarrow \infty$. The lifetime of the resonance is given by W^{-1}. Consider the lowest angular momentum subspace $K = \pm(d-1)/2$ and $E < 0$, the most general solution with no outgoing component is given by

$$\begin{pmatrix}\Psi_1(r)\\ \Psi_2(r)\end{pmatrix} \approx r^{-\frac{d-1}{2}} \left((2iEr)^{\sqrt{\beta_c^2-\beta^2}} \left(\begin{pmatrix}a_{11}\\ a_{12}\end{pmatrix} + O(|E|r) \right) \right.$$
$$\left. + \, (2iEr)^{-\sqrt{\beta_c^2-\beta^2}} \left(\begin{pmatrix}a_{21}\\ a_{22}\end{pmatrix} + O(|E|r) \right) \right) \tag{3.7}$$

where $\beta_c \equiv (d-1)/2$[2] and a is a 2×2 energy independent coefficient matrix.

As in the case of H_S above it is necessary at this point to remove the singularity of the $1/r$ potential by introducing a radial short distance cutoff L and imposing a boundary condition. To identify this explicitly, consider the case where $E = i$ in (3.7). Since (3.7) is (asymptotically) an ingoing e^{iEr} plane wave solution if $E \in \mathbb{R}$, it decays exponentially for $E = i$ and $r \rightarrow \infty$. If additionally $\beta^2 > \beta_c^2 - 1/4$, then (3.7) is a normalizable eigenfunction with a complex valued eignvalue which renders H_D not self-adjoint.

The equivalent mixed boundary condition of (3.3) can be written as follows [90]

$$h = \frac{\Psi_2(L)}{\Psi_1(L)} \tag{3.8}$$

where $h \in \mathbb{R}$ is determined by the short range physics. Equipped with this condition H_D is now a well defined self-adjoint operator on the interval $L < r < \infty$. The spectrum of H_D exhibits two distinct pictures in the low energy $|E|L \ll 1$ regime. For $\beta < \beta_c \equiv (d-1)/2$, the expression of $\Psi_2(L)/\Psi_1(L)$ as given from (3.7) is independent of E to leading order in $|E|L$. As a result, Eq. (3.8) does hold for a

[2]For higher angular momentum channels β_c is larger and given by $|K|$ where K is defined as in (3.5).

general choice of h. For $\beta > \beta_c$, the insertion of (3.7) and (3.8) reduces into

$$(2iEL)^{2i\sqrt{\beta^2-\beta_c^2}} = z_0 \tag{3.9}$$

where

$$z_0\,(h,\beta) \equiv \frac{h\,a_{21} - a_{22}}{a_{12} - h\,a_{11}} \tag{3.10}$$

is a complex valued number[3] (the explicit exporession for z_0, which can be found in [69], is not important for the purpose of this section). The solution of (3.9) yields a set of quasi-bound energies at

$$E_n = E_0 e^{-\frac{\pi n}{\sqrt{\beta^2-\beta_c^2}}} \tag{3.11}$$

where $n \in \mathbb{Z}$, such that $|E_n|\,L \ll 1$ and $E_0 \equiv \mathrm{Re}\left(\frac{1}{2iL} z_0^{\frac{1}{2i\sqrt{\beta^2-\beta_c^2}}}\right)$. It can be directly verified that $E_R = \mathrm{Re}\,E_n < 0$ and $W = -2\mathrm{Im}\,E_n > 0$ [69].

Thus, in complete analogy with the $-\lambda/r^2$ inverse squared potential described in Sect. 2, for $\beta < \beta_c \equiv (d-1)/2$, the spectrum contains a CSI phase with no quasi-bound states close to $E = 0$. As β exceeds β_c, an infinite series of quasi-bound states appears which arrange in a DSI geometric series such that

$$E_{n+1}/E_n = e^{-\frac{\pi}{\sqrt{\beta^2-\beta_c^2}}}. \tag{3.12}$$

As seen explicitly in (3.7), the characteristic behavior of the eigenstates for $|E|\,r \ll 1$ manifests an abrupt transition from real to complex valued exponents at $\beta = \beta_c$. The characteristics of this transition are independent of the values of L, h which enter only into the overall factor E_0 in (3.9). Thus, under a proper trasformation between λ and β, Table 1 represents a valid and consistent description of the massless Dirac Coulomb system as well.

3.2 *Distinct Features Associated with Spin* 1/2

On top of the similarities emphasized above, an interesting difference in the quantum phase transition exhibited by H_S and H_D results from the distinct spin of the associated Schrödinger and Dirac wave functions. Unlike the scalar

[3]Here z_0 is not a phase like in (2.6), a reflection of the fact that the solutions for E would have an imaginary component corresponding to a finite lifetime.

Schrödinger case, the lowest angular momentum subspace of H_D contains two channels corresponding to $K = \pm(d-1)/2$. As a result, not one but two copies of geometric ladders of the form (3.11) appear at $\beta = \beta_c$ (see Fig. 1). These two ladders may be degenerate or intertwined depending on the choice of boundary condition in (3.8).

The breaking of the degeneracy between the ladders is directly related to the breaking of a symmetry. To understand this point more explicitly consider the case where $d = 2$. There, in a basis where $\gamma^0 = \sigma_z$, $\gamma^1 = i\sigma_1$, $\gamma^2 = -i\sigma_2$, H_D is given by

$$H_D = \sigma_i p_i - \beta/r. \tag{3.13}$$

From (3.13) it is seen that H_D is symmetric under the following parity transformation

$$x \to -x,\ y \to y,\ H_D \to \sigma_2 H_D \sigma_2, \tag{3.14}$$

which in terms of $\Psi_1(r)$, $\Psi_2(r)$ is equivalent to [69]

$$\Psi_1(r) \to \Psi_2(r),\ \Psi_2(r) \to -\Psi_1(r),\ m \to -m-1 \tag{3.15}$$

where m is the orbital angular momentum. Consequently, the Dirac equation (3.4) is invariant under (3.15), however, the boundary condition (3.8) can break (3.15). Typical choices of boundary conditions are

1. Continuously connected constant potential $V(r < L) = -\beta/L$ [70] corresponding to $h = J_{m+1}(\beta + EL)/J_m(\beta + EL)$, where $J_n(x)$ is Bessel's function.
2. Zero wavefunction of one of the spinor components [81] corresponding to $h = 0$ or $h = \infty$.
3. Infinite mass term on boundary [71] corresponding to $h = 1$.
4. Chiral boundary conditions [68]

$$h = \begin{cases} 0 & m \geq 1 \\ \infty & m \leq 0 \end{cases} \tag{3.16}$$

inducing a zero mode localized at the boundary.

Under (3.15), a solution of the Dirac equation with angular momentum m obeying boundary condition (3.8) will transform into a different solution with angular momentum $-m-1$ obeying (3.8) with $h \to -h^{-1}$. Thus, the boundary condition respects parity if and only if

$$h_m = -h^{-1}_{-m-1}. \tag{3.17}$$

Thus case 1 above preserves parity while 2–4 break parity. If (3.17) holds, transformation (3.15) links between the $m \leftrightarrow -m-1$ eigenspace solutions. The lowest angular momentum subspaces correspond to orbital angular momentum $m = 0, -1$. If (3.17) holds, then the two geometric ladders (3.11) associated with $m = 0, -1$ are degenerate. The reason is that, as seen in (3.7), under (3.15)

$$a_{11} \to a_{12},\ a_{12} \to -a_{11},\ a_{21} \to a_{22},\ a_{22} \to -a_{21}$$
$$h \to -h^{-1} \tag{3.18}$$

which render z_0 in (3.10) and consequently E_0 invariant. Thus $E_{0,m=\pm 1/2}$ are identical in this case. If (3.17) does not hold, this symmetry is not enforced and the degeneracy between the ladders is broken.

The visualization of parity breaking is displayed in Fig. 1 where the density of states $\rho(E)$ of H_D is plotted for the $m = 0, -1$ channels and $\beta > \beta_c$. The boundary

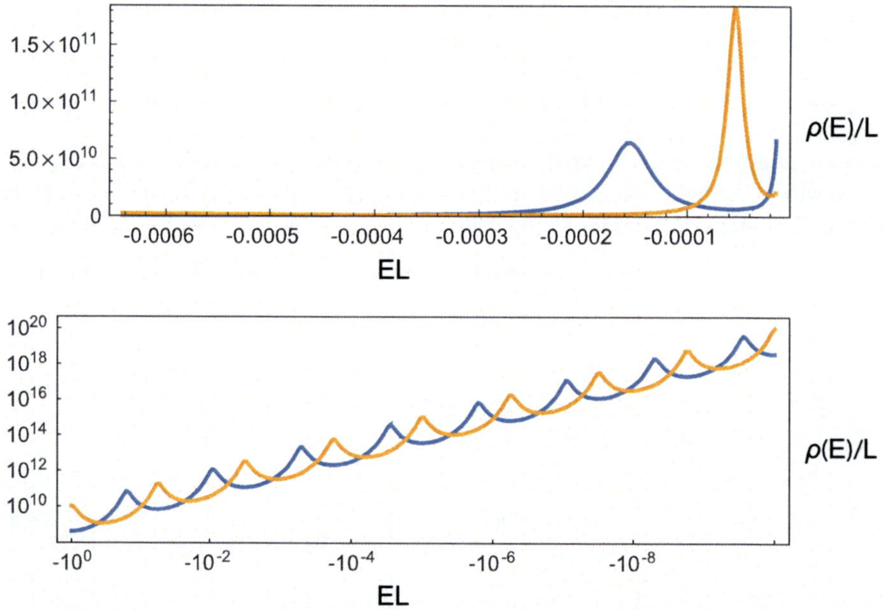

Fig. 1 Density of states $\rho(E)$ of H_D for $d = 2$, $\beta = 1.2 > \beta_c$ and different angular momentum eigenstates. The yellow and blue curves correspond to the $m = 0, -1$ angular momentum channels respectively. The boundary condition h used here is the chiral boundary condition (3.16) which breaks parity. The parameter L is the short distance cutoff taken here to be 0.195 nm. The numeric values on both axis are in units of $\hbar c = 0.197\,\text{eV}\,\mu\text{m}$. The set of pronounced peaks in both curves describes the quasi-bound spectrum in the overcritical regime $\beta > \beta_c$ as calculated in (3.11). The lower panel displays the detailed structure of the infinite geometric ladders of the quasi-bound states in a logarithmic scale. The $m = 0, -1$ ladders are intertwined, indicative of the breaking of parity by the boundary condition. These results are independent of the specific choice of L or h (provided that it breaks parity)

condition that was used in Fig. 1 is the chiral boundary condition (3.16) which breaks parity. Both curves exhibit an identical set of pronounced peaks condensing near $E = 0^-$. These peaks describe quasi-bound states (3.11) and, accordingly, are arranged in a set of two geometric ladders. The separation between the ladder is a distinct signal of parity breaking.

3.3 Experimental Realization

The CSI to DSI transition has recently received further validity and interest due to a detailed experimental observation in graphene [69]. In what follows, we summarize the results of this observation and emphasize its most significant features.

Graphene is a particularly interesting condensed matter system where H_D is relevant (for $d = 2$). The basic reason for this argument is that low energy excitations in graphene behave as a massless Dirac fermion field with a linear dispersion $E = \pm v_F |p|$ where the Fermi velocity $v_F \approx 10^6$ m/s appears instead of c [48]. These characteristics have been extensively exploited to make graphene a useful platform to emulate specific features of quantum field theory, topology and quantum electrodynamics (QED) [49, 60, 80–82, 89, 91], since an effective fine structure constant $\alpha_G = e^2/\hbar v_F$ of order unity is obtained by replacing the velocity of light c by v_F.

It has been shown that single-atom vacancies in graphene can host a local and stable charge [55, 57, 69]. This charge can be modified and measured at the vacancy site by means of scanning tunneling spectroscopy and Landau level spectroscopy [57]. The presence of massless Dirac excitations in the vicinity of the vacancy charge motivates the assumption that these will interact in a way that can be described by a massless Dirac Coulomb system. Particularly, the low energy spectral features of the charged vacancy would be the same as that of a tunable Coulomb source. The experimental results of [69] provide confirmation of this hypothesis as will be detailed below.

The measurements and data analysis presented below were carried out as follows: positive charges are gradually increased into an initially prepared single atom vacancy in graphene. Using a scanning tunneling microscope (STM) the differential conductance $dI/dV\,(V)$ through the STM tip is measured at each charge increment at the vacancy site. The conductance $dI/dV\,(V)$ is expected to be proportional to the local density of states of the system [2, 69]. Thus, quasi-bound states should also appear as pronounced peaks in the dI/dV curves.

For low enough values of the charge, the differential conductance displayed in Fig. 2b, shows the existence of a single quasi-bound state resonance whose distance from the Dirac point increases with charge. The behaviour close to the Dirac point, is very similar to the theoretical prediction of the under-critical regime $\beta < \beta_c$ displayed in Fig. 2a. The β value associated with the data of Fig. 2b is obtained from matching the position of the quasi-bound state with the theoretical model where the cutoff L and the boundary condition h are fixed model parameters that will

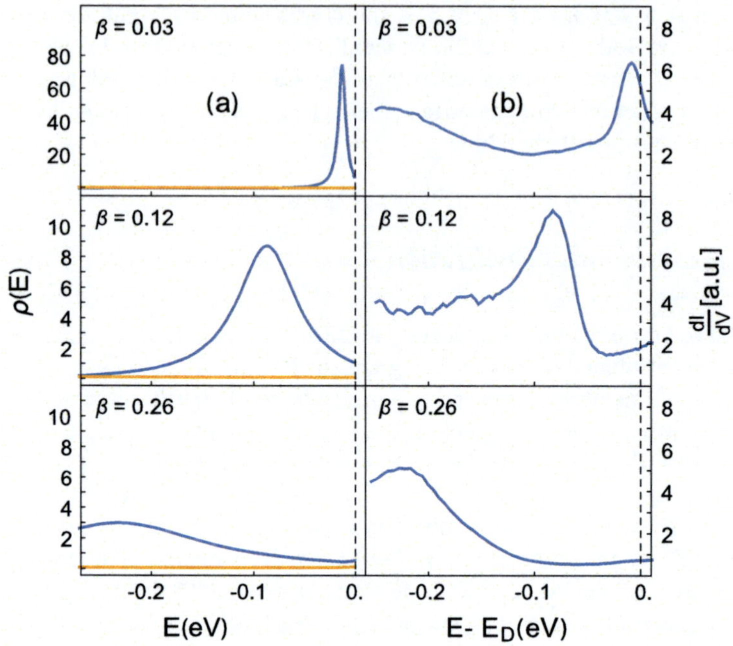

Fig. 2 Experimental and theoretical picture in the undercritical regime. (**a**) Theoretical behaviour of the density of states $\rho(E)$ of the Dirac Hamiltonian H_D in (3.1) with $d = 2$, $c \to v_F = 0.003c$ and angular momentum channels $m = -1$ (blue) and $m = 0$ (yellow). The cutoff and boundary conditions are assigned here with the optimized values $L = 0.195\,\text{nm}$, $h = -0.85(m+1)$ as explained in the text. The $m = -1$ (blue) branch contains a single peak and the $m = 0$ (yellow) branch shows no peak. While increasing β, the resonance shifts to lower energy and becomes broader. (**b**) The conductance dI/dV measured at a single vacancy in graphene using STM as a function of the applied voltage V. The determination of the parameter β is obtained from matching the position of the peak in the dI/dV curve with the theoretical model where the cutoff L and the boundary condition h are fixed model parameters

be given later. The theoretical position of the under-critical quasi-bound state as a function of β is displayed in Fig. 4 along with the positions of the peak extracted from measurements. The existence of a quasi-bound state does not contradict CSI of the undercritical phase since the absence of any states occurs only in the low energy limit.

At the point where the build up charge exceeds a certain value, three additional resonances emerge out of the Dirac point. These resonances are interpreted as the lowest overcritical ($\beta > 1/2$) resonances which we denote E_1, E_1', E_2 respectively. The corresponding theoretical and experimental behaviours displayed in Figs. 1 and 3, show a very good qualitative agreement. To achieve a quantitative comparison solely based on the massless Dirac Coulomb Hamiltonian (3.13), the theoretical β values corresponding to the respective positions of the lowest overcritical experimental resonance E_1 (as demonstrated in Fig. 3) are deduced for fixed L and

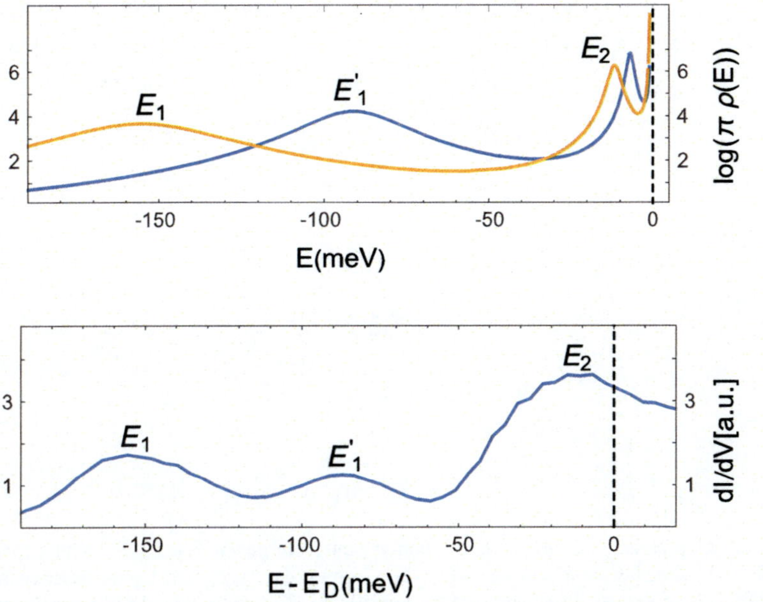

Fig. 3 Experimental and theoretical picture in the overcritical regime. Upper plot: Theoretical behaviour of the low energy density of states $\rho(E)$ for overcritical $\beta = 1.33$. The Blue (Yellow) line corresponds to $m = -1$ ($m = 0$) orbital angular momentum. The peaks on the vertical scale describe the first quasi-bound states with two (Blue and Yellow) infinite geometric towers of states. Lower plot: Experimental values of the tunnelling conductance measured at the charged vacancy site in graphene

the boundary condition h (as before). This allows to determine the lowest branch $E_1(\beta)$ represented in Fig. 4. Then, the experimental points E_1', E_2 are now free points to be directly compared to their corresponding theoretical branch as seen in Fig. 4. Parameters L and h, are determined according to the ansatz $h = a(m+1)$, and correspond to optimal values of $L = 0.195$ nm, $a \simeq -0.85$. The comparison of the experimental E_2/E_1 ratio with the universal prediction $E_{n+1}/E_n = e^{-\pi/\sqrt{\beta^2-1/4}}$ is given in Fig. 5. A trend-line of the form $e^{-b/\sqrt{\beta^2-1/4}}$ is fitted to the ratios E_2/E_1 yielding a statistical value of $b = 3.145$ with standard error of $\Delta b = 0.06$ consistent with the predicted value π. An error of $\pm 1meV$ is assumed for the position of the energy resonances.

A few further comments are appropriate:

1. The points on the $E_2(\beta)$ curve follow very closely the theoretical prediction $E_{n+1}/E_n = e^{-\pi/\sqrt{\beta^2-1/4}}$. This result is relatively insensitive to the choice of h, L.
2. In contrast, the correspondence between the $E_1'(\beta)$ points and the theoretical branch is sensitive to the choice of h, L. This reflects the fact that while each geometric ladder is of the form (3.11), the energy scale E_0 is different between

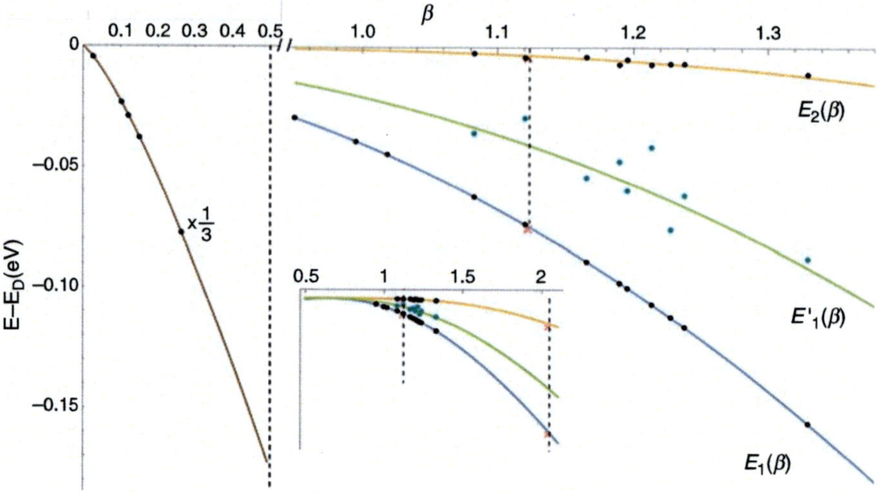

Fig. 4 Comparison of lowest quasi bound state energy curves $E_n(\beta)$ with experimentally measured tunneling conductance peaks. The curves $E_1(\beta)$, $E_1'(\beta)$, $E_2(\beta)$ describe resonances extracted from the density of states of the $m = 0$ (E_1, E_2) and $m = -1$ (E_1') angular momentum channels. E_1, E_2 and E_1' are the first quasi bound states appearing for $\beta > 1/2$ in the $m = 0, -1$ channels respectively. The brown curve is the position of the single under-critical quasi bound state as a function of $\beta < \beta_c$ scaled by a factor of 1/3 in the vertical axis. The black and cyan dots correspond to the positions of the tunneling conductance peaks as measured in graphene. The determination of the β value associated with these points is obtained from matching the position of the single under-critical peak and first over-critical peak (E_1) in the dI/dV curves with the theoretical model where the cutoff L and boundary condition h are fixed parameters. The two pink x's are the values of Efimov energy states as measured in Caesium atoms [42, 51] and rescaled by an appropriate overall factor. These points corresponds to the (overcritical) fixed Efimov value $\beta_E = 1.1236$. Similarly, additional experimental points obtained in [72, 87] are displayed in the inset

the E_1 (β) and E_1' (β) channels thus leading to a shifted relative position. The ansatz taken for h is phenomenological, however, in order to get reasonable correspondence to theory, the explicit dependence on m is needed. More importantly, it is necessary to use a parity breaking boundary condition (see Sect. 3.2) to describe the $E_1'(\beta)$ points, otherwise, both angular momentum channels E_1 (β), E_1' (β) will become degenerate and there would be no theoretical line to describe the E_1' (β) points. The existence of the experimental E_1' (β) branch is therefore a distinct signal that parity symmetry in the corresponding Dirac description is broken. In graphene, exchanging the triangular sub-lattices is equivalent to a parity transformation. Creating a vacancy breaks the symmetry between the two sub-lattices and is therefore at the origin of broken parity in the Dirac model.

3. The optimal value obtained for the short distance cutoff $L = 0.195$ nm is fully consistent with the low energy requirement $E_1 L/\hbar v_F \simeq 0.03 \ll 1$ necessary to be in the regime relevant to observe the β-driven QPT. Furthermore, it is quite close the lattice spacing of graphene (≈ 0.15 nm)

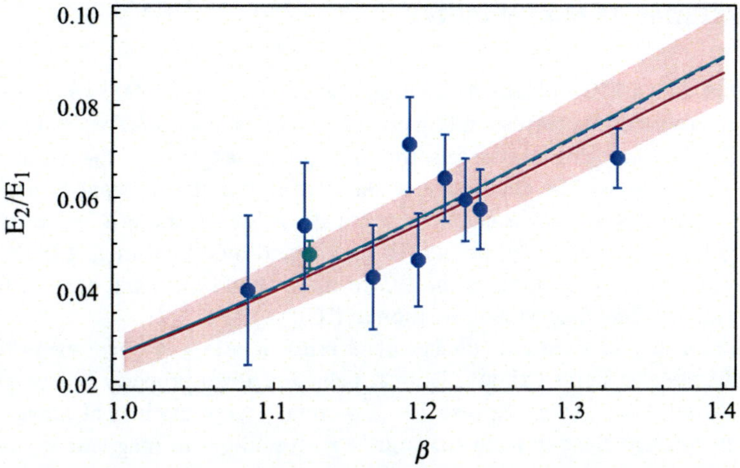

Fig. 5 Comparison between the experimentally obtained E_2/E_1 ratio and the universal factor $e^{-\pi/\sqrt{\beta^2-1/4}}$. Blue points: the ratio E_2/E_1 obtained from the position of the points in Fig. 4. Green point: Universal Efimov energy ratio as measured in Caesium atoms [42, 51]. Blue line (dashed): the corresponding optimized curve, fitted according to the model $e^{-b/\sqrt{\beta^2-1/4}}$ and corresponding to $b = 3.145$ with standard error of $\Delta b = 0.06$ consistent with the predicted value π. The shaded pink region is the $\pm 2\Delta b$ confidence interval of the curve. Cyan line: universal low energy factor $e^{-\pi/\sqrt{\beta^2-1/4}}$. Purple line: theoretical ratio E_2/E_1 obtained from the exact solution of the Dirac equation. As $\beta \to 0.5$, $|E_n|$ becomes smaller therefore the green and purple curves coincide for low β. The error bar on the resonance energies is ± 1 meV

One of the most interesting features of observed quasi bound states is their similarity with the Efimov spectrum. As discussed in Sect. 2.2, Efimov states are a geometric tower of states with a fixed geometric factor which is derived from an effective Schrödinger equation with a $V = -\lambda/r^2$ potential (as in (2.1)) and overcritical potential strength $\lambda = s_0 + 1/4$, $s_0 \approx 1.00624$. To emphasize the similarities between the Dirac quasi bound spectrum and the Efimov spectrum or, more generally, between the CSI to DSI transition in the Dirac and Schrödinger Hamiltonians H_D, H_S, two additional experimental points (pink x's) are presented in Fig. 4. These points are the values of Efimov energy states measured in Caesium atoms [42, 51] and scaled with an appropriate overall factor. The points are placed at the (overcritical) fixed Efimov value $\beta_{\mathrm{E}} = 1.1236$ corresponding to the geometric factor of Efimov states. The universality of the transition is thereby emphasized in Fig. 4 in which curves calculated from a massless Dirac Hamiltonian, energy positions of tunneling conductance peaks in graphene and resonances of a gas of Caesium atoms are combined in a meaningful context.

4 Relation to Universality

In Sects. 2 and 3 we obtained the properties of the CSI to DSI transition from a direct analysis of the corresponding eigenstates of each system. In what follows, we describe the same physics, but this time through the language of the renormalization group (RG). As will be detailed next, the description of this phenomenon in a RG picture provides a notable example of a case in which there is universality even in the absence of any fixed points. To understand this point more clearly, we first recall the physical meaning of the RG formalism and the usual context for which universality is understood with relation to RG.

Universality is a central concept of physics. It refers to phenomena for which very different systems exhibit identical behavior when properly coarse-grained to large distance (or low energy) scales. Important representatives of universality are systems that are close to a critical point, e.g., liquid-gas or magnetic systems. Near the critical point, these systems exhibit continuous scale invariance (as in (1.1)) where the free energy and correlation length vary as a power of the temperature (or some other control parameter). The exponents of these functions are real valued and are identical for a set of different systems thereby constituting a "universality class".

The contemporary understanding of university in critical phenomena is provided by the tools of RG and effective theory. In the framework of the later, low energy physics is described by a Hamiltonian H with a series of interaction terms $g_n O_n$ constrained by symmetries. Intrinsic to this description is an ultraviolet cutoff Λ reflecting the conceptual idea that H is obtained from some microscopic Hamiltonian H_0 by integrating out degrees of freedom with length scale shorter than $1/\Lambda$. The dependence of $\vec{g} \equiv (g_1, g_2 \ldots)$ on Λ defines the RG space of parameters $\vec{g}(\Lambda)$ which represent a large set of Hamiltonians $H(\vec{g}(\Lambda))$. Within this picture, the scale invariant character of critical phenomena is attributed to the case where $H_0(\vec{g}_0)$ flows in the infrared limit, $\Lambda \to 0$, to $H(\vec{g}_*)$ where $\vec{g}_*$ is a fixed point. Additionally, universality classes arise since trajectories starting at distinct positions on RG space can flow to the same fixed point for $\Lambda \to 0$. The role of RG fixed points in the description of universality, effective theory and scale invariance is central and extends throughout broad sub-fields in physics.

4.1 Renormalization Group Formalism for the Schrödinger $1/r^2$ Potential

The RG picture which describes the low energy physics of the Schrödinger $-\lambda/r^2$ potential in the $\lambda > \lambda_c$ regime cannot be associated with a fixed point because of the absence of CSI. However, even without fixed points, we expect universality to appear in this regime since the geometric series factor $E_{n+1}/E_n = \exp\left(-2\pi/\sqrt{\lambda - \lambda_c}\right)$ is independent of the short distance parameters associated with the cutoff L and the boundary condition g.

To see this explicitly [5, 10, 47, 50, 63], consider the radial Schrödinger equation for H_S given by

$$-\left(\frac{d^2}{dr^2}+\frac{d-1}{r}\frac{d}{dr}-\frac{l(l+d-2)}{r^2}\right)-\frac{\lambda}{r^s}\psi(r)=E\psi(r), \quad L<r<\infty \tag{4.1}$$

where $\psi(r)$ is the radial wavefunction, l the orbital angular momentum, d the space dimension, L a short distance cutoff and $s = 2$ but remains implicit for a reason that will be clear shortly. A well defined eigenstate of (4.1) is obtained by imposing a boundary condition at $r = L$

$$L\frac{\psi'(L)}{\psi(L)}=g, \tag{4.2}$$

$g \in \mathbb{R}$, which encodes the short-distance physics. To initiate a RG transformation we transform

$$L \to L+dL \equiv \epsilon L; \quad 0<\epsilon-1 \ll 1 \tag{4.3}$$

and obtain an equivalent effective description with the short distance cut-off ϵL and correspondingly, a new boundary condition at $r = \epsilon L$:

$$\epsilon L\frac{\psi'(\epsilon L)}{\psi(\epsilon L)}=g(\epsilon L). \tag{4.4}$$

As a result of (4.3), Eq. (4.1) is now defined in the range $\epsilon L \le r < \infty$ with the same functional form. With the help of the rescaling $r' \equiv \epsilon^{-1}r$, $E' \equiv \epsilon^2 E$, Eq. (4.1) is modified to the equivalent form

$$-\left(\frac{d^2}{dr'^2}+\frac{d-1}{r'}\frac{d}{dr'}-\frac{l(l+d-2)}{r'^2}\right)-\frac{\lambda\epsilon^{2-s}}{r'^s}\psi(r')=E'\psi(r') \quad L<r'<\infty. \tag{4.5}$$

Thus, transformation (4.3) is accounted in (4.1) by $\lambda \to \lambda\epsilon^{2-s}$ and using (4.3) leads to the infinitesimal form

$$L\frac{d\lambda}{dL}=(2-s)\lambda. \tag{4.6}$$

Similarly, $g(\epsilon L)$ in (4.4) can be related to $g(L)$ as follows. The series expansion of $g(\epsilon L)$ in $\epsilon - 1$ is

$$g(\epsilon L)=L\frac{\psi'(L)}{\psi(L)}+(\epsilon-1)\left(L\frac{\psi'(L)}{\psi(L)}-L^2\left(\frac{\psi'(L)}{\psi(L)}\right)^2+L^2\frac{\psi''(L)}{\psi(L)}\right)+O(\epsilon-1)^2. \tag{4.7}$$

Manipulation of (4.7) by insertion of the radial Schrödinger equation (4.1) and the definition of $g(L)$ yield

$$g(\epsilon L) = g(L) + (\epsilon - 1)\Big((2-d)\,g(L) - g(L)^2 - \lambda L^{2-s} + l(l+d-2) - L^2 E\Big) \tag{4.8}$$

where terms of order $(\epsilon - 1)^2$ or higher were eliminated. The equivalent differential form is thus

$$L\frac{dg}{dL} = (2-d)\,g - g^2 - \lambda L^{2-s} + l(l+d-2) - L^2 E. \tag{4.9}$$

In the low energy regime

$$L^2\,|E| \ll |\lambda - l(l+d-2)| \tag{4.10}$$

Eq. (4.9) reduces to

$$L\frac{dg}{dL} = (2-d)\,g - g^2 - \lambda \tag{4.11}$$

where the orbital angular momentum was taken to be $l = 0$ and s set to $s = 2$ for brevity. Finally, the combination of (4.6) and (4.11) constitutes the RG equations

$$\begin{aligned} \beta(\lambda) &\equiv L\frac{d\lambda}{dL} = (2-s)\,\lambda \\ \beta(g) &\equiv L\frac{dg}{dL} = -(g - g_+)(g - g_-) \end{aligned} \tag{4.12}$$

where

$$g_\pm = \frac{2-d}{2} \pm \sqrt{\lambda_c - \lambda} \tag{4.13}$$

and $\lambda_c = (d-2)^2/4$.

Since $\beta(\lambda) = 0$ for $s = 2$, $\lambda(L)$ remains unchanged under the RG transformation. In contrary, the function $\beta(g)$ is not trivial and has two roots $g_\pm$. For $\lambda < \lambda_c$, the two roots correspond to two fixed points, g_- unstable and g_+ stable. However, as λ increases, the two fixed points get closer and merge for $\lambda = \lambda_c$. For $\lambda > \lambda_c$, $g_\pm$ become complex valued and the two fixed points vanish as can be seen in Fig. 6a. The solution for $g(L)$ in this regime is given explicitly by (see Fig. 6b)

$$g(L) = \frac{2-d}{2} - \sqrt{\lambda - \lambda_c}\tan\left[\sqrt{\lambda - \lambda_c}\ln(L/L_0) - \phi_g\right] \tag{4.14}$$

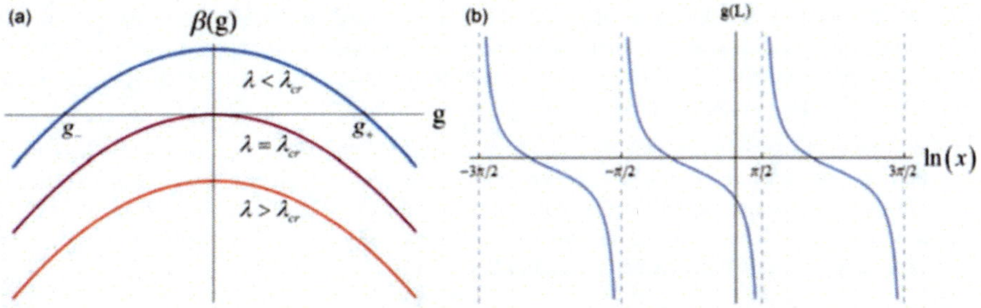

Fig. 6 Visualization of the renormalization group picture associated with the boundary condition $g(L)$ at the short distance cutoff $r = L$ for the case of the Schrödinger $V(r) = -\lambda/r^2$ potential H_S. **(a)** The $\beta(g)$ function in the over-critical and under-critical regimes. For $\lambda < \lambda_c$, $\beta(g)$ has two roots correspond to two fixed points, g_- unstable and g_+ stable. The point $\lambda = \lambda_c$ is a transition point where the roots merge into a single fixed point. For $\lambda > \lambda_c$ there are no real fixed points. **(b)** The behaviour of the boundary condition $g(L)$ in the overcritical regime $\lambda > \lambda_c$ and $d = 3$ as a function of $\ln(x)$ with $x \equiv \sqrt{\lambda - \lambda_c}\ln(L/L_0) - \phi_g$, $\phi_g \equiv \arctan\left(\frac{g_0 - \frac{2-d}{2}}{\sqrt{\lambda-\lambda_c}}\right)$. Independent of the initial condition $g_0(L_0)$, $g(L)$ is a log-periodic function of L which, as shown in (1.3), is a generic feature of DSI

where $\phi_g \equiv \arctan\left(\frac{g_0 - \frac{2-d}{2}}{\sqrt{\lambda-\lambda_c}}\right)$. Unlike the case of a fixed point, the flow of $g(L)$ in (4.14) does not terminate at any specific point but rather oscillate periodically in $\log L$ with period $L \to e^{\pi/\sqrt{\lambda-\lambda_c}}L$ independent of the initial condition $g(L_0) = g_0$.

The appearance of two fixed points for $\lambda < \lambda_c$, which annihilate at λ_c and give rise to a log-periodic flow for $\lambda > \lambda_c$ is the transcription of the CSI to DSI transition in the RG picture. The periodicity $e^{\pi/\sqrt{\lambda-\lambda_c}}$, being independent on the initial conditions, $g(L_0) = g_0$, represents a universal content even in the absence of fixed points.

An analogue of the RG equations (4.12) can be derived for the boundary condition $h(L)$ in (3.8) of the massless Dirac Coulomb system described in Sect. 3 [33].

5 Discussion

The similarities between the Dirac and Schrödinger system H_S, H_D

$$H_S = p^2 - \lambda/r^2 \tag{5.1}$$

$$H_D = \gamma^0\gamma^j p_j - \beta/r \tag{5.2}$$

presented in Sects. 2 and 3 motivate the study of whether a similar transition from CSI to DSI is possible for a generic class of systems and, if so, what are the common

Table 2 Comparison between the various cases discussed in the text for which a transition between a continuous scale invariant phase and a discrete scale invariant phase occurs. In the DSI regime each system is characterized by the sudden appearance of a geometric tower of modes with the universal form $O_n = O_0 \exp\left(-b\frac{\pi n}{\sqrt{x-x_c}}\right)$. In lines 1–3 of the table, O_n are one body bound states. Lines 4–5 describe many body quantum systems where O_n are fermion masses and 3-body bound states respectively. Line 6 provides a comparison with the Berezinskii–Kosterlitz–Thouless phase transition where the analog quantity for O_n is the free energy F for $T \gtrsim T_c$. In line 4, α_N is a N dependent real number whose exact value can be found in [15]. In line 5, $c_\pm$ are d dependent real positive number defined in Sect. 5.3

System	O_n	x	x_c	b
$H_S = p^2 - \lambda/r^2$	E_n	λ	$(d-2)^2/4$	2
$H_D = \gamma^0\gamma^j p_j - \beta/r$	E_n	β^2	$(d-1)^2/4$	1
$H_L = \left(-\frac{d^2}{dx^2}\right)^N - \frac{\lambda_L}{x^{2N}}$	E_n	λ_L	$\left(\frac{(2N-1)!!}{2^N}\right)^2$	$N\alpha_N$
QED3 with N massless flavours	m_n	N^{-1}	$\pi^2/32$	$\pi/\sqrt{8}$
Efimov effect in d dimensions	E_n	d	2.3	$1/c_\pm(d)$
BKT	F	T	T_c	System dependent

ingredients within this class. Below we briefly survey some other setups which interestingly give rise to a CSI to DSI transition. The relation between all these cases is summarized in Table 2.

5.1 Lifshitz Scaling Symmetry

Since H_S and H_D share the property that the power law form of the corresponding potential matches the order of the kinetic term, it is interesting to examine whether this property is a sufficient ingredient by considering a generalized class of one dimensional Hamiltonians,

$$H_L = \left(-\frac{d^2}{dx^2}\right)^N - \frac{\lambda_L}{x^{2N}}, \tag{5.3}$$

where N is a natural integer and λ_L a real valued coupling. The Hamiltonian H_L describes a quantum system with non-quadratic anisotropic scaling between space and time for $N > 1$. This so called "Lifshitz scaling symmetry" [6], manifest in (5.3), can be seen for example at the finite temperature multicritical points of certain materials [34, 41] or in strongly correlated electron systems [9, 30, 88]. Quartic dispersion relations $E \sim p^4$ can also be found in graphene bilayers [58] and heavy fermion metals [74]. It may also have applications in particle physics [6], cosmology [64] and quantum gravity [39, 40, 46].

The detailed solution of the corresponding Schrödinger equation $H_L\psi = E\psi$ [15] confirms that a transition from CSI to DSI occurs at $\lambda_{L,c} = (2N-1)!!^2/2^{2N}$, $\forall N \geq 1$. The CSI phase contains no low energy, $|E|^{1/2N}\, L \ll 1$ (L is a short distance cutoff), bound states and the DSI phase is characterized by an infinite set of bound states forming the geometric series

$$E_n = -E_0 e^{-\frac{N\alpha_N \pi n}{\sqrt{\lambda_L - \lambda_{L,c}}}}, \; 0 < \lambda_L - \lambda_{L,c} \ll 1 \tag{5.4}$$

where $E_0 > 0$ and α_N is an N dependent real number. For $\lambda_L - \lambda_{L,c} \to 0^+$, the analytic behavior of the spectrum is characteristic of the Berezinskii–Kosterlitz–Thouless (BKT) scaling in analogy with the $N = 1$ case. However, unlike the $N = 1$ case, the BKT scaling appears only for $\lambda_L - \lambda_{L,c} \to 0^+$. Deeper in the overcritical regime, the dependence on $\left(\lambda_L - \lambda_{L,c}\right)^{1/2}$ in (5.4) is replaced by a more complicated function of $\lambda_L - \lambda_{L,c}$ [15]. The transition as well as the value of $\lambda_{L,c}$ is independent of the short distance physics characterized by the boundary conditions and cutoff L.

Since H_L is a high order differential operator it requires the specification of several $x = L$ boundary condition parameters (unlike the one parameter g in Sect. 2.1) in order to render it as a well defined self-adjoint operator on the interval $L < x < \infty$. The most general choice of boundary conditions is parameterized by a unitary $N \times N$ matrix. Accordingly, the corresponding N^2 dimensional RG space is characterized by fixed points in the under-critical regime $\lambda_L < \lambda_{L,c}$. Interestingly, the DSI over-critical regime $\lambda_L > \lambda_{L,c}$ is not filled with an infinite number of cyclic flows such as represented in Fig. 6b. Instead, there is a ‘limit cycle’ [85], i.e., an isolated closed trajectory at which flows terminate [14] (see Fig. 7).

5.2 QED in 2 + 1 Dimensions and N Fermionic Flavors

The study of dynamical fermion mass generation in 2 + 1 dimensional quantum electrodynamics (QED) [8, 38] provides an interesting many body instance of the CSI to DSI transition. Consider the 2 + 1 dimensional QED Lagrangian

$$\mathcal{L} = i\bar{\Psi}\gamma^{\mu}\left(\partial_{\mu} - ieA_{\mu}\right)\Psi - \frac{1}{4}F_{\mu\nu}F^{\mu\nu} \tag{5.5}$$

where Ψ is a vector of N identical types of fermion fields with zero mass. In this theory, e^2 or alternatively $\alpha \equiv Ne^2/8$, is a dimension-full coupling. Analogous to the short distance cutoff L of Sects. 2, 3, and 5.1, α constitutes the only energy scale of the theory. Consequently, the low energy regime $E \ll \alpha$ can be shown to exhibit CSI.

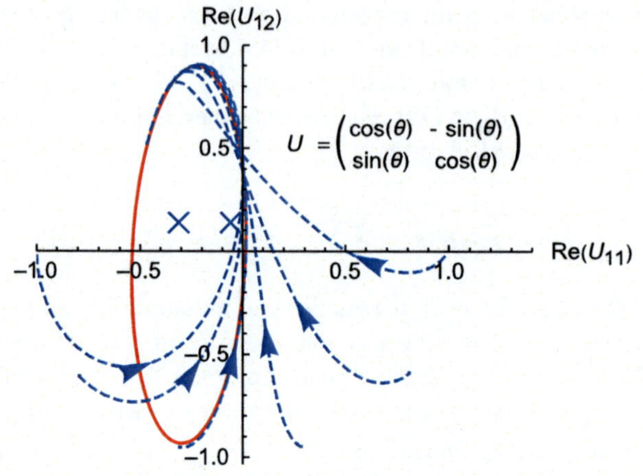

Fig. 7 A two dimensional projection of the (four dimensional) RG picture of the system $H = d_x^4 - 2/x^4$. The four boundary conditions at $x = L$ are parametrized by a unitary 2×2 matrix U. The initial conditions for the dashed blue flows are specified by choosing $\theta = -\pi, \ldots, -\pi/10, 0$ for the U matrix as displayed. All the trajectories flow towards a limit cycle. There exists a non-unitary fixed point, denoted by the blue cross, which is enclosed by the cycle when projected down onto any two dimensional subspace

To understand whether or not fermion mass appears as a result of quantum fluctuations it is required to calculate the fermion propagator, specifically, the self-energy $\Sigma(p)$. Under a particular (non-perturbative) approximation scheme [8], the expression for $\Sigma(p)$ can be extracted from the solution of the following differential equation

$$-\Sigma''(p) - \frac{2}{p}\Sigma'(p) - \frac{\lambda_Q}{p^2 + \Sigma(p)^2}\Sigma(p) = 0, \quad 0 < p < \alpha \tag{5.6}$$

with boundary condition

$$\alpha\frac{\Sigma'(\alpha)}{\Sigma(\alpha)} = -1 \tag{5.7}$$

where $\lambda_Q \equiv 8/(\pi^2 N)$. Close to a transition point the fermion mass and thereby $\Sigma(p)$ are non-zero but arbitrarily small such that $\Sigma(p) \ll p < \alpha$. As a result, (5.6) can be further approximated by assuming $\Sigma(p)^2$ in the denominator is a constant which we define as $\Sigma(p)^2 \to m^2/\lambda_Q$. Expanding to order m^2 yields

$$-\Sigma''(p) - \frac{2}{p}\Sigma'(p) - \frac{\lambda_Q}{p^2}\Sigma(p) = -\frac{m^2}{p^4}\Sigma(p). \tag{5.8}$$

A closer look on Eqs. (5.7) and (5.8) reveals that they are the same as the radial form of the Schrödinger equation with a $V = -\lambda/r^2$ potential

$$-\left(\frac{d^2}{dr^2}+\frac{d-1}{r}\frac{d}{dr}-\frac{l(l+d-2)}{r^2}\right)-\frac{\lambda}{r^2}\psi(r) = -k^2\psi(r),\quad L<r<\infty \tag{5.9}$$

$$L\frac{\psi'(L)}{\psi(L)} = g \tag{5.10}$$

where $k = \sqrt{-E}$ as described in Sect. 2 and in Eqs. (4.1) and (4.2). To see this explicitly, we rewrite (5.7) and (5.8) in terms of $r \equiv 1/p$, $\psi(r) \equiv \Sigma(p)$, $L \equiv 1/\alpha$ which then yields

$$-\psi''(r) - \frac{\lambda_Q}{r^2}\psi(r) = -m^2\psi(r),\quad L<r<\infty \tag{5.11}$$

$$L\frac{\psi'(L)}{\psi(L)} = 1. \tag{5.12}$$

Thus, the appearance of a non-vanishing fermion self energy constitutes a system of the form (5.9) and (5.10) with $d = 1$, $\lambda = \lambda_Q$ and $g = 1$. The resulting implication is that a transition from a CSI to DSI occurs at $\lambda_{Q,c} = 1/4$. For $\lambda_Q < 1/4$ there will be no $\Sigma(p) \neq 0$ solution for the self-energy in the $m/\alpha \ll 1$ regime. However, once λ_Q exceeds $\lambda_{Q,c} = 1/4$ an infinite geometric tower of possible non-trivial self-energy solutions appears with eigenvalues

$$m_{n+1}/m_n = e^{-\frac{\pi}{\sqrt{\lambda_Q-\lambda_{Q,c}}}}. \tag{5.13}$$

The critical point $\lambda_{Q,c} = 1/4$ corresponds to a critical fermion number $N_c = 32/\pi^2$ for which the DSI regime is $N < N_c$. In these term, (5.13) reduces to

$$m_{n+1}/m_n = e^{-\frac{\pi}{\sqrt{\frac{1}{N}-\frac{\pi^2}{32}}}\pi/\sqrt{8}}. \tag{5.14}$$

5.3 *Efimov Effect in d Dimensions*

As described in Sect. 2.2, the Efimov effect [12, 23, 24] is a remarkable phenomenon in which three particles form an infinite geometric ladder of low energy bound states. The effect occurs when at least two of the three pairs interact with a range that is small compared to the scattering length. It can be shown that the Efimov effect is possible only for space dimensions $2.3 < d < 3.76$ [66] which essentially limits the phenomenon to 3 dimensions. Interestingly, in the case where d is allowed to be tuned continuously, two CSI to DSI transitions are initiated at the critical dimensions

$d_- = 2.3$, $d_+ = 3.76$ [61]. In what follows we outline the main features of this result.

Low energy 3-body observables of locally interacting identical bosons can be described by an effective field theory with Lagrangian

$$\mathcal{L} = \psi^\dagger \left(i\frac{\partial}{\partial t} + \frac{1}{2}\nabla^2 \right) \psi + \frac{g_2}{4}\left(\Delta^\dagger \Delta\right) - \frac{g_2}{4}\left(\Delta^\dagger \psi^2 + \psi^{\dagger 2}\Delta\right) - \frac{g_3}{36}\left(\Delta^\dagger \Delta \psi^\dagger \psi\right) \tag{5.15}$$

where ψ is a non-relativistic bosonic atom field, Δ is a non dynamical 'diatom' field annihilating two atoms at one point and g_2, g_3 are bare 2-body and 3-body couplings respectively. With the diatom field and these interaction terms, it is possible to reproduce the physics of the Efimov effect [11]. The main ingredient of this procedure is the diagramatic calculation of the atom-diatom scattering amplitude as shown in Fig. 8. The self-consistent equations described in Fig. 8 leads to the following approximate relation for the s-wave atom-diatom amplitude A_s

$$A_s(p) = -\left(\frac{4}{3}\right)^{\frac{d-2}{2}} \frac{4\sin\left(\frac{d}{2}\pi\right)}{\pi} \int_0^\infty dq \frac{q}{p^2+q^2}\, {}_2F_1\left(\begin{matrix}\frac{1}{2} & 1 \\ \frac{d}{2} & \end{matrix}; \frac{p^2q^2}{p^2+q^2}\right) A_s(q) \tag{5.16}$$

Since there are no dimension-full parameters in (5.16) we are, once again, faced with a CSI equation, in analogy with the characteristics of equations (2.2), (3.3), (5.3), and (5.8). By inserting the ansatz $A_s(p) = p^{s-1}$, two possible solutions for (5.16) are obtained

$$A_s \approx a_1 p^{\sqrt{s^2}-1} + a_2 p^{-\sqrt{s^2}-1} \tag{5.17}$$

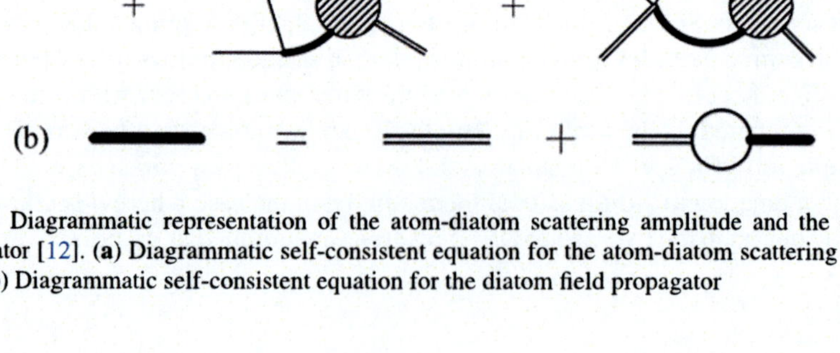

Fig. 8 Diagrammatic representation of the atom-diatom scattering amplitude and the diatom propagator [12]. (**a**) Diagrammatic self-consistent equation for the atom-diatom scattering amplitude. (**b**) Diagrammatic self-consistent equation for the diatom field propagator

where $s^2(d)$ is a solution of the $s \to -s$ invariant equation

$$2\sin\left(\frac{d\pi}{2}\right) {}_2F_1\left(\begin{matrix}\frac{d-1+s}{2} & \frac{d-1-s}{2} \\ & \frac{d}{2}\end{matrix}; \frac{1}{4}\right) + \cos\left(\frac{s}{2}\pi\right) = 0. \tag{5.18}$$

The numerical solution $s^2(d)$ of (5.18) shows that near $d = d_\pm$, $s^2(d_\pm) = 0$, $\partial_d s^2(d_-) < 0$, $\partial_d s^2(d_+) > 0$ and it is analytic. Consequently, near the critical dimensions $d_\pm$

$$s^2(d) = \pm c_\pm^2 (d - d_\pm) + O(d - d_\pm) \tag{5.19}$$

with $c_\pm > 0$. The insertion of (5.19) into (5.17) imply a CSI to DSI transition from real to complex valued power law behaviour of A_s. The DSI regime $d_- < d < d_+$ is consistent with the strip within which the Efimov states appear. Consequently, close to the critical points $d = d_\pm$, $A_s(p)$ in (5.17) obeys the following DSI scaling relation (as in (1.1))

$$A_s\left(e^{\frac{\pi n}{c_\pm\sqrt{|d-d_\pm|}}} p\right) = e^{-\frac{\pi n}{c_\pm\sqrt{|d-d_\pm|}}} A_s(p). \tag{5.20}$$

The corresponding RG equation for the couplings is

$$\Lambda\frac{d}{d\Lambda}G = \frac{1-s^2(d)}{2}(G - G_-)(G - G_+) \tag{5.21}$$

where $G(\Lambda) \equiv \Lambda^2 g_3(\Lambda)/9g_2(\Lambda)^2$, Λ is a UV cutoff and

$$G_\pm \equiv -\left(1 \pm \sqrt{s^2(d)}\right) / \left(1 \mp \sqrt{s^2(d)}\right). \tag{5.22}$$

In accordance with the RG picture detailed in Sect. 4.1, the insertion of (5.19) shows that the β-function of G contains two fixed points outside the strip $d_- < d < d_+$ which annihilate at $d = d_\pm$.

6 Summary

The breaking of continuous scale invariance (CSI) into discrete scale invariance (DSI) is a rich phenomenon with roots in multiple fields in physics. Theoretically, this transition plays an important role in various fundamental quantum systems such as the inverse-squared potential (Sect. 2), the massless hydrogen atom (Sect. 3), 2+1 dimensional quantum electrodynamics (Sect. 5.2) and the Efimov effect (Sects. 2.2 and 5.3). This CSI to DSI transition constitutes a quantum phase transition which appears for single body and strongly coupled many body systems and extends through non-relativistic, relativistic and Lifshitz dispersion relations. In a RG picture the transition describes universal low energy physics without fixed points and constitutes a physical realization of a limit cycle. Remarkably, the features of this transition have been measured recently in various systems such as cold atoms, graphene and Fermi gases [20]. In the DSI phase, the dependence of the geometric ladder of states on the control parameter (see Table 2) is in the class of Berezinskii–Kosterlitz–Thouless transitions. This provides an interesting, yet to be studied, bridge between DSI and two dimensional systems associated with BKT physics.

The characteristics described above provide the motivation to further study the ingredients associated with CSI to DSI transitions and we expect that these transitions will have an increasingly important role across the physics community in the future.

More generally, it will be interesting to understand how this phenomenon relate to recent realization of fractal structure in theories of quantum gravitation [75–78] or to the characteristics of systems that are placed on an explicit fractal space [1, 3, 4, 7, 18, 22, 25, 26, 83, 84, 86].

Acknowledgments This work was supported by the Israel Science Foundation Grant No. 924/09 and by the Pazy Foundation.

References

1. Akkermans, E.: Statistical mechanics and quantum fields on fractals. In: D. Carfí, M.L. Lapidus, E.P.J. Pearse, M. van Frankenhuijsen (eds.) Fractal Geometry and Dynamical Systems in Pure and Applied Mathematics II: Fractals in Applied Mathematics, vol. 601, pp. 1–21. American Mathematical Society (AMS), Providence (2013). http://dx.doi.org/10.1090/conm/601/11962
2. Akkermans, E., Montambaux, G.: Mesoscopic Physics of Electrons and Photons, chap. 7. Cambridge University Press, Cambridge (2007). https://doi.org/10.1017/CBO9780511618833
3. Akkermans, E., Dunne, G.V., Teplyaev, A.: Thermodynamics of photons on fractals. Phys. Rev. Lett. **105**, 230,407 (2010). https://link.aps.org/doi/10.1103/PhysRevLett.105.230407
4. Akkermans, E., Dunne, G., Levy, E.: Wave propagation in one-dimension. In: Optics of Aperiodic Structures: Fundamentals and Device Applications, pp. 407–449. Pan Stanford Publishing, New York (2014)
5. Albeverio, S., Høegh-Krohn, R., Wu, T.T.: A class of exactly solvable three-body quantum mechanical problems and the universal low energy behavior. Phys. Lett. A **83**(3), 105–109 (1981). http://www.sciencedirect.com/science/article/pii/0375960181905077
6. Alexandre, J.: Lifshitz-type quantum field theories in particle physics. Int. J. Mod. Phys. **A26**, 4523–4541 (2011). https://doi.org/10.1142/S0217751X11054656
7. Alonso Ruiz, P.: Explicit formulas for heat kernels on diamond fractals. Commun. Math. Phys. **364**(3), 1305–1326 (2018). https://doi.org/10.1007/s00220-018-3221-x
8. Appelquist, T., Nash, D., Wijewardhana, L.C.R.: Critical behavior in (2+1)-dimensional QED. Phys. Rev. Lett. **60**, 2575–2578 (1988). http://link.aps.org/doi/10.1103/PhysRevLett.60.2575
9. Ardonne, E., Fendley, P., Fradkin, E.: Topological order and conformal quantum critical points. Ann. Phys. **310**, 493–551 (2004). https://doi.org/10.1016/j.aop.2004.01.004

10. Beane, S.R., Bedaque, P.F., Childress, L., Kryjevski, A., McGuire, J., van Kolck, U.: Singular potentials and limit cycles. Phys. Rev. A **64**, 042103 (2001). http://link.aps.org/doi/10.1103/PhysRevA.64.042103
11. Bedaque, P.F., Hammer, H.W., van Kolck, U.: Renormalization of the three-body system with short-range interactions. Phys. Rev. Lett. **82**, 463–467 (1999). https://link.aps.org/doi/10.1103/PhysRevLett.82.463
12. Braaten, E., Hammer, H.W.: Universality in few-body systems with large scattering length. Phys. Rep. **428**(5), 259–390 (2006). https://doi.org/10.1016/j.physrep.2006.03.001
13. Braaten, E., Phillips, D.: Renormalization-group limit cycle for the $1/r^2$ potential. Phys. Rev. A **70**, 052111 (2004). https://doi.org/10.1103/PhysRevA.70.052111
14. Brattan, D.K., Ovdat, O., Akkermans, E.: On the landscape of scale invariance in quantum mechanics. J. Phys. A Math. Theor. **51**(43), 435401 (2018). http://stacks.iop.org/1751-8121/51/i=43/a=435401
15. Brattan, D.K., Ovdat, O., Akkermans, E.: Scale anomaly of a lifshitz scalar: a universal quantum phase transition to discrete scale invariance. Phys. Rev. D **97**, 061701 (2018). https://doi.org/10.1103/PhysRevD.97.061701
16. Camblong, H.E., Epele, L.N., Fanchiotti, H., Garcia Canal, C.A.: Quantum anomaly in molecular physics. Phys. Rev. Lett. **87**, 220402 (2001). https://doi.org/10.1103/PhysRevLett.87.220402
17. Case, K.M.: Singular potentials. Phys. Rev. **80**(5), 797–806 (1950). https://doi.org/10.1103/PhysRev.80.797
18. Chen, J.P., Molchanov, S., Teplyaev, A.: Spectral dimension and bohr's formula for schrödinger operators on unbounded fractal spaces. J. Phys. A Math. Theor. **48**(39), 395203 (2015). https://doi.org/10.1088/1751-8113/48/39/395203
19. De Martino, A., Klöpfer, D., Matrasulov, D., Egger, R.: Electric-dipole-induced universality for dirac fermions in graphene. Phys. Rev. Lett. **112**, 186603 (2014). https://doi.org/10.1103/PhysRevLett.112.186603
20. Deng, S., Shi, Z.Y., Diao, P., Yu, Q., Zhai, H., Qi, R., Wu, H.: Observation of the efimovian expansion in scale-invariant fermi gases. Science **353**(6297), 371–374 (2016). https://doi.org/10.1126/science.aaf0666
21. Dong, S.H.: Wave Equations in Higher Dimensions. Springer, Berlin (2011)
22. Dunne, G.: Heat kernels and zeta functions on fractals. J. Phys. A Math. Theor. **45** 374016 (2012). https://doi.org/10.1088/1751-8113/45/37/374016
23. Efimov, V.: Energy levels arising from resonant two-body forces in a three-body system. Phys. Lett. B **33**(8), 563–564 (1970). https://doi.org/10.1016/0370-2693(70)90349-7
24. Efimov, V.: Weakly-bound states of three resonantly-interacting particles. Sov. J. Nucl. Phys **12**, 589–595 (1971). https://www.uibk.ac.at/exphys/ultracold/projects/levt/FourBodies/SovJNucPhys12.589.efimov.pdf
25. Fan, E., Khandker, Z., Strichartz, R.S.: Harmonic oscillators on infinite sierpinski gaskets. Commun. Math. Phys. **287**(1), 351–382 (2009). https://doi.org/10.1007/s00220-008-0633-z
26. Fernandes, R.M., Schmalian, J.: Complex critical exponents for percolation transitions in josephson-junction arrays, antiferromagnets, and interacting bosons. Phys. Rev. Lett. **106**, 067,004 (2011). https://doi.org/10.1103/PhysRevLett.106.067004
27. Fomin, P., Miransky, V.: On the dynamical vacuum rearrangement and the problem of fermion mass generation. Phys. Lett. B **64**(2), 166–168 (1976). https://doi.org/10.1016/0370-2693(76)90321-X
28. Fomin, P., Gusynin, V., Miransky, V.: Vacuum instability of massless electrodynamics and the gell-mann-low eigenvalue condition for the bare coupling constant. Phys. Lett. B **78**(1), 136–139 (1978). https://doi.org/10.1016/0370-2693(78)90366-0
29. Fomin, P.I., Gusynin, V.P., Miransky, V.A., Sitenko, Y.A.: Dynamical symmetry breaking and particle mass generation in gauge field theories. La Rivista del Nuovo Cimento (1978–1999) **6**(5), 1–90 (1983). https://doi.org/10.1007/BF02740014

30. Fradkin, E., Huse, D.A., Moessner, R., Oganesyan, V., Sondhi, S.L.: Bipartite Rokhsar–Kivelson points and Cantor deconfinement. Phys. Rev. B **69**, 224415 (2004). https://doi.org/10.1103/PhysRevB.69.224415
31. Friedrich, H.: Scattering Theory. Springer, Berlin (2013). https://doi.org/10.1007/978-3-662-48526-2
32. Gitman, D.M., Tyutin, I., Voronov, B.L.: Self-Adjoint Extensions in Quantum Mechanics: General Theory and Applications to Schrödinger and Dirac Equations with Singular Potentials, vol. 62. Springer, Berlin (2012)
33. Gorsky, A., Popov, F.: Atomic collapse in graphene and cyclic renormalization group flow. Phys. Rev. D **89**, 061702 (2014). https://doi.org/10.1103/PhysRevD.89.061702
34. Grinstein, G.: Anisotropic sine-gordon model and infinite-order phase transitions in three dimensions. Phys. Rev. B **23**, 4615–4630 (1981). https://link.aps.org/doi/10.1103/PhysRevB.23.4615
35. Gross, N., Shotan, Z., Kokkelmans, S., Khaykovich, L.: Observation of universality in ultracold ^{7}Li three-body recombination. Phys. Rev. Lett. **103**, 163202 (2009). https://doi.org/10.1103/PhysRevLett.103.163202
36. Gusynin, V.P., Schreiber, A.W., Sizer, T., Williams, A.G.: Chiral symmetry breaking in dimensionally regularized nonperturbative quenched QED. Phys. Rev. D **60**, 065007 (1999). https://link.aps.org/doi/10.1103/PhysRevD.60.065007
37. Hammer, H.W., Swingle, B.G.: On the limit cycle for the $1/r^2$ potential in momentum space. Ann. Phys. **321**(2), 306–317 (2006). https://doi.org/10.1016/j.aop.2005.04.017
38. Herbut, I.F.: Chiral symmetry breaking in three-dimensional quantum electrodynamics as fixed point annihilation. Phys. Rev. D **94**, 025036 (2016). https://link.aps.org/doi/10.1103/PhysRevD.94.025036
39. Horava, P.: Quantum gravity at a lifshitz point. Phys. Rev. **D79**, 084008 (2009). https://doi.org/10.1103/PhysRevD.79.084008
40. Horava, P.: Spectral dimension of the universe in quantum gravity at a Lifshitz point. Phys. Rev. Lett. **102**, 161301 (2009). https://doi.org/10.1103/PhysRevLett.102.161301
41. Hornreich, R.M., Luban, M., Shtrikman, S.: Critical behavior at the onset of $\vec{k}$-space instability on the λ line. Phys. Rev. Lett. **35**, 1678–1681 (1975). https://link.aps.org/doi/10.1103/PhysRevLett.35.1678
42. Huang, B., Sidorenkov, L.A., Grimm, R., Hutson, J.M.: Observation of the second triatomic resonance in Efimov's scenario. Phys. Rev. Lett. **112**, 190401 (2014). http://link.aps.org/doi/10.1103/PhysRevLett.112.190401
43. Jackiw, R.W.: Diverse Topics in Theoretical and Mathematical Physics. World Scientific, Singapore (1995)
44. Jensen, K.: Semi-holographic quantum criticality. Phys. Rev. Lett. **107**, 231601 (2011). https://doi.org/10.1103/PhysRevLett.107.231601
45. Jensen, K., Karch, A., Son, D.T., Thompson, E.G.: Holographic Berezinskii-Kosterlitz-Thouless transitions. Phys. Rev. Lett. **105**, 041601 (2010). https://doi.org/10.1103/PhysRevLett.105.041601
46. Kachru, S., Liu, X., Mulligan, M.: Gravity duals of Lifshitz-like fixed points. Phys. Rev. **D78**, 106005 (2008). https://doi.org/10.1103/PhysRevD.78.106005
47. Kaplan, D.B., Lee, J.W., Son, D.T., Stephanov, M.A.: Conformality lost. Phys. Rev. D **80**, 125005 (2009). http://link.aps.org/doi/10.1103/PhysRevD.80.125005
48. Katsnelson, M.I.: Graphene: Carbon in Two Dimensions. Cambridge University Press, New York (2012)
49. Katsnelson, M.I., Novoselov, K.S., Geim, A.K.: Chiral tunnelling and the Klein paradox in graphene. Nat. Phys. **2**(9), 620–625 (2006). http://dx.doi.org/10.1038/nphys384
50. Kolomeisky, E.B., Straley, J.P.: Universality classes for line-depinning transitions. Phys. Rev. B **46**, 12664–12674 (1992). http://link.aps.org/doi/10.1103/PhysRevB.46.12664

51. Kraemer, T., Mark, M., Waldburger, P., Danzl, J., Chin, C., Engeser, B., Lange, A., Pilch, K., Jaakkola, A., Nägerl, H.C., et al.: Evidence for Efimov quantum states in an ultracold gas of Caesium atoms. Nature **440**(7082), 315–318 (2006). https://doi.org/10.1038/nature04626
52. Kunitski, M., Zeller, S., Voigtsberger, J., Kalinin, A., Schmidt, L.P.H., Schöffler, M., Czasch, A., Schöllkopf, W., Grisenti, R.E., Jahnke, T., Blume, D., Dörner, R.: Observation of the Efimov state of the Helium trimer. Science **348**(6234), 551–555 (2015). https://doi.org/10.1126/science.aaa5601
53. Landau, L.D.: Quantum Mechanics: Non-relativistic Theory. Butterworth-Heinemann, Oxford (1991)
54. Lévy-Leblond, J.M.: Electron capture by polar molecules. Phys. Rev. **153**, 1–4 (1967). http://link.aps.org/doi/10.1103/PhysRev.153.1
55. Liu, Y., Weinert, M., Li, L.: Determining charge state of graphene vacancy by noncontact atomic force microscopy and first-principles calculations. Nanotechnology **26**(3), 035702 (2015)
56. Lompe, T., Ottenstein, T.B., Serwane, F., Wenz, A.N., Zürn, G., Jochim, S.: Radio-frequency association of Efimov trimers. Science **330**(6006), 940–944 (2010). https://doi.org/10.1126/science.1193148
57. Mao, J., Jiang, Y., Moldovan, D., Li, G., Watanabe, K., Taniguchi, T., Masir, M.R., Peeters, F.M., Andrei, E.Y.: Realization of a tunable artificial atom at a supercritically charged vacancy in graphene. Nat. Phys. **12**(6), 545–549 (2016). http://dx.doi.org/10.1038/nphys3665
58. McCann, E., Koshino, M.: The electronic properties of bilayer graphene. Rep. Prog. Phys. **76**(5), 056503 (2013). http://stacks.iop.org/0034-4885/76/i=5/a=056503
59. Meetz, K.: Singular potentials in nonrelativistic quantum mechanics. Il Nuovo Cimento (1955-1965) **34**(3), 690–708 (1964). http://dx.doi.org/10.1007/BF02750010
60. Miransky, V.: Dynamic mass generation and renormalizations in quantum field theories. Phys. Lett. B **91**(3), 421–424 (1980). https://doi.org/10.1016/0370-2693(80)91011-4
61. Mohapatra, A., Braaten, E.: Conformality lost in efimov physics. Phys. Rev. A **98**, 013633 (2018). https://link.aps.org/doi/10.1103/PhysRevA.98.013633
62. Moroz, S., Schmidt, R.: Nonrelativistic inverse square potential, scale anomaly, and complex extension. Ann. Phys. **325**(2), 491–513 (2010). https://doi.org/10.1016/j.aop.2009.10.002
63. Mueller, E.J., Ho, T.L.: Renormalization group limit cycles in quantum mechanical problems (2004). arXiv cond-mat/0403283. https://arxiv.org/pdf/cond-mat/0403283.pdf
64. Mukohyama, S.: Horava-Lifshitz cosmology: a review. Class. Quant. Grav. **27**, 223101 (2010). https://doi.org/10.1088/0264-9381/27/22/223101
65. Nakajima, S., Horikoshi, M., Mukaiyama, T., Naidon, P., Ueda, M.: Measurement of an Efimov trimer binding energy in a three-component mixture of ^{6}Li. Phys. Rev. Lett. **106**, 143201 (2011). http://link.aps.org/doi/10.1103/PhysRevLett.106.143201
66. Nielsen, E., Fedorov, D., Jensen, A., Garrido, E.: The three-body problem with short-range interactions. Phys. Rep. **347**(5), 373–459 (2001). https://doi.org/10.1016/S0370-1573(00)00107-1
67. Nisoli, C., Bishop, A.R.: Attractive inverse square potential, $U(1)$ gauge, and winding transitions. Phys. Rev. Lett. **112**, 070401 (2014). https://doi.org/10.1103/PhysRevLett.112.070401
68. Ovdat, O., Don, Y., Akkermans, E.: Vacancies in graphene: dirac physics and fractional vacuum charges (2018). arXiv:1807.10297. https://arxiv.org/pdf/1807.10297.pdf
69. Ovdat, O., Mao, J., Jiang, Y., Andrei, E.Y., Akkermans, E.: Observing a scale anomaly and a universal quantum phase transition in graphene. Nat. Commun. **8**(1), 507 (2017). https://doi.org/10.1038/s41467-017-00591-8
70. Pereira, V.M., Kotov, V.N., Castro Neto, A.H.: Supercritical Coulomb impurities in gapped graphene. Phys. Rev. B **78**, 085101 (2008). http://link.aps.org/doi/10.1103/PhysRevB.78.085101
71. Pereira, V.M., Nilsson, J., Castro Neto, A.H.: Coulomb impurity problem in graphene. Phys. Rev. Lett. **99**, 166802 (2007). http://link.aps.org/doi/10.1103/PhysRevLett.99.166802

72. Pires, R., Ulmanis, J., Häfner, S., Repp, M., Arias, A., Kuhnle, E.D., Weidemüller, M.: Observation of Efimov resonances in a mixture with extreme mass imbalance. Phys. Rev. Lett. **112**, 250404 (2014). http://link.aps.org/doi/10.1103/PhysRevLett.112.250404
73. Pollack, S.E., Dries, D., Hulet, R.G.: Universality in three-and four-body bound states of ultracold atoms. Science **326**(5960), 1683–1685 (2009). https://doi.org/10.1126/science.1182840
74. Ramires, A., Coleman, P., Nevidomskyy, A.H., Tsvelik, A.M.: β-YbAlB$_4$: a critical nodal metal. Phys. Rev. Lett. **109**(17), 176404 (2012). https://doi.org/10.1103/PhysRevLett.109.176404
75. Reuter, M., Saueressig, F.: Renormalization group flow of quantum gravity in the Einstein-Hilbert truncation. Phys. Rev. D **65**(6), 065016 (2002)
76. Reuter, M., Saueressig, F.: Fractal space-times under the microscope: a renormalization group view on monte carlo data. J. High Energy Phys. **2011**(12), 12 (2011). https://doi.org/10.1007/JHEP12(2011)012
77. Reuter, M., Saueressig, F.: Quantum Einstein gravity. New J. Phys. **14**, 055022 (2012). https://doi.org/10.1088/1367-2630/14/5/055022
78. Reuter, M., Saueressig, F.: Asymptotic safety, fractals, and cosmology. In: Quantum Gravity and Quantum Cosmology, pp. 185–226. Springer, Berlin (2013)
79. Scopa, S., Karevski, D.: One-dimensional bose gas driven by a slow time-dependent harmonic trap. J. Phys. A Math. Theor. **50**(42), 425301 (2017). https://doi.org/10.1088/1751-8121/aa890f
80. Shytov, A.V., Katsnelson, M.I., Levitov, L.S.: Atomic collapse and Quasi-Rydberg states in graphene. Phys. Rev. Lett. **99**, 246802 (2007). https://doi.org/10.1103/PhysRevLett.99.246802
81. Shytov, A.V., Katsnelson, M.I., Levitov, L.S.: Vacuum polarization and screening of supercritical impurities in graphene. Phys. Rev. Lett. **99**, 236801 (2007). http://link.aps.org/doi/10.1103/PhysRevLett.99.236801
82. Stander, N., Huard, B., Goldhaber-Gordon, D.: Evidence for Klein tunneling in graphene $p-n$ junctions. Phys. Rev. Lett. **102**, 026807 (2009). https://link.aps.org/doi/10.1103/PhysRevLett.102.026807
83. Steinhurst, B.A., Teplyaev, A.: Existence of a meromorphic extension of spectral zeta functions on fractals. Lett. Math. Phys. **103**(12), 1377–1388 (2013). https://doi.org/10.1007/s11005-013-0649-y
84. Strichartz, R.: A fractal quantum mechanical model with coulomb potential. Commun. Pure Appl. Anal. **8**, 743–755 (2008). https://doi.org/10.3934/cpaa.2009.8.743
85. Strogatz, S.H.: Nonlinear Dynamics and Chaos: With Applications to Physics, Biology, Chemistry, and Engineering. Westview Press, Boulder (2014)
86. Teplyaev, A.: Spectral zeta functions of fractals and the complex dynamics of polynomials. Trans. Am. Math. Soc. **359**(9), 4339–4358 (2007)
87. Tung, S.K., Jiménez-García, K., Johansen, J., Parker, C.V., Chin, C.: Geometric scaling of Efimov states in a ^{6}Li-^{133}Cs mixture. Phys. Rev. Lett. **113**, 240402 (2014). https://doi.org/10.1103/PhysRevLett.113.240402
88. Vishwanath, A., Balents, L., Senthil, T.: Quantum criticality and deconfinement in phase transitions between valence bond solids. Phys. Rev. B **69**, 224416 (2004). https://link.aps.org/doi/10.1103/PhysRevB.69.224416
89. Wang, Y., Wong, D., Shytov, A.V., Brar, V.W., Choi, S., Wu, Q., Tsai, H.Z., Regan, W., Zettl, A., Kawakami, R.K., Louie, S.G., Levitov, L.S., Crommie, M.F.: Observing atomic collapse resonances in artificial nuclei on graphene. Science **340**(6133), 734–737 (2013). http://science.sciencemag.org/content/340/6133/734
90. Yang, C.N.: Generalization of Sturm-Liouville theory to a system of ordinary differential equations with Dirac type spectrum. Commun. Math. Phys. **112**(1), 205–216 (1987). http://projecteuclid.org/euclid.cmp/1104159815
91. Zhang, Y., Tan, Y.W., Stormer, H.L., Kim, P.: Experimental observation of the quantum Hall effect and Berry's phase in graphene. Nature **438**(7065), 201–204 (2005). http://dx.doi.org/10.1038/nature04235

The Random Conductance Model with Heavy Tails on Nested Fractal Graphs

David A. Croydon

Abstract Recently, Kigami's resistance form framework has been applied to provide a general approach for deriving the scaling limits of random walks on graphs with a fractal scaling limit (Croydon, Ann Inst Henri Poincaré Probab Stat 54(4):1939–1968, 2018; Croydon et al., Electron J Probab 22, paper no.82, 41, 2017). As an illustrative example, this article describes an application to the random conductance model with heavy tails on nested fractal graphs.

Keywords Nested fractal · Random conductance model · Scaling limit · FIN diffusion

Mathematics Subject Classifications (2010) Primary: 28A80; Secondary: 60K37

1 Introduction

One of the early motivations for the study of stochastic processes on fractals came from physics, where there was an interest in understanding the dynamical properties of disordered media. Specifically, certain examples of the latter were modelled by critical percolation, which is believed to exhibit large scale fractal structure. (See [15] for background.) The initial response from the mathematics community was to construct Brownian motion on idealised fractals, such as the Sierpiński gasket [27, 34]. Since then, the technology has developed to the point where it can engage with some of the original questions about critical percolation. For instance, recent work in this direction underlines that the notion of a resistance form, as introduced by Kigami to provide a broad framework for studying analysis on fractals [30, 31], is useful for understanding the scaling limits of various models of random walks on random graphs in critical regimes [20, 21]. We highlight that resistance forms

D. A. Croydon (✉)
Research Institute for Mathematical Sciences, Kyoto University, Kyoto, Japan
e-mail: croydon@kurims.kyoto-u.ac.jp

U. Freiberg et al. (eds.), *Fractal Geometry and Stochastics VI*,
Progress in Probability 76, https://doi.org/10.1007/978-3-030-59649-1_10

are only really applicable in low-dimensional settings, with the stochastic processes constructed from them typically being point recurrent (note that in the case of the standard Brownian motion on $\mathbb{R}^d$, the latter property holds only when $d = 1$, and this is indeed the only dimension in which the Brownian motion can be described by a resistance form). A brief introductory survey of the work of [20, 21] already appears in [19], where a number of applications to random graphs are listed (see also [4, 5] for some further ones that have appeared more recently), and a conjecture for critical percolation is made. Here, the aim will be to introduce the general resistance form results of [20, 21] specifically to an audience that has some familiarity with analysis on self-similar fractals by presenting in detail an example from [21] which is of interest in its own right: the random conductance model with heavy tails on nested fractal graphs.

The nested fractals were originally introduced in [35], and are a class of self-similar fractals that are finitely-ramified, embedded into Euclidean space and admit a high degree of symmetry. In the next section we will introduce sequences of graphs associated with nested fractals, but to keep the presentation concise here, we focus for the moment on a concrete example of a nested fractal, the Sierpiński gasket in two dimensions. Let $V_0 := \{x_0, x_1, x_2\} \subseteq \mathbb{R}^2$ consist of the vertices of an equilateral triangle of side length 1. Write $\psi_i(x) := |x + x_i|/2$ for $i = 0, 1, 2$. Then there exists a unique compact set F such that $F = \cup_{i=0}^2 \psi_i(F)$; this is the Sierpiński gasket. We define the associated Sierpiński gasket graphs $(G_n)_{n\geq 0}$ by setting the vertex set $V(G_n) := V_n$, where $V_n := \cup_{i=0}^2 \psi_i(V_{n-1})$ for $n \geq 1$, (note that V_0 was already defined,) and defining the edge set $E(G_n)$ to be the collection of pairs of elements of V_n at a Euclidean distance 2^{-n} apart. (The first three graphs in this sequence are shown in Fig. 1.) For each n, we associate a stochastic process $X^n = (X_t^n)_{t\geq 0}$ by supposing X^n is the continuous time Markov chain that has exponential holding times of unit mean, and at jump times moves to a neighbour of the current location with uniform probability amongst the possibilities. If we moreover assume that $X_0^n = x_0$ for each n, then, from the seminal early works in the area [12, 27, 34, 35] it is known that

$$\left(X^n_{5^n t}\right)_{t\geq 0} \to \left(X_t^{SG}\right)_{t\geq 0} \tag{1.1}$$

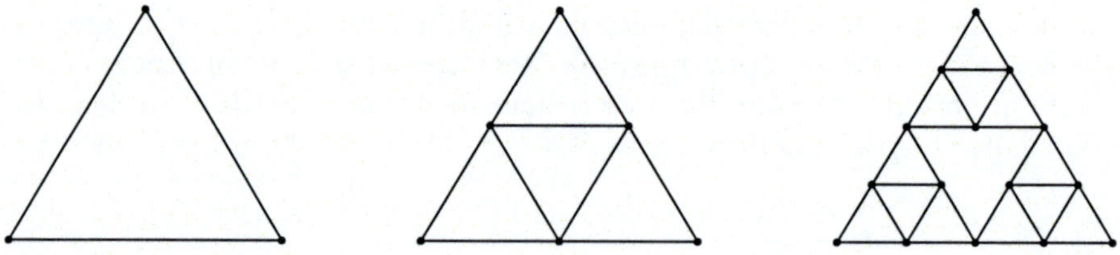

Fig. 1 The Sierpiński gasket graphs G_0, G_1, G_2

in distribution in $D([0,\infty),\mathbb{R}^2)$ (that is, the space of cadlag processes on $\mathbb{R}^2$, i.e. those that are right-continuous and have left-hand limits, equipped with the usual Skorohod J_1-topology—for elementary introductions to this framework, see [16, Chapter 3] or [39, Chapter 3], for example), where X^{SG} is a strong Markov diffusion—the so-called Brownian motion on the Sierpiński gasket, started from x_0. We remark that the terminology 'Brownian motion' reflects the fact that X^{SG} is apparently the most natural stochastic process on the Sierpiński gasket—apart from being a strong Markov diffusion that arises as a scaling limit of random walks on approximating lattices, it has a distribution that is invariant under the symmetries of the underlying space, and also satisfies natural scale invariance properties. Given this, as in other settings, it is natural to ask how robust a result such as (1.1) is to perturbations in the environment in which the process X^n is based.

One simple, canonical way in which to introduce disorder into the situation is in terms of the random conductance model. Specifically, let $G=(V_G,E_G)$ be a locally finite, connected graph. Let $\omega=(\omega_e)_{e\in E_G}$ be a collection of independent and identically distributed (i.i.d.) strictly-positive random variables built on a probability space with probability measure $\mathbf{P}$; these are the so-called *random conductances*. (Actually, for our model of self-similar fractals, we will later allow some local dependence.) Conditional on ω, we define the *variable speed random walk (VSRW)* $X^V=(X^V_t)_{t\geq 0}$ to be the continuous-time V_G-valued Markov chain with jump rate from x to y given by ω_{xy} if $\{x,y\}\in E_G$, and jump rate 0 otherwise. We obtain the associated *constant speed random walk (CSRW)* $X^C=(X^C_t)_{t\geq 0}$ by setting the jump rate along edge x to y to be $\omega_{xy}/\nu(\{x\})$, where

$$\nu(\{x\}):=\sum_{e\in E_G:\,x\in e}\omega_e; \tag{1.2}$$

note that the latter process has unit mean holding times at each vertex, and so X^n as described in the previous paragraph is simply the CSRW when G_n is equipped with unit conductances $\omega_e\equiv 1$.

An important observation is that the VSRW and CSRW experience different trapping behaviour on edges of large conductance. In particular, if we have an edge of conductance $\omega_e\gg 1$ (surrounded by other edges of conductance close to 1), then both the VSRW and CSRW cross the edge order ω_e times before escaping. However, each crossing only takes the VSRW a time of $1/\omega_e$, meaning that it is only trapped for a time of order 1, whereas each crossing for the CSRW takes a time of order 1, and so the latter process is trapped for a total time of order ω_e. In particular, when the weights are bounded away from 0, but not bounded above, we might expect the VSRW of the random conductance model to behave like the VSRW on the unweighted graph. For the CSRW, however, we would expect the trapping to be more significant, potentially leading to anomalous scaling if the weights are suitably inhomogeneous.

The random conductance model has been studied in a range of settings, via which the intuition of the previous paragraph has been shown to reflect the actual behaviour of the VSRW and CSRW. In the case of $\mathbb{Z}^d$ with $d \geq 2$, for example, it has been established that if the weights are bounded away from 0, then the VSRW always scales diffusively to a Brownian motion [11]. On the other hand, for the CSRW this is only true when the weights also have a finite first moment [11]. (In fact, both these results also apply when $d = 1$, cf. remarks in [17, 21]. See also [2] for the case when the weights are unbounded below, and [3] for results beyond the case of i.i.d. conductances.) For weights whose tail no longer has a first moment, but is in the normal domain of attraction of an α-stable random variable, namely there exists a constant $c \in (0, \infty)$ such that

$$u^\alpha \mathbf{P}(\omega_e > u) \to c \tag{1.3}$$

as $u \to \infty$, one instead sees as a scaling limit for the CSRW the fractional kinetics process—this is a Brownian motion subordinated by an α-stable process, which is subdiffusive [13, 38]. The subordination here reflects that in its first n jumps, the random walk visits Cn sites, and the time spent in these grows like a sum of n i.i.d. α-stable random variables, so is of order $n^{1/\alpha} \gg n$ (there are logarithmic corrections needed when $d = 2$ [38]). In $d = 1$, the simple random walk revisits sites more often, and so although it is also true that the CSRW is subdiffusive when the weights satisfy (1.3), the nature of the process is different. Rather, the limiting process, is a Brownian motion time-changed by the Poisson random measure

$$\nu(dx) = \sum_i v_i \delta_{x_i}(dx), \tag{1.4}$$

where $(v_i, x_i)_{i \in \mathbb{N}}$ is a Poisson point process with intensity $\alpha v^{-1-\alpha} dv dx$, and δ_{x_i} is the probability measure placing all its mass at x_i; this random measure can be viewed as the scaling limit of the random trapping environment [38]. After its introduction in [25] as a scaling limit for a random walk with strongly inhomogeneous random jump rates, the Brownian motion time-changed by ν is called the Fontes–Isopi–Newman diffusion.

For fractals, the random conductance model has previously been studied in [32, 33], where homogenisation was shown for certain classes of fractal graphs when the weights were bounded uniformly below and above. Here, we explain the progress of [21], in which a framework was developed that allowed unbounded weights, and particularly weights satisfying (1.3) to be considered. For the particular case of nested fractals (the precise definition of which is recalled in the next section), one knows that diffusions on such spaces are point recurrent, and so it is natural to conjecture that the nature of the random conductance model is likely to be more closely related to the one-dimensional Euclidean picture than the higher dimensional situation. The aim of this article is to explain that this is indeed the case, with the main result being stated as Theorem 4.5. We note that, although we restrict to nested fractals here, in [21], the slightly more general setting of uniformly finitely ramified fractals was considered. Moreover, we also remark that heat kernel estimates for the limiting processes are given in [22].

The remainder of the article is organised as follows. After introducing the model in Sect. 2, we go on to study the renormalisation and homogenisation of associated resistance metrics in Sect. 3, and then present the main scaling result in Sect. 4.

2 Random Conductance Model on Nested Fractal Graphs

In this section, we introduce precisely the model that will be of interest in the remainder of the article, starting with the notion of a nested fractal. For $\beta > 1$ and $I = \{1, 2, \cdots, N\}$, let $(\psi_i)_{i\in I}$ be a family of contraction maps on $\mathbb{R}^d$ such that $\psi_i(x) = \beta^{-1}U_i x + \gamma_i$ for $x \in \mathbb{R}^d$, where U_i is a unitary map and $\gamma_i \in \mathbb{R}^d$. As $(\psi_i)_{i\in I}$ is a family of contraction maps, there exists a unique non-void compact set F such that $F = \cup_{i\in I}\psi_i(F)$. We assume the following.

Open set condition There is a non-empty, bounded open set W such that the sets $(\psi_i(W))_{i\in I}$ are disjoint and $\cup_{i\in I}\psi_i(W) \subseteq W$.

The maps $(\psi_i)_{i\in I}$ have unique fixed points, and we denote the set of these by Fix. A point $x \in Fix$ is called an *essential fixed point* if there exist $i, j \in I,\ i \neq j$ and $y \in Fix$ such that $\psi_i(x) = \psi_j(y)$. We write V_0 for the set of essential fixed points. Denoting $\psi_{i_1,\dots,i_n} = \psi_{i_1} \circ \cdots \circ \psi_{i_n}$ for each $n \geq 0$ and $i_1, \cdots, i_n \in I$, we call a set of the form $\psi_{i_1,\cdots,i_n}(V_0)$ an n-cell. The further assumptions we make are the following.

Connectivity For any 1-cells C and C', there is a sequence $C = C_0, C_1, \dots, C_n = C'$ of 1-cells such that $C_{i-1} \cap C_i \neq \emptyset$ for $i = 1, \dots, n$.

Symmetry For any $x, y \in \mathbb{R}^d$ with $x \neq y$, let H_{xy} denote the hyperplane perpendicularly bisecting x and y, and U_{xy} denote reflection with respect to H_{xy}. If $x, y \in V_0$ and $x \neq y$, then U_{xy} maps n-cells to n-cells, and maps any n-cell which contains elements on both sides of H_{xy} to itself for each $n \geq 0$.

Nesting/Finite ramification If $n \geq 1$ and if $(i_1, \cdots, i_n)$ and $(j_1, \cdots, j_n)$ are distinct elements of I^n, then

$$\psi_{i_1,\dots,i_n}(F) \bigcap \psi_{j_1,\dots,j_n}(F) = \psi_{i_1,\dots,i_n}(V_0) \bigcap \psi_{j_1,\dots,j_n}(V_0).$$

A *nested fractal* F is a set determined by $(\psi_i)_{i\in I}$ satisfying the above assumptions with $|V_0| \geq 2$. Throughout, we assume without loss of generality that $\psi_1(x) = \beta^{-1}x$ and 0 belongs to V_0. We observe that the class of nested fractals was introduced in [35], and is included in the class of uniformly finitely ramified fractals, first introduced in [28] (and upon which the random conductance model was studied in [21]), and the latter collection is included in the class of post-critically finite self-similar sets [30]. We note that the Sierpiński gasket is a nested fractal, other examples include the Vicsek set, and Lindstrøm's snowflake. Some discussion about the restrictiveness of the axioms for nested fractals appears in [8, Remark 5.25].

Related to the nested fractal itself, we now introduce a sequence of nested fractal graphs $(G_n)_{n\geq 0}$. As in the case of the Sierpiński gasket described in the introduction, the G_n has vertex set V_n given by $\cup_{i=1}^N \psi_i(V_{n-1})$, where V_0 is as defined above. Moreover, for each n, the edge set E_n of G_n consists of the collection of pairs of vertices that are contained in the same n-cell. We let μ_n be the counting measure on V_n (placing mass one on each vertex).

Finally for this section, let us describe the version of the random conductance model that is of interest here. For each $n \geq 1$, let $\omega^n = (\omega_e^n)_{e\in E_n}$ be a collection of strictly-positive random variables built on a probability space with probability measure $\mathbf{P}$. We assume the following conditions on the weights.

Independence Weights within each n-cell are independent copies of ω^0.

Uniform lower bound There exists a deterministic constant $c > 0$ such that, $\mathbf{P}$-a.s.,

$$\omega_e^0 \geq c.$$

α-stable tail decay There exist constants $\alpha \in (0, 1)$ and $c \in (0, \infty)$ such that the random conductance distribution satisfies

$$u^\alpha \mathbf{P}\left(\sum_{e\in E_0} \omega_e^0 > u\right) \to c \tag{2.1}$$

as $u \to \infty$.

Given a realisation of weights satisfying these assumptions, we define the variable speed random walk $X^{n,V}$ and constant speed random walk $X^{n,C}$ on G_n, as per the conventions in the introduction. Specifically, both have jump chains given by the simple random walk on the graph G_n. The process $X^{n,V}$ has exponential holding times, with the mean of the holding times at vertex $x \in V_n$ being given by $1/\nu_n(\{x\})$, where, similarly to (1.2),

$$\nu_n(\{x\}) := \sum_{e\in E_n:\, x\in e} \omega_e^n; \tag{2.2}$$

the process $X^{n,C}$ has unit mean exponential holding times. The so-called quenched, i.e. conditional on the conductances, laws of $X^{n,V}$ and $X^{n,C}$ started from a vertex $x \in V_n$ will be denoted $P_x^{n,V}$ and $P_x^{n,C}$, respectively. The corresponding averaged/annealed laws are then given by

$$\mathbb{P}_x^{n,V} := \int P_x^{n,V}(\cdot)\, d\mathbf{P}, \qquad \mathbb{P}_x^{n,C} := \int P_x^{n,C}(\cdot)\, d\mathbf{P}.$$

The aim of this article is to describe scaling limits for both $X^{n,V}$ and $X^{n,C}$ under their annealed laws; the main result is stated as Theorem 4.5. Some discussion as to why we consider the annealed laws, rather than the quenched laws, is given in Remark 4.8.

3 Homogenisation of Resistance

In this section, we will briefly recall the now classical construction of a resistance metric on a nested fractal via graphical approximations. Following this, we explain what is perhaps the main result of [21] concerning self-similar fractals, which is that the same resistance metric arises from the random conductance model defined in the previous section, i.e. homogenisation of the resistance occurs. Roughly speaking this can be interpreted as meaning that, apart from normalisation by a deterministic constant, the randomness of the conductances is insignificant on large scales. Intuitively, this might be expected since, whilst the tail decay at (2.1) leads to the occasional exceptionally large edge conductance, or equivalently the occasional exceptionally small edge resistance, as we rescale, neighbouring points are anyway close in terms of resistance, and so this does not lead to large scale distortions.

Before getting to resistance metrics, however, we introduce the canonical Dirichlet form and Brownian motion on a nested fractal. In Lindstrøm's original work on nested fractals [35], transition probabilities $(q_{x,y})_{x,y\in V_0}$ satisfying $q_{x,x}=0$ and $\sum_{y\in V_0} q_{x,y}=1$ for $x\in V_0$, and also $q_{x,y}=q_{y,x}>0$ for $x\neq y\in V_0$ were introduced. Importantly, it was further established that the quantities $(q_{x,y})_{x,y\in V_0}$ could be chosen to be invariant under renormalisation in the sense we now describe. Specifically, define a quadratic form by setting

$$\mathcal{E}_0(f,f)=\frac{1}{2}\sum_{x,y\in V_0} q_{x,y}\left(f(x)-f(y)\right)^2$$

for $f\in\mathcal{F}_0:=\{f:V_0\to\mathbb{R}\}$. One obtains a further quadratic form on the same space by defining

$$\tilde{\mathcal{E}}_0(f,f)=\inf\left\{\sum_{i\in I}\mathcal{E}_0\left(g\circ\psi_i,g\circ\psi_i\right):\ g:V_1\to\mathbb{R},\ g|_{V_0}=f\right\}$$

for $f\in\mathcal{F}_0$. The invariance under renormalisation of [35, Theorem V.5] then has the equivalent statement that there exists a constant $\rho>1$ such that $\mathcal{E}_0=\rho\tilde{\mathcal{E}}_0$. Moreover, it is now known that the latter condition, together with the assumption that q are the entries of a stochastic matrix, ensure the uniqueness of $(q_{x,y})_{x,y\in V_0}$ (see [37, Theorem 6.8] and [33, Corollary 3.5]). Given $(q_{x,y})_{x,y\in V_0}$ and ρ, for $n\geq 1$ we then let

$$\mathcal{E}_n(f,f)=\rho^n\sum_{i_1,\dots,i_n\in I}\mathcal{E}_0\left(f\circ\psi_{i_1,\dots,i_n},f\circ\psi_{i_1,\dots,i_n}\right)$$

for $f \in \mathcal{F}_n := \{f : V_n \to \mathbb{R}\}$. One then obtains a canonical quadratic form on F by setting

$$\mathcal{E}(f, f) := \lim_{n\to\infty} \mathcal{E}_n(f|_{V_n}, f|_{V_n})$$

for any $f \in \mathcal{F} := \{f \in C(F, \mathbb{R}) : \lim_{n\to\infty} \mathcal{E}_n(f|_{V_n}, f|_{V_n}) < \infty\}$. Importantly, the resulting quadratic form $(\mathcal{E}, \mathcal{F})$ turns out to be a Dirichlet form on $L^2(F, \mu)$, where μ is the unique self-similar probability measure on F, that is, the only probability measure satisfying

$$\mu = \frac{1}{N} \sum_{i\in I} \mu \circ \psi_i^{-1}.$$

As a consequence, standard machinery from probability theory (see [26], for example) yields that there exists a corresponding Markov process $X^F = (X_t^F)_{t\geq 0}$, which is now commonly called the Brownian motion on the nested fractal F.

We next describe the parallel construction of the resistance metric on F. To start with one possible definition, we observe that from the quadratic form $(\mathcal{E}, \mathcal{F})$, one obtains a metric on F by defining

$$R(x, y)^{-1} := \inf\{\mathcal{E}(f, f) : f \in \mathcal{F},\ f(x) = 1,\ f(y) = 0\}, \qquad x, y \in F,\ x \neq y; \tag{3.1}$$

this is the resistance metric on F. In fact, the above description of R yields a one-to-one relationship between a class of quadratic forms called resistance forms (of which $(\mathcal{E}, \mathcal{F})$ is one) and a class of metrics called resistance metrics (see [30, Theorems 2.3.4, 2.3.6], for example). An alternative definition of R is via resistance metrics on the finite graphs. Specifically, suppose R_n is the resistance metric on V_n induced by placing conductances according to $(\rho^{-n} q_{x,y})_{x,y\in V_0}$ along edges of n-cells, i.e. setting the conductance from $\psi_{i_1,\dots,i_n}(x)$ to $\psi_{i_1,\dots,i_n}(y)$ to be $\rho^{-n} q_{x,y}$; alternatively, R_n can be defined from $(\mathcal{E}_n, \mathcal{F}_n)$ analogously to (3.1). From the invariance under renormalisation of $\mathcal{E}_0$, one can check that

$$R_n = R_m|_{V_n}, \qquad \forall m \geq n.$$

From this it readily follows that we have $R = \lim_{n\to\infty} R_n(x, y)$ on $V_* = \cup_{n\geq 0} V_n$. In particular, $R|_{V_n} = R_n$. With some additional work to check that (F, R) is the completion of (V_*, R), we obtain that V_n converges to F with respect to Hausdorff topology on compact subsets of (F, R). (See [29] for proofs of these claims.)

It transpires that one obtains the limit described in the preceding paragraph if the deterministic conductances characterised by $(q_{x,y})_{x,y\in V_0}$ are replaced by the random conductances of the previous section. That is, suppose R_n^ω is the resistance metric on V_n induced by placing conductances according to $(c\rho^{-n}\omega_e^n)_{e\in E_n}$ along edges of the graph, where $c \in (0, \infty)$ is a deterministic constant that depends on the law of the conductances; this is the metric given by (3.1) for the following quadratic form

$$\frac{1}{2c}\rho^n \sum_{i_1,\dots,i_n\in I}\sum_{x,y\in V_0} \omega^n_{\psi_{i_1,\dots,i_n}(x),\psi_{i_1,\dots,i_n}(y)} \left(f\circ\psi_{i_1,\dots,i_n}(x) - f\circ\psi_{i_1,\dots,i_n}(y)\right)^2,$$

which is defined for $f\in\mathcal{F}_n$. From [21, Theorem 6.11], we then have that, in **P**-probability,

$$\left(R^\omega_n(x,y)\right)_{x,y\in V_0} \to (R(x,y))_{x,y\in V_0}, \tag{3.2}$$

where we note that the constant c is determined by this result. The proof in [21], which can heuristically be understood as establishing contractivity of a renormalisation map, resembles that of the corresponding results in [32, 33]. However, the lack of a uniform upper bound on the conductances leads to significant technical challenges, particularly in checking that certain quantities are integrable, as is required for the argument to work. From (3.2) and the trivial bound that $R^\omega_n \le CR_n$, (which follows from the fact that the conductances are bounded away from 0,) we readily obtain the following proposition.

Proposition 3.1 ([21, Lemma 6.14]) *In* **P***-probability,*

$$\sup_{x,y\in V_n}\left|R^\omega_n(x,y) - R(x,y)\right| \to 0.$$

Since (V_n, R^ω_n) can not in general be isometrically embedded into (F, R), then the usual Hausdorff topology on (F, R) is not the right topology with which to discuss convergence. However, one can instead conclude from the previous result (and some small additional technical work again depending on the bound $R^\omega_n \le CR_n$) that (V_n, R^ω_n) converges to (F, R) with respect to the Gromov–Hausdorff topology, that is, all the spaces in question can be isometrically embedded into a common metric space so that the V_n converges to F with respect to the usual Hausdorff metric on this space (see [18, Chapter 7] for background on the Gromov–Hausdorff topology).

4 Random Walk Scaling Limits

Proposition 3.1 is the main ingredient to proving scaling limits for the variable speed random walk $X^{n,V}$ and the constant speed random walk $X^{n,C}$. Indeed, the only additional input required is the convergence under scaling of the counting measure μ_n and the measure ν_n defined in terms of conductances at (2.2), which is straightforward to prove. The machinery that allows us to proceed with this program is the main result of [20] (which gives a more general version of the result of [21]).

To introduce the abstract result we appeal to precisely, let us fix the framework. In particular, we write $\mathbb{F}_c^*$ for the collection of quintuples of the form $(K, R_K, \mu_K, \rho_K, \phi_K)$, where: K is a non-empty set; R_K is a resistance metric on K such that (K, R_K) is compact; μ_K is a locally finite Borel regular measure of full support on (K, R_K); ρ_K is a marked point in K, and ϕ_K is a continuous map from K to some fixed metric space (M, d_M). From the point of view of metric geometry, there is a natural notion of convergence of such spaces which gives rise to the marked spatial Gromov–Hausdorff–Prohorov topology. Specifically, convergence of some sequence in $\mathbb{F}_c^*$ means that all the spaces can be isometrically embedded into a common metric space $(\mathcal{M}, d_{\mathcal{M}})$ in such a way that: the embedded sets converge with respect to the Hausdorff distance, the embedded measures converge weakly, the embedded marked points converge, and the image of the continuous map is close in M for points that are close in $\mathcal{M}$. We note that such Gromov–Hausdorff-type topologies have proved useful for studying various kinds of random metric spaces; see [18] for an introduction to the classical theory. More specifically, the marked spatial Gromov–Hausdorff–Prohorov topology was introduced in [14], building on the notions of the Gromov–Hausdorff–Prohorov/Gromov–Hausdorff-vague topologies of [1, 7, 24, 36] and the topology for spatial trees of [23] (cf. the spectral Gromov–Hausdorff topology of [21]).

Importantly, that the elements $(K, R_K, \mu_K, \rho_K, \phi_K)$ of $\mathbb{F}_c^*$ incorporate a resistance metric means that there is a naturally associated stochastic process. For, it is a result of Kigami that the corresponding resistance form, characterised via (3.1), is a regular Dirichlet form on $L^2(K, \mu_K)$, and so naturally associated with a Markov process (see [31, Chapter 9], for example). The following result establishes that, if the convergence described in the previous paragraph occurs, then we also obtain convergence of stochastic processes.

Theorem 4.1 ([20, Theorem 7.2]) *Suppose that* $(K_n, R_{K_n}, \mu_{K_n}, \rho_{K_n}, \phi_{K_n})_{n\geq 1}$ *is a sequence in* $\mathbb{F}_c^*$ *satisfying*

$$\left(K_n, R_{K_n}, \mu_{K_n}, \rho_{K_n}, \phi_{K_n}\right) \to (K, R_K, \mu_K, \rho_K, \phi_K) \tag{4.1}$$

in the marked spatial Gromov–Hausdorff–Prohorov topology for some element $(K, R_K, \mu_K, \rho_K, \phi_K) \in \mathbb{F}_c^*$. *It then holds that*

$$P^n_{\rho_{K_n}}\left(\left(\phi_{K_n}(X^n_t)\right)_{t\geq 0} \in \cdot\right) \to P_{\rho_K}\left((\phi_K(X_t))_{t\geq 0} \in \cdot\right)$$

weakly as probability measures on $D(\mathbb{R}_+, M)$, *where* $((X^n_t)_{t\geq 0}, (P^n_x)_{x\in K_n})$ *is the Markov process corresponding to* $(K_n, R_{K_n}, \mu_{K_n}, \rho_{K_n})$, *and* $((X_t)_{t\geq 0}, (P_x)_{x\in K})$ *is the Markov process corresponding to* (K, R_K, μ_K, ρ_K).

Remark 4.2 The key to the proof of the above result in [20] is the observation that for a process associated with a resistance metric, it is possible to explicitly express the associated resolvent kernel in terms of the resistance metric. (This was also the basis of the corresponding argument for trees from [6].) Specifically, if

$((X_t)_{t\geq 0}, (P_x)_{x\in K})$ is the Markov process associated with $(K, R_K, \mu_K, \rho_K, \phi_K) \in \mathbb{F}_c^*$, define the resolvent of X killed on hitting x by

$$G_x f(y) = E_y \int_0^{\sigma_x} f(X_s) ds,$$

where E_y is the expectation under P_y, and $\sigma_x := \inf\{t \geq 0 : X_t = x\}$ is the hitting time of x by X. (NB. Processes associated with resistance forms hit points; the above expression is well-defined and finite.) One can then write

$$G_x f(y) = \int_K g_x(y,z) f(z) \mu_K(dz),$$

where the resolvent kernel is given by

$$g_x(y,z) = \frac{R_K(x,y) + R_K(x,z) - R_K(y,z)}{2}.$$

(See [31, Theorem 4.3].) Appealing to this formula, the metric measure convergence at (4.1) enables one to check the convergence of resolvents in a certain sense. One can then use more standard machinery from probability theory to establish semi-group convergence, and moreover convergence of finite dimensional distributions. To complete the proof, one is also required to check tightness of the processes (see [16, Chapter 16]), but again this can be deduced from the above resolvent density formula (or, more precisely, a slight generalisation thereof). See [20] for details.

Remark 4.3 Whilst Theorem 4.1 has an appealingly concise statement, checking the assumption at (4.1) is by no means trivial. Indeed, beyond the case of graph trees (or graphs that are close to trees), where the resistance metric corresponds to (or is close to, respectively) a shortest path metric, or certain finitely ramified self-similar fractals, where the resistance metric can be studied by using the particular structure of the space, understanding detailed properties of the resistance metric remains a challenge. To give just one example of an open problem from the world of self-similar fractals, it is still not known how to compute the value of the resistance exponent for graphs based on the two-dimensional Sierpiński carpet, see [9] for some work in this direction, and the discussion in [10, Example 4] concerning the graphical Sierpiński carpet in particular.

We will apply Theorem 4.1 with $K_n = V_n$, $R_{K_n} = R_n^\omega$, $\mu_{K_n} = \mu_n$ or $\mu_{K_n} = \nu_n$, $\rho_{K_n} = 0$, and $\phi_{K_n} := I_n$, where I_n is the identity map from K_n into $\mathbb{R}^d$. The following lemma gives us the scaling limits of the measures. To state the result, we introduce a Poisson random measure on F by setting

$$\nu(dx) = \sum_i v_i \delta_{x_i}(dx),$$

where $(v_i, x_i)_{i\in\mathbb{N}}$ is a Poisson point process with intensity $\alpha v^{-1-\alpha} dv \mu(dx)$, and δ_{x_i} is the probability measure placing all its mass at x_i. (This is the analogue of the measure defined at (1.4) in the present setting.) Note that the exponent α is given by the tail of the conductance distribution (2.1).

Lemma 4.4 *It holds that $N^{-n}\mu_n \to \mu$, and also there exists a deterministic constant $c_0 \in (0, \infty)$ such that $c_0^{-1} N^{-n/\alpha} \nu_n \to \nu$ in distribution, in both cases with respect to the weak topology for finite measures on $\mathbb{R}^d$.*

Combining Proposition 3.1 and Lemma 4.4, we readily obtain that

$$\left(V_n, R_n^\omega, N^{-n}\mu_n, 0, I_n\right) \to (F, R, \mu, 0, I), \tag{4.2}$$

in **P**-probability, and

$$\left(V_n, R_n^\omega, c_0^{-1} N^{-n/\alpha} \nu_n, 0, I_n\right) \to (F, R, \nu, 0, I),$$

in distribution under **P** with respect to the marked spatial Gromov–Hausdorff–Prohorov topology, where I is the identity map from F into $\mathbb{R}^d$. Since $X^{n,V}$ is the process associated with $(V_n, c^{-1}\rho^n R_n^\omega, \mu_n, 0, I_n)$, and $X^{n,C}$ is the process naturally associated with $(V_n, c^{-1}\rho^n R_n^\omega, \nu_n, 0, I_n)$, we are consequently able to apply Theorem 4.1 to deduce a scaling limit for these processes. (By considering the generators of the relevant Markov processes, it is readily checked how the resistance and mass scaling factors can be interpreted in terms of time scaling.) As for the limiting processes, we note that the Brownian motion X^F is the process associated with $(F, R, \mu, 0)$—we write the law of this process started from 0 as P_0. Moreover, the process associated with $(F, R, \nu, 0)$ is the time-change of X^F according to ν, that is, defining an additive functional

$$A_t := \int_0^t L_t(x)\nu(dx),$$

where $(L_t(x))_{x\in F, t>0}$ are the jointly continuous local times of X^F (with respect to μ), and its right-continuous inverse $\tau(t) := \inf\{s > 0 : A_s > t\}$, we set

$$X_t^{F,\nu} := X_{\tau(t)}^F;$$

following the definition of the corresponding one-dimensional process in [25], we call this the FIN diffusion on F. The averaged/annealed law of the FIN diffusion on F, started from 0, will be denoted

$$\mathbb{P}_0^{\mathrm{FIN}} := \int P_0\left(X^{F,\nu} \in \cdot\right) d\mathbf{P},$$

i.e. one chooses ν according to $\mathbf{P}$, and then the law of $X^{F,\nu}$ is determined by the law of X^F under P_0.

Theorem 4.5 *There exists a deterministic constant* $c_1 \in (0, \infty)$ *such that*

$$\mathbb{P}_0^{n,V}\left(\left(X^{n,V}_{c_1 t(\rho N)^n}\right)_{t\geq 0} \in \cdot\right) \to P_0\left(\left(X^F_t\right)_{t\geq 0} \in \cdot\right)$$

weakly as probability measures on $D(\mathbb{R}_+, \mathbb{R}^d)$*. Moreover, there exists a deterministic constant* $c_2 \in (0, \infty)$ *such that*

$$\mathbb{P}_0^{n,C}\left(\left(X^{n,C}_{c_2 t(\rho N^{1/\alpha})^n}\right)_{t\geq 0} \in \cdot\right) \to \mathbb{P}_\rho^{\mathrm{FIN}}\left(\left(X^{F,\nu}_t\right)_{t\geq 0} \in \cdot\right)$$

weakly as probability measures on $D(\mathbb{R}_+, \mathbb{R}^d)$*.*

Remark 4.6 To state the result for the Sierpiński gasket explicitly, note that in this case we have $N = 3$ and $\rho = 5/3$, so that

$$\mathbb{P}_0^{n,V}\left(\left(X^{n,V}_{c_1 t 5^n}\right)_{t\geq 0} \in \cdot\right) \to P_0\left(\left(X^F_t\right)_{t\geq 0} \in \cdot\right),$$

and we also have

$$\mathbb{P}_0^{n,C}\left(\left(X^{n,C}_{c_2 t 5^n (3^{\frac{1}{\alpha}-1})^n}\right)_{t\geq 0} \in \cdot\right) \to \mathbb{P}_0^{\mathrm{FIN}}\left(\left(X^{F,\nu}_t\right)_{t\geq 0} \in \cdot\right).$$

In particular, the scaling regime for the variable speed random walk matches that of the simple random walks on the unweighted graphs, as stated at (1.1); and since $\alpha < 1$, the constant speed random walk (or limiting diffusion) moves through the relevant graph more slowly than the unweighted simple random walk (or Brownian motion, respectively). Together with known results for simple random walks on nested fractal graphs, Theorem 4.5 implies that these qualitative comments apply to nested fractal graphs in general.

Remark 4.7 When $\mathbf{E}\omega_e^0 < \infty$ for each $e \in E_0$, one obtains in place of the second claim of Lemma 4.4 that there exists a constant c_0 such that $c_0^{-1} N^{-n} \nu_n \to \mu$. Consequently, if (2.1) is replaced by the assumption of finite first moments, then one can check the annealed limit of $X^{n,C}$ is Brownian motion, rather than the FIN diffusion that appears in the second statement of Theorem 4.5.

Remark 4.8 A stronger notion of convergence than convergence with respect to the annealed law is convergence with respect to the quenched law for $\mathbf{P}$-a.e. realisation of the conductances. Typically, one might hope to be able to prove such a quenched convergence statement in the case where the conductances homogenise, as has been established when the underlying graph is a Euclidean lattice (see [2, 3, 11], for example). In particular, it would be natural to conjecture that for the example described in this article, the quenched law of the VSRW $X^{n,V}$ converges as $n \to \infty$ for typical realisations of the environment. To do this, it would be sufficient to replace the weak (i.e. in probability) statement of (4.2) with a strong (i.e. **P**-a.s.) one. However, the techniques of [21] are not sufficient to yield such a result. As for the CSRW $X^{n,C}$, the typical fluctuations of the conductance environment as n varies will be too large to permit a quenched limit statement (cf. the law of the iterated logarithm for simple random walk on $\mathbb{Z}$, which implies that individual sample paths can not be rescaled to a realisation of Brownian motion on $\mathbb{R}$, even though the discrete paths have the latter process as a distributional limit).

Acknowledgments The author is grateful to the organisers of the conference Fractal Geometry and Stochastics 6 for arranging a wonderful meeting, where he was given the chance to present the work of [20, 21], and for inviting him to produce this article. His attendance at the latter event was partially supported by the JSPS Grant-in-Aid for Research Activity Start-up, 18H05832. The author is grateful to a referee for their careful reading of an earlier version of the article, which led to a number of improvements being made.

References

1. Abraham, R., Delmas, J.-F., Hoscheit, P.: A note on the Gromov-Hausdorff-Prokhorov distance between (locally) compact metric measure spaces. Electron. J. Probab. **18**(14), 21 (2013)
2. Andres, S., Barlow, M.T., Deuschel, J.-D., Hambly, B.M.: Invariance principle for the random conductance model. Probab. Theory Relat. Fields **156**(3–4), 535–580 (2013)
3. Andres, S., Deuschel, J.-D., Slowik, M.: Invariance principle for the random conductance model in a degenerate ergodic environment. Ann. Probab. **43**(4), 1866–1891 (2015)
4. Andriopoulos, G.: Invariance principles for random walks in random environment on trees (2018). Preprint. arXiv:1812.10197
5. Archer, E.: Brownian motion on stable looptrees (2019). Preprint. arXiv:1902.01713.
6. Athreya, S., Löhr, W., Winter, A.: Invariance principle for variable speed random walks on trees. Ann. Probab. **45**(2), 625–667 (2017).
7. Athreya, S., Löhr, W., Winter, A.: The gap between Gromov-vague and Gromov-Hausdorff-vague topology. Stoch. Process. Appl. **126**(9), 2527–2553 (2016)
8. Barlow, M.T.: Diffusions on fractals. In: Lectures on Probability Theory and Statistics (Saint-Flour, 1995). Lecture Notes in Mathematics, vol. 1690, pp. 1–121. Springer, Berlin (1998)
9. Barlow, M.T., Bass, R.F.: On the resistance of the Sierpiński carpet. Proc. Roy. Soc. Lond. Ser. A **431**(1882), 345–360 (1990)
10. Barlow, M.T., Coulhon, T., Kumagai, T.: Characterization of sub-Gaussian heat kernel estimates on strongly recurrent graphs. Commun. Pure Appl. Math. **58**(12), 1642–1677 (2005)
11. Barlow, M.T., Deuschel, J.-D.: Invariance principle for the random conductance model with unbounded conductances. Ann. Probab. **38**(1), 234–276 (2010)
12. Barlow, M.T., Perkins, E.A.: Brownian motion on the Sierpiński gasket. Probab. Theory Relat. Fields **79**(4), 543–623 (1988)
13. Barlow, M.T., Černý, J.: Convergence to fractional kinetics for random walks associated with unbounded conductances. Probab. Theory Relat. Fields **149**(3–4), 639–673 (2011)
14. Barlow, M.T., Croydon, D.A., Kumagai, T.: Subsequential scaling limits of simple random walk on the two-dimensional uniform spanning tree. Ann. Probab. **45**(1), 4–55 (2017)

15. Ben Avraham, D., Havlin, S.: Diffusion and Reactions in Fractals and Disordered Systems. Cambridge University Press, Cambridge (2000)
16. Billingsley, P.: Convergence of Probability Measures. Wiley Series in Probability and Statistics: Probability and Statistics, 2nd edn. Wiley, New York (1999). A Wiley-Interscience Publication
17. Biskup, M., Prescott, T.M.: Functional CLT for random walk among bounded random conductances. Electron. J. Probab. **12**(49), 1323–1348 (2007)
18. Burago, D., Burago, Y., Ivanov, S.: A Course in Metric Geometry. Graduate Studies in Mathematics, vol. 33. American Mathematical Society, Providence (2001)
19. Croydon, D.A.: An introduction to stochastic processed associated with resistance forms and their scaling limits. RIMS Kôkyûroku **2030**, 1–8 (2018)
20. Croydon, D.A.: Scaling limits of stochastic processes associated with resistance forms. Ann. Inst. Henri Poincaré Probab. Stat. **54**(4), 1939–1968 (2018)
21. Croydon, D.A., Hambly, B.M., Kumagai, T.: Time-changes of stochastic processes associated 481 with resistance forms. Electron. J. Probab. **22**, paper no. 82, 41 (2017)
22. Croydon, D.A., Hambly, B.M., Kumagai, T.: Heat kernel estimates for FIN processes associated with resistance forms. Stoch. Process. Appl. **129**(9), 2991–3017 (2019)
23. Duquesne, T., Le Gall, J.-F.: Probabilistic and fractal aspects of Lévy trees. Probab. Theory Relat. Fields **131**(4), 553–603 (2005)
24. Evans, S.N.: Probability and Real Trees. Lecture Notes in Mathematics, vol. 1920. Springer, Berlin (2008). Lectures from the 35th Summer School on Probability Theory held in Saint-Flour, July 6–23, 2005
25. Fontes, L.R.G., Isopi, M., Newman, C.M.: Random walks with strongly inhomogeneous rates and singular diffusions: convergence, localization and aging in one dimension. Ann. Probab. **30**(2), 579–604 (2002)
26. Fukushima, M., Oshima, Y., Takeda, M.: Dirichlet Forms and Symmetric Markov Processes. de Gruyter Studies in Mathematics, extended edn., vol. 19. Walter de Gruyter, Berlin (2011)
27. Goldstein, S.: Random walks and diffusions on fractals. In: Percolation Theory and Ergodic Theory of Infinite Particle Systems (Minneapolis, 1984–1985). The IMA Volumes in Mathematics and its Applications, vol. 8, pp. 121–129. Springer, New York (1987)
28. Hambly, B.M., Kumagai, T.: Heat kernel estimates for symmetric random walks on a class of fractal graphs and stability under rough isometries. Fractal Geometry and Applications: A Jubilee of Benoît Mandelbrot, Part 2. Proceedings of Symposia in Pure Mathematics, vol. 72, pp. 233–259. American Mathematical Society, Providence (2004)
29. Kigami, J.: Effective resistances for harmonic structures on P.C.F. self-similar sets. Math. Proc. Cambridge Philos. Soc. **115**(2), 291–303 (1994)
30. Kigami, J.: Analysis on Fractals. Cambridge Tracts in Mathematics, vol. 143. Cambridge University Press, Cambridge (2001)
31. Kigami, J.: Resistance forms, quasisymmetric maps and heat kernel estimates. Mem. Am. Math. Soc. **216**(1015), vi+132 (2012)
32. Kumagai, T.: Homogenization on finitely ramified fractals. In: Stochastic Analysis and Related Topics in Kyoto. Advanced Studies in Pure Mathematics, vol. 41, pp. 189–207. Mathematics Society, Tokyo (2004)
33. Kumagai, T., Kusuoka, S.: Homogenization on nested fractals. Probab. Theory Relat. Fields **104**(3), 375–398 (1996)
34. Kusuoka, S.: A diffusion process on a fractal. Probabilistic Methods in Mathematical Physics (Katata/Kyoto, 1985), pp. 251–274. Academic Press, Boston (1987)
35. Lindstrøm, T.: Brownian motion on nested fractals. Mem. Am. Math. Soc. **83**(420), iv+128 (1990)
36. Miermont, G.: Tessellations of random maps of arbitrary genus. Ann. Sci. Éc. Norm. Supér. **42**(5), 725–781 (2009)

37. Sabot, C.: Existence and uniqueness of diffusions on finitely ramified self-similar fractals. Ann. Sci. École Norm. Sup. **30**(5), 605–673 (1997)
38. Černý, J.: On two-dimensional random walk among heavy-tailed conductances. Electron. J. Probab. **16**(10), 293–313 (2011)
39. Whitt, W.: Stochastic-Process Limits. Springer Series in Operations Research. Springer, New York (2002). An introduction to stochastic-process limits and their application to queues

Space-Time Duality for Semi-Fractional Diffusions

Peter Kern and Svenja Lage

Abstract Almost 60 years ago Zolotarev proved a duality result which relates an α-stable density for $\alpha \in (1, 2)$ to the density of a $\frac{1}{\alpha}$-stable distribution on the positive real line. In recent years Zolotarev duality was the key to show space-time duality for fractional diffusions stating that certain heat-type fractional equations with a negative fractional derivative of order α in space are equivalent to corresponding time-fractional differential equations of order $\frac{1}{\alpha}$. The point source solutions of the former are given by negatively skewed α-stable densities, whereas the latter are solved by densities of corresponding inverse $\frac{1}{\alpha}$-stable subordinators. We review this space-time duality and take it as a recipe for a previously unknown generalization from the stable to the semistable situation.

Keywords Zolotarev duality · Fractional diffusion · Semi-fractional derivative · Semistable Lévy process · Subordinator · Hitting-time

Mathematics Subject Classifications (2010) Primary: 35R11, Secondary: 26A33, 60G18, 60G22, 60G51, 82C31

1 Introduction

The objects of our study are one-dimensional Lévy processes with a certain self-similarity property. A Lévy process $X = (X_t)_{t \geq 0}$ on $\mathbb{R}$ is a stochastic process starting in $X_0 = 0$ with the following properties:

- X has independent increments, i.e. $(X_{t(k)} - X_{t(k-1)})_{k=1,\dots,n}$ are independent random variables for finitely many time points $0 = t(0) < t(1) < \cdots < t(n)$.
- X has stationary increments, i.e. $X_t - X_s \stackrel{d}{=} X_{t-s}$ for all $0 \leq s \leq t$, where $\stackrel{d}{=}$ denotes equality in distribution.

P. Kern (✉) · S. Lage
Mathematical Institute, Heinrich-Heine-University Düsseldorf, Düsseldorf, Germany
e-mail: kern@hhu.de; Svenja.Lage@uni-duesseldorf.de

U. Freiberg et al. (eds.), *Fractal Geometry and Stochastics VI*,
Progress in Probability 76, https://doi.org/10.1007/978-3-030-59649-1_11

- X is stochastically continuous, i.e. $P\{|X_t - X_s| > \varepsilon\} \to 0$ as $|t-s| \to 0$ for all $\varepsilon > 0$.

We will further assume that the process is strictly self-similar in the statistical sense that

$$(X_{ct})_{t\geq 0} \overset{\text{fd}}{=} (c^{1/\alpha} X_t)_{t\geq 0} \quad \text{for all } c > 0, \tag{1.1}$$

where $\overset{\text{fd}}{=}$ denotes equality of all joint distributions for finitely many time points. Then necessarily X is a stable Lévy process with parameter $\alpha \in (0, 2]$, where we exclude the degenerate case $X_t = \mu t$ for some $\mu \in \mathbb{R}$, corresponding to $\alpha = 1$. In the case of $\alpha = 2$ we have Brownian motion with a certain variance parameter $\sigma^2 > 0$. Brownian motion is exceptional among the α-stable Lévy processes, since its sample paths $t \mapsto X_t(\omega)$ are continuous for almost all $\omega \in \boldsymbol{\Omega}$ of the underlying probability space $(\boldsymbol{\Omega}, \mathcal{A}, P)$, whereas for $0 < \alpha < 2$ the sample paths of an α-stable process are almost surely càdlàg functions (right-continuous with left limits) with jumps as illustrated in Fig. 1. For all $\alpha \in (0, 2]$ the paths of an α-stable process can be considered as random fractals which in many aspects almost surely share the same fractal behavior. E.g. the Hausdorff dimension of the range, the graph or multiple points only depend on α and the space dimension for multivariate stable Lévy processes. Classical results on the fractal behavior of Brownian paths [7, 21, 44] were later extended to the case of multivariate stable processes in [4, 11, 13, 34, 36, 37, 45, 46] to mention just a few striking results. For an excellent overview on fractal path behavior we refer to the survey article [48].

It is further known that X_t has a smooth probability density $x \mapsto p(x, t)$ for every $t > 0$, in particular these are $C^\infty(\mathbb{R})$-functions such that the density itself and all its derivatives belong to $C_0(\mathbb{R}) \cap L^1(\mathbb{R})$. However, no closed form solution of stable densities are known besides $\alpha = 2$ (Gaussian), $\alpha = 1$ (Cauchy) and a certain density for $\alpha = \frac{1}{2}$ called Lévy density, which becomes important later on. In the following we will exclude the often exceptional cases $\alpha = 2$ (Brownian motion)

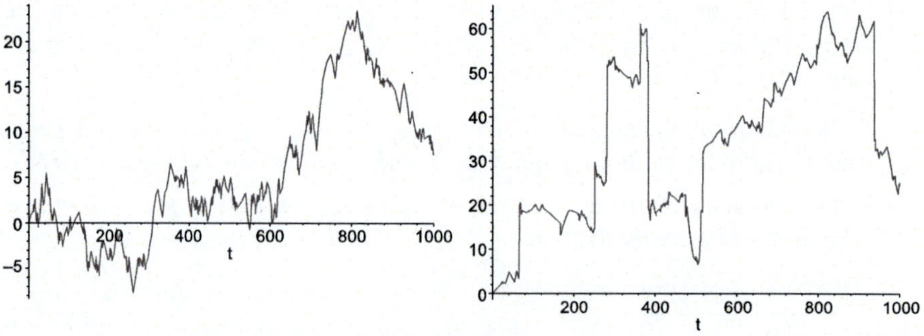

Fig. 1 Sample paths of a Brownian motion for $\alpha = 2$ are continuous (left), whereas sample paths of a stable Lévy process with $\alpha \in (0, 2)$ have jumps (right)

and $\alpha = 1$ (Cauchy and degenerate process). Since the description of stable Lévy processes by their probability densities are not easily accessible, the processes are best characterized by their Fourier transforms (FT) in terms of the Lévy–Khintchine formula

$$\mathbb{E}[\exp(ik X_t)] = \widehat{p}(k,t) = \exp(t\psi(k))$$

with log-characteristic function

$$\psi(k) = i\mu k + \int_{\mathbb{R}\setminus\{0\}} \left(e^{ikx} - 1 - \frac{ikx}{1+x^2}\right) d\phi(x) \tag{1.2}$$

for some unique drift parameter $\mu \in \mathbb{R}$ and a unique Lévy measure

$$d\phi(x) = D\left(p \cdot x^{-\alpha-1} 1_{\{x>0\}} + q \cdot |x|^{-\alpha-1} 1_{\{x<0\}}\right) dx, \tag{1.3}$$

where $D > 0$ and $p, q \geq 0$ with $p + q = 1$. Thus it is sufficient to describe the distribution of X_1 for which three additional parameters $D > 0$, $p \in [0, 1]$ and $\mu \in \mathbb{R}$ are needed besides the parameter $\alpha \in (0, 2) \setminus \{1\}$. According to [39], there is an alternative parametrization of the probability density $p(x, 1) = g(x; \alpha, \beta, \sigma, v)$ as the unique function with FT

$$\widehat{g}(k; \alpha, \beta, \sigma, v) = \widehat{p}(k, 1) = \exp\left(ivk - \sigma^{\alpha}|k|^{\alpha}\left(1 - i\beta\, \mathrm{sign}(k) \tan(\alpha\tfrac{\pi}{2})\right)\right), \tag{1.4}$$

where $\beta = p - q \in [-1, 1]$ is a skewness parameter, $\sigma = (D \cdot |\cos(\alpha\frac{\pi}{2})|)^{1/\alpha} > 0$ is a scale parameter, and $v = \mu - \int_{\mathbb{R}\setminus\{0\}}(\frac{x}{1+x^2} - x\, 1_{\{\alpha>1\}})d\phi(x)$ is a centering parameter. In particular, the strict self-similarity (1.1) holds iff $v = 0$. To visualize the impact of the parameters α and β on ranges that will become important later in this article, various stable densities for constant $\nu = 0$ and $\sigma > 0$ are plotted in Fig. 2 using Fourier inversion techniques.

A further description of stable Lévy processes comes from the fact that for suitable functions f the operators $T_t f(x) = \mathbb{E}[f(x - X_t)]$, $t \geq 0$, determine a C_0-semigroup with generator

$$Lf(x) = -\mu f'(x) + \int_{\mathbb{R}\setminus\{0\}} \left(f(x-y) - f(x) + \frac{y f'(x)}{1+y^2}\right) d\phi(x) \tag{1.5}$$

and FT $\widehat{Lf}(k) = \psi(k) \cdot \widehat{f}(k)$ with ψ from (1.2) and ϕ as in (1.3). For a comprehensive overview on (stable) Lévy processes we refer to the monographs [39, 40].

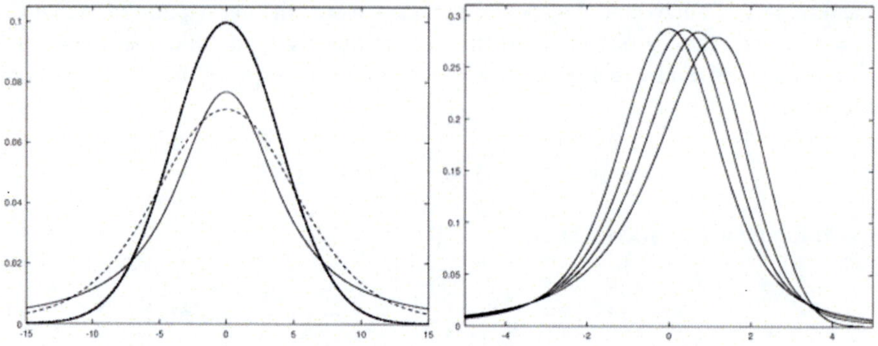

Fig. 2 Stable densities $g(x; \alpha, 0, 4, 0)$ for $\alpha = 2$, $\alpha = 1.7$ and $\alpha = 1.1$ (left) and $g(x; 1.5, \beta, 1, 0)$ for $\beta = 0$, $\beta = -0.3$, $\beta = -0.6$ and $\beta = -1$ (right)

It is well known that for $\alpha = 2$ the Brownian motion with variance parameter $\sigma^2 > 0$ has probability density

$$p(x,t) = \frac{1}{\sqrt{2\pi\sigma^2 t}} \exp\left(-\frac{1}{2}\frac{x^2}{\sigma^2 t}\right) \qquad \text{for } x \in \mathbb{R} \text{ and } t > 0$$

which is a solution to the one-dimensional heat equation

$$\frac{\partial}{\partial t} p(x,t) = \frac{\sigma^2}{2}\frac{\partial^2}{\partial x^2} p(x,t)$$

with initial point source $p(x, 0) = \delta(x)$. As laid out in [32], for $\alpha \in (0, 2) \setminus \{1\}$ the stable densities are point source solutions to the fractional diffusion equation

$$\frac{\partial}{\partial t} p(x,t) = -v\frac{\partial}{\partial x} p(x,t) + C\left(\frac{1+\beta}{2}\frac{\partial^\alpha}{\partial x^\alpha} p(x,t) + \frac{1-\beta}{2}\frac{\partial^\alpha}{\partial(-x)^\alpha} p(x,t)\right), \tag{1.6}$$

where $C > 0$ if $\alpha \in (1, 2)$, $C < 0$ if $\alpha \in (0, 1)$, $v \in \mathbb{R}$ is a velocity (centering) parameter, and $\beta \in [-1, 1]$ is the skewness parameter. Here $\frac{\partial^\alpha}{\partial x^\alpha} f(x)$ and $\frac{\partial^\alpha}{\partial(-x)^\alpha} f(x)$ denote the positive and negative Riemann–Liouville fractional derivatives defined for suitable functions f as the unique functions with FT $(-ik)^\alpha \widehat{f}(k)$, respectively $(ik)^\alpha \widehat{f}(k)$. Due to $\widehat{Lf} = \psi \cdot \widehat{f}$ these can also be defined by means of the generators of α-stable Lévy processes with $\nu = 0$ and skewness parameter $\beta = 1$, respectively $\beta = -1$. Formally, for integers $\alpha \in \mathbb{N}$ the FT of the Riemann–Liouville fractional derivative coincides with $\int_{\mathbb{R}} e^{\pm ikx} f^{(\alpha)}(x)\,dx = \widehat{f^{(\alpha)}}(\pm k)$ and thus fractional derivatives generalize integer order derivatives. For details on fractional calculus we refer to the monographs [22, 38].

Since stable Lévy processes contain both fractal behavior of sample paths and probability densities solving a fractional pde, they contribute to an ongoing discussion on the connection of fractal geometry and fractional calculus [6, 23, 43]. The fractional pde (1.6) is our starting point towards space-time duality for fractional diffusions. In Sect. 2 we will review this fractional pde approach and a remarkable connection to Zolotarev duality. In the special case of a negatively skewed stable Lévy process with $\alpha \in (1,2)$, the fractional diffusion equation is known to be equivalent to a time-fractional pde with an ordinary first-order derivative in space, which is called space-time duality [1, 14]. This perfectly reflects Zolotarev duality for the related stable densities. From a physical point of view this space-time duality has an important impact. Since fractional derivatives are non-local operators, the fractional diffusion equation lacks a meaningful physical interpretation. As mentioned by Hilfer [12], due to non-locality in space, experimentally a closed system cannot be separated from its outer environment, whereas non-locality in time does not violate physical principles if one accepts long memory effects.

We will further consider Lévy processes with a discrete scaling property such that (1.1) only holds for some $c > 1$ and thus for all integer powers of c, but not necessarily for all $c > 0$:

$$(X_{c^m t})_{t\geq 0} \stackrel{\text{fd}}{=} (c^{m/\alpha} X_t)_{t\geq 0} \quad \text{for some } c > 1 \text{ and all } m \in \mathbb{Z}.$$

These processes are called semistable Lévy processes and are determined by log-periodic perturbations of the tails of the Lévy measure, i.e. instead of (1.3) we have for all $x > 0$

$$\phi(x,\infty) = x^{-\alpha}\theta_+(\log x) \quad \text{and} \quad \phi(-\infty,-x) = x^{-\alpha}\theta_-(\log x), \tag{1.7}$$

where $\theta_\pm$ are non-negative, $\log(c^{1/\alpha})$-periodic functions such that $x \mapsto x^{-\alpha}\theta_\pm(\log x)$ are non-increasing, which we call admissable. Replacing the Lévy measure (1.3) by (1.7), the formulas (1.2) for the log-characteristic function and (1.5) for the generator remain valid for a semistable Lévy process. For details on semistable distributions and Lévy processes we refer to the monographs [30, 40]. Log-periodic disturbances of power law behavior frequently appears in a variety of physical applications [41, 49] and also in finance [42]. The most prominent example of a semistable Lévy process which is not stable is the limiting process for the normalized gain in successive St. Petersburg games derived in [8, 29]. Here, the Lévy measure ϕ is concentrated on $2^{\mathbb{Z}}$ with $\phi(\{2^m\}) = 2^{-m}$ for all $m \in \mathbb{Z}$ such that $c = 2$, $\alpha = 1$, $\theta_- \equiv 0$ and $\theta_+(x) = 2^{\langle \frac{x}{\log 2}\rangle}$, where $\langle x\rangle = x - \lfloor x\rfloor$ denotes the fractional part of $x \in \mathbb{R}$. For details see [17] where also fractal path properties of this particular semistable Lévy process are investigated. However, since $\alpha = 1$ the example is outside the scope of this article. In recent years the fractal path behavior of general semistable Lévy processes has been investigated, complementing the above mentioned classical results for their stable counterparts. It turned out that in terms of Hausdorff dimension the range, the graph and multiple points of the sample paths almost surely are not affected by the log-periodic perturbations [16, 19, 28, 47] even in terms of exact Hausdorff measure [18]. Nevertheless, semistable Lévy processes show a different behavior when turning to probability densities which are known to be of class $C^\infty(\mathbb{R})$ again with all its derivatives belonging to $C_0(\mathbb{R}) \cap L^1(\mathbb{R})$. Recently, semi-fractional derivatives have been introduced in [20] such that densities of semistable Lévy processes solve corresponding semi-fractional diffusion equations. This new class of fractional derivatives can be seen as a special case of general fractional derivatives as in [24, 25]. In Sect. 3 we ask for a new duality result concerning the more general class of semistable Lévy processes. The approach allows us to develop a novel dual equation with a semi-fractional derivative in time in which the log-periodic disturbances cause an additional inhomogeneity and thus shows a significantly different behavior compared to their stable counterpart. Finally, proofs of our new results are given in Sect. 4.

2 Fractional Diffusions and Zolotarev Duality

In this section we follow the arguments laid out in [14, 32] to derive the probabilistic solution to certain fractional diffusion equations by stable densities, and the approach in [14] to space-time duality in the negatively skewed case. This is best suited to our desired generalization towards the semistable setting in Sect. 3.

We will frequently make use of the following transforms of our densities for $k \in \mathbb{R}$ and $s > 0$.

$$\text{Fourier transform (FT):} \quad \widehat{p}(k,t) = \int_{\mathbb{R}} e^{ikx} p(x,t)\,dx$$

$$\text{Laplace transform (LT):} \quad \widetilde{p}(x,s) = \int_0^\infty e^{-st} p(x,t)\,dt$$

$$\text{Fourier–Laplace transform (FLT):} \quad \bar{p}(k,s) = \int_0^\infty \int_{\mathbb{R}} e^{-st+ikx} p(x,t)\,dx\,dt$$

Turning to the FT on both sides of (1.6) yields

$$\begin{aligned} \frac{\partial}{\partial t}\widehat{p}(k,t) &= v\,ik\widehat{p}(k,t) + C\left(\frac{1+\beta}{2}(-ik)^\alpha + \frac{1-\beta}{2}(ik)^\alpha\right)\widehat{p}(k,t) \\ &= v\,ik\widehat{p}(k,t) - \sigma^\alpha |k|^\alpha \left(1 - i\beta\,\mathrm{sign}(k)\tan(\alpha\tfrac{\pi}{2})\right)\widehat{p}(k,t), \end{aligned} \tag{2.1}$$

where the last equality follows after a short calculation with the scale parameter $\sigma = (-C\cos(\alpha\frac{\pi}{2}))^{1/\alpha} > 0$; see equations (5.5) and (5.6) in [32] for details. With the initial conditions $\widehat{p}(0,t) = 1$ for a probability density, and $\widehat{p}(k,0) = 1$ corresponding to the point source $p(x,0) = \delta(x)$, using (1.4) the unique solution to the ode (2.1) is given by $\widehat{p}(k,t) = \widehat{g}(k;\alpha,\beta,\sigma t^{1/\alpha}, vt)$, showing that the stable densities $p(x,t) = g(x;\alpha,\beta,\sigma t^{1/\alpha}, vt)$ solve (1.6).

We now restrict our considerations to the negatively skewed case $\beta = -1$ with $\alpha \in (1,2)$, $v = 0$ and $C = 1$. The corresponding fractional diffusion equation

$$\frac{\partial}{\partial t} p(x,t) = \frac{\partial^\alpha}{\partial(-x)^\alpha} p(x,t) \tag{2.2}$$

is solved by the stable densities

$$p(x,t) = g\left(x;\alpha,-1,\left(\left|\cos\left(\alpha\tfrac{\pi}{2}\right)\right| t\right)^{1/\alpha},0\right). \tag{2.3}$$

Applying FLT to both sides of (2.2) yields $s\,\bar{p}(k,s) - 1 = (ik)^\alpha \bar{p}(k,s)$ for the point source fulfilling $\widehat{p}(k,0) = 1$ with solution

$$\bar{p}(k,s) = \frac{1}{s-(ik)^\alpha} = \frac{1}{s-\psi(k)}, \tag{2.4}$$

where ψ is as in (1.2) for the Lévy measure ϕ concentrated on the negative axis with $\phi(-\infty,-x) = x^{-\alpha}\frac{\alpha-1}{\Gamma(2-\alpha)}$ and $\mu = \int_{-\infty}^{0}(\frac{x}{1+x^2} - x)\,d\phi(x)$. Note that $\bar{p}$ has a simple pole at $k = -i\,s^{1/\alpha}$. Inverting the FT with the help of Cauchy's residue theorem (details are given in Sect. 4), for $x > 0$ this leads to

$$\widetilde{p}(x,s) = \frac{1}{\alpha} s^{-1+1/\alpha} \exp\left(-x\,s^{1/\alpha}\right) = \frac{1}{\alpha}\widetilde{h}(x,s) \tag{2.5}$$

for the Laplace transform (LT) $\widetilde{p}(x,s) = \int_0^\infty e^{-st} p(x,t)\,dt$ as shown in [14], where $\widetilde{h}$ is the LT of the inverse $\frac{1}{\alpha}$-stable subordinator (see Remark 2.3) with $\frac{1}{\alpha} \in (\frac{1}{2},1)$ and density

$$h(x,t) = \alpha t\,x^{-1-\alpha} g\left(t\,x^{-\alpha};\tfrac{1}{\alpha},1,\left|\cos\left(\tfrac{1}{\alpha}\tfrac{\pi}{2}\right)\right|^\alpha,0\right) \tag{2.6}$$

for $x > 0$; see [31] or equation (4.47) in [32]. Combining (2.3), (2.5), and (2.6) directly leads to Zolotarev's duality result relating negatively skewed α-stable densities for $\alpha \in (1,2)$ with positively skewed $\frac{1}{\alpha}$-stable densities:

Theorem 2.1 ([50, Theorem 1]) *For $\alpha \in (1,2)$ and stable densities g parametrized as in (1.4) we have for all $x > 0$ and $t > 0$*

$$g\left(x;\alpha,-1,\left(\left|\cos\left(\alpha\tfrac{\pi}{2}\right)\right| t\right)^{1/\alpha},0\right) = t\,x^{-1-\alpha} g\left(t\,x^{-\alpha};\tfrac{1}{\alpha},1,\left|\cos\left(\tfrac{1}{\alpha}\tfrac{\pi}{2}\right)\right|^\alpha,0\right).$$

Note that Zolotarev uses a different parametrization which can be transferred to the above parametrization (1.4) as described in [1]. Zolotarev proved this result in [50] by transforming the FT of the α-stable density using complex contour integrals; cf.

also Theorem 2.3.1 in [51]. Lukacs [27, Theorem 3.3] gave a different proof using a series representation of stable densities independently obtained by Bergström [3] and Feller [10]. In this work of Feller the α-stable density is also shown to be a solution to a fractional diffusion equation with a fractional integral operator of negative order $-\alpha$. It is worth mentioning that Zolotarev duality also holds for arbitrary values of the skewness parameter β, but then the below interpretation as a solution of a time-fractional pde fails. Zolotarev's result further holds for $\alpha = 2$ which leads to a closed form expression of a positively skewed $\frac{1}{2}$-stable density, the only closed form expression known besides the Gaussian and the Cauchy density. This density is frequently called Lévy density due to its appearance in [26], but according to section 3.7 in [9] it was already observed by Heavyside in 1871. The fractional pde connection for the case $\alpha = 2$ can be found in [2].

Coming back to duality, we now want to show that (2.6) is related to a time-fractional pde. Therefore, applying FT for $x > 0$ to (2.5) yields $\bar{h}(k,s) = \frac{s^{-1+1/\alpha}}{s^{1/\alpha}-ik}$ which leads to the equation

$$s^{1/\alpha}\bar{h}(k,s) - s^{-1+1/\alpha} = ik\,\bar{h}(k,s).$$

Inverting the FT on both sides gives

$$s^{1/\alpha}\widetilde{h}(x,s) - s^{-1+1/\alpha}\delta(x) = -\frac{\partial}{\partial x}\widetilde{h}(x,s). \tag{2.7}$$

For suitable functions f and $t \geq 0$ denote by $(\frac{\partial}{\partial t})^{\gamma} f(t)$ the Caputo fractional derivative of order $\gamma \in (0,1)$ which is the unique function with LT $s^{\gamma}\widetilde{f}(s) - s^{\gamma-1}f(0)$, whereas the Riemann–Liouville fractional derivative $\frac{\partial^{\gamma}}{\partial t^{\gamma}}$ of order $\gamma \in (0,1)$ is the unique function with LT $s^{\gamma}\widetilde{f}(s)$. Then Laplace inversion on both sides of (2.7) yields

$$\left(\frac{\partial}{\partial t}\right)^{1/\alpha} h(x,t) = -\frac{\partial}{\partial x}h(x,t) \tag{2.8}$$

for $x > 0$ and $t > 0$. Since $p(x,t) = \alpha^{-1}h(x,t)$ by (2.5), the original α-stable density p also solves the time-fractional pde (2.8) under point source initial condition $p(x,0) = \delta(x)$ leading directly to space-time duality for fractional diffusions:

Theorem 2.2 ([1, 14]) *For $x > 0$ and $t > 0$ the point source solutions $p(x,t)$ of the fractional diffusion equation (2.2) of order $\alpha \in (1,2)$ and $h(x,t)$ of the time-fractional pde (2.8) of order $\frac{1}{\alpha} \in (\frac{1}{2},1)$ are equivalent, i.e. they are proportional to each other: $p(x,t) = \alpha^{-1}h(x,t)$ for all $x > 0$ and $t > 0$.*

The proof in [1] directly uses Zolotarev duality, whereas the above arguments from [14] only use FLT techniques and gives the partial result on Zolotarev duality stated in Theorem 2.1 as a byproduct. In the semistable setup corresponding duality results

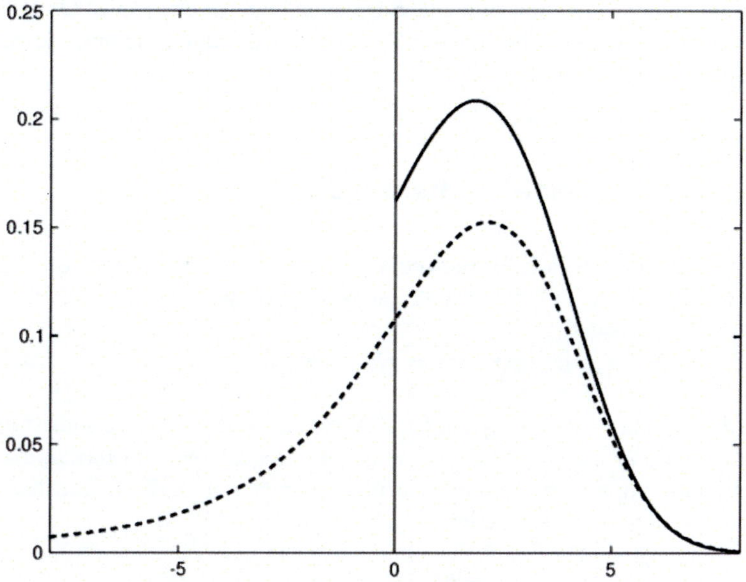

Fig. 3 Solutions $p(x, t_0)$ (dashed line) of the fractional diffusion equation (2.2) and $h(x, t_0)$ (solid line) of the time-fractional pde (2.8) for fixed $t_0 = 3.5$ and $\alpha = 1.5$

are not known in the literature and the above FLT method is our preferable choice in Sect. 3 to derive a corresponding semistable duality result.

To illustrate Theorem 2.2 we plotted numerical solutions $p(x, t)$ of the fractional diffusion equation (2.2) and $h(x, t)$ of the time-fractional pde (2.8) for fixed $t_0 = 3.5$ and $\alpha = 1.5$ in Fig. 3. For the stable density $p(x, t_0)$ in (2.3) we use a Fourier inversion technique together with the representation (1.4), whereas $h(x, t_0)$ was approximated from (2.8) by a finite difference method [33] involving Grünwald–Letnikov differences for the time-fractional derivative. Note that in Fig. 1 the ratio $h(x, t_0)/p(x, t_0)$ decreases from the true value $\alpha = 1.5$ at $x = 0$ almost linearly to 1.2 at $x = 4$ which is an effect of the rather weak approximation by Grünwald–Letnikov differences for which the error increases with the distance from the origin.

Remark 2.3 The time-fractional equation (2.8) has the following probabilistic interpretation. If $(D_t)_{t\geq 0}$ is a $\frac{1}{\alpha}$-stable subordinator, i.e. a $\frac{1}{\alpha}$-stable Lévy process with almost surely strictly increasing sample paths, then its hitting-time process $(E_t := \inf\{u > 0 : D_u > t\})_{t\geq 0}$ which is also called an inverse stable subordinator, has a smooth probability density $x \mapsto h(x, t)$ which solves (2.8) with initial point source condition; see [31, 32] for details. The space-time duality in Theorem 2.2 does not cover the full range $\frac{1}{\alpha} \in (0, 1)$ for $\frac{1}{\alpha}$-stable subordinators. Extending Theorem 2.2 for $\frac{1}{\alpha} \in (0, \frac{1}{2})$ would lead to an equivalent space-fractional pde of order $\alpha > 2$ for which in its full generality no meaningful stochastic solution exists. A first result towards this direction is given in [15] for $\frac{1}{\alpha} \in (\frac{1}{3}, \frac{1}{2})$ leading to a probabilistic interpretation of a space-fractional pde of order $\alpha \in (2,3)$ by means of an inverse $\frac{1}{\alpha}$-stable subordinator. This stochastic solution is much stronger than the higher order approach in [5].

3 Duality for Semi-Fractional Diffusions

We now turn to a negatively skewed semistable distribution for $\alpha \in (1,2)$ with a Lévy measure ϕ as in (1.7) concentrated on the negative axis

$$\phi(-\infty,-x) = x^{-\alpha}\theta(\log x), \qquad x > 0.$$

Here θ is an admissable function, i.e. θ is a positive, $\log(c^{1/\alpha})$-periodic function for some $c > 1$ and $x \mapsto x^{-\alpha}\theta(\log x)$ is non-increasing. We will further assume that θ is smooth, i.e. θ is continuous and piecewise continuously differentiable, hence representable by a Fourier series

$$\theta(x) = \sum_{n\in\mathbb{Z}} c_n\, e^{in\widetilde{c}x} \quad \text{with} \quad \widetilde{c} = \frac{2\pi\alpha}{\log c}.$$

In the special case of constant $\theta \equiv c_0 = \frac{\alpha-1}{\Gamma(2-\alpha)}$ and $\mu = \int_{-\infty}^{0}(\frac{x}{1+x^2} - x)\, d\phi(x)$ in (1.2) this reduces to the stable distribution corresponding to the fractional diffusion equation (2.2). For the more general semistable distribution with the same drift parameter μ the corresponding semi-fractional diffusion equation is given by

$$\frac{\partial}{\partial t} p(x,t) = \frac{\partial^\alpha}{\partial_{c,\theta}(-x)^\alpha} p(x,t). \tag{3.1}$$

Here, for suitable functions f the negative semi-fractional derivative of order $\alpha \in (1,2)$ was recently introduced in [20] by its generator form

$$\begin{aligned}\frac{\partial^\alpha}{\partial_{c,\theta}(-x)^\alpha} f(x) = Lf(x) &= \int_{-\infty}^{0} \big(f(x-y) - f(x) + yf'(x)\big)\, d\phi(y)\\ &= \int_0^\infty \big(f'(x+y) - f'(x)\big)\, y^{-\alpha}\theta(\log y)\, dy,\end{aligned} \tag{3.2}$$

where the last equality follows from reflection and integration by parts. As shown in [20], with this definition the negatively skewed semistable densities $x \mapsto p(x,t)$ are a solution to (3.1). Moreover, it was shown in [20] that the corresponding log-characteristic function admits the series representation

$$\psi(k) = -\sum_{n\in\mathbb{Z}} c_n\, \Gamma(in\widetilde{c} - \alpha + 1)(ik)^{\alpha - in\widetilde{c}} \tag{3.3}$$

which for the stable case $\theta \equiv c_0 = \frac{\alpha-1}{\Gamma(2-\alpha)} = -\frac{1}{\Gamma(1-\alpha)}$ reduces to $\psi(k) = (ik)^\alpha$ and gives back the negative Riemann–Liouville fractional derivative of order $\alpha \in (1, 2)$. Applying the FLT on both sides of (3.1) again yields $\bar{p}(k, s) = \frac{1}{s-\psi(k)}$ as in (2.4) for the corresponding semistable densities, but now with ψ from (3.3). We will show in Lemma 4.1 that the FLT $\bar{p}$ has again a single pole at some $k = -i\xi(s)$ on the negative imaginary axis which enables us to invert the FT with the help of Cauchy's residue theorem to come to:

Theorem 3.1 *For $\alpha \in (1, 2)$ the LT with respect to time of the semistable densities corresponding to the the semi-fractional diffusion equation* (3.1) *takes the form*

$$\widetilde{p}(x, s) = \frac{1}{\alpha} \frac{s^{1/\alpha} g(\log s) \exp\left(-x\, s^{1/\alpha} g(\log s)\right)}{s + f(s)} =: \frac{1}{\alpha} \widetilde{h}(x, s), \tag{3.4}$$

where g is a continuously differentiable, $\log(c)$*-periodic function and f is some specific function such that $s + f(s) > 0$. Moreover, f and g only depend on $c > 1$, $\alpha \in (1, 2)$ and the admissible function θ.*

The proof of Theorem 3.1 is given in Sect. 4. As in Sect. 2 we now calculate the FT of $\widetilde{h}$ on the right-hand side of (3.4) and then apply FLT inversion which also justifies the LT notation $\widetilde{h}(x, s)$ in Theorem 3.1. Writing $\xi(s) = s^{1/\alpha} g(\log s)$ to simplify notation (it turns out that this is indeed the location of the pole of $\bar{p}(k, s)$ on the negative imaginary axis stated above) and applying FT for $x > 0$ to (3.4) yields

$$\begin{aligned}
\bar{h}(k, s) &= \frac{\xi(s)}{s + f(s)} \int_0^\infty \exp\left(-x\,(\xi(s) - ik)\right) dx \\
&= \frac{\xi(s)}{s + f(s)} \frac{1}{\xi(s) - ik} = \left(\frac{1}{s} - \frac{1}{s} \frac{f(s)}{s + f(s)}\right) \frac{\xi(s)}{\xi(s) - ik},
\end{aligned}$$

which leads to the equation

$$\xi(s)\bar{h}(k, s) - s^{-1}\xi(s) - ik\bar{h}(k, s) = -\frac{1}{s} \frac{f(s)}{s + f(s)} \xi(s) =: \frac{1}{s} s^{1/\alpha} \gamma(\log s).$$

Inverting the FT on both sides gives

$$\xi(s)\widetilde{h}(x, s) - s^{-1}\xi(s)\delta(x) + \frac{\partial}{\partial x}\widetilde{h}(x, s) = \frac{1}{s} s^{1/\alpha} \gamma(\log s)\delta(x). \tag{3.5}$$

We will show in Lemma 4.3 that γ is a smooth $\log(c)$-periodic function and thus γ and g from Theorem 3.1 both admit a Fourier series representation

$$g(x) = \sum_{n \in \mathbb{Z}} d_n\, e^{-in\widetilde{d}x} \quad \text{and} \quad \gamma(x) = \sum_{n \in \mathbb{Z}} h_n\, e^{-in\widetilde{d}x} \tag{3.6}$$

with $\widetilde{d} = \frac{2\pi}{\log c} = \frac{2\pi \frac{1}{\alpha}}{\log d}$ for $d = c^{1/\alpha} > 1$. Let us define the functions

$$\tau(x) = \sum_{n\in\mathbb{Z}} \frac{d_n}{\Gamma(in\widetilde{d} - \frac{1}{\alpha} + 1)} e^{in\widetilde{d}x} \quad \text{and} \quad \rho(x) = \sum_{n\in\mathbb{Z}} \frac{h_n}{\Gamma(in\widetilde{d} - \frac{1}{\alpha} + 1)} e^{in\widetilde{d}x} \tag{3.7}$$

which clearly are $\log(d^\alpha)$-periodic functions. Note that formally $\tau(-\log s)$ and $\rho(-\log s)$ are related to $\xi(s) = s^{1/\alpha} g(\log s)$ and $s^{1/\alpha}\gamma(\log s)$ in the same manner than $\theta(-\log(ik))$ is related to $-\psi(k)$ in (3.3), simply by multiplying the Fourier coefficients with appropriate values of the gamma function depending on the admissability parameters. We conjecture that τ and ρ are admissable with respect to the parameters $d > 1$ and $\frac{1}{\alpha} \in (\frac{1}{2}, 1)$. If so, then for suitable functions f and $t \geq 0$ we may formally introduce the Riemann–Liouville and the Caputo semi-fractional derivative by LT inversion in analogy to time-fractional derivatives:

$$\frac{\partial^{1/\alpha}}{\partial_{d,\tau}\, t^{1/\alpha}} f(t) = r(t) \iff \widetilde{r}(s) = \xi(s)\widetilde{f}(s),$$

$$\left(\frac{\partial}{\partial_{d,\tau} t}\right)^{1/\alpha} f(t) = r(t) \iff \widetilde{r}(s) = \xi(s)\widetilde{f}(s) - s^{-1}\xi(s) f(0).$$

Remark 3.2 It is worth mentioning that this formal introduction of semi-fractional derivatives for functions on the positive real line can be strengthened from a probabilistic perspective. In fact the densities $h(x,t)$ of an inverse $\frac{1}{\alpha}$-semistable subordinator with a $\log(d^\alpha)$-periodic admissable function τ in the positive tail of the Lévy measure solve the semi-fractional pde

$$\left(\frac{\partial}{\partial_{d,\tau} t}\right)^{1/\alpha} h(x,t) = -\frac{\partial}{\partial x} h(x,t)$$

in analogy to (2.8) for the densities of an inverse $\frac{1}{\alpha}$-stable subordinator. This fact is outside the scope of this article and will be published elsewhere.

Finally, since $\frac{1}{s} = \int_0^\infty e^{-st}\, dt$ is the LT of the function $1_{(0,\infty)}(t)$, we may now rewrite (3.5) as

$$\left(\frac{\partial}{\partial_{d,\tau} t}\right)^{1/\alpha} h(x,t) + \frac{\partial}{\partial x} h(x,t) = \delta(x) \frac{\partial^{1/\alpha}}{\partial_{d,\rho}\, t^{1/\alpha}} 1_{(0,\infty)}(t). \tag{3.8}$$

Similar to (3.2), for suitable functions f the semi-fractional Caputo derivative of order $\frac{1}{\alpha} \in (0,1)$ (here we have $\frac{1}{\alpha} \in (\frac{1}{2}, 1)$) with respect to $d > 1$ and the admissable function ρ is given in [20] by

$$\left(\frac{\partial}{\partial_{d,\rho} t}\right)^{1/\alpha} f(t) = \int_0^\infty f'(t-s) s^{-1/\alpha} \rho(\log s)\, ds \tag{3.9}$$

and the corresponding Riemann–Liouville derivative is obtained by interchanging differentiation and integration on the right-hand side of (3.9). Hence, on the right-hand side of (3.8) we get

$$\begin{aligned}\frac{\partial^{1/\alpha}}{\partial_{d,\rho}\,t^{1/\alpha}}1_{(0,\infty)}(t) &= \frac{d}{dt}\int_0^\infty 1_{(0,\infty)}(t-s)\,s^{-1/\alpha}\rho(\log s)\,ds\\ &= \frac{d}{dt}\int_0^t s^{-1/\alpha}\rho(\log s)\,ds = t^{-1/\alpha}\rho(\log t)\end{aligned}$$

which yields

$$\left(\frac{\partial}{\partial_{d,\tau}t}\right)^{1/\alpha}h(x,t)+\frac{\partial}{\partial x}h(x,t)=\delta(x)\,t^{-1/\alpha}\rho(\log t). \tag{3.10}$$

Thus we have shown space-time duality for semi-fractional diffusions:

Theorem 3.3 *Assume that τ and ρ in (3.7) are admissable functions with respect to the parameters $d=c^{1/\alpha}>1$ and $\frac{1}{\alpha}\in(\frac{1}{2},1)$. Then for $x>0$ and $t>0$ the point source solutions $p(x,t)$ of the semi-fractional diffusion equation (3.1) of order $\alpha\in(1,2)$ in space and $h(x,t)$ of the semi-fractional pde (3.10) of order $\frac{1}{\alpha}\in(\frac{1}{2},1)$ in time are equivalent, i.e. $p(x,t)=\alpha^{-1}h(x,t)$ for all $x>0$ and $t>0$.*

Note that with f and g also τ and ρ do only depend on $c>1$, $\alpha\in(1,2)$ and the admissable function θ of the underlying semistable distribution.

4 Proofs for Sect. 3

For simplicity, we write $\omega_n=-c_n\Gamma(in\tilde{c}-\alpha+1)$ for the coefficients in (3.3). Extending ψ for $z\in\mathbb{C}$ shows that

$$\psi(z)=\sum_{n\in\mathbb{Z}}\omega_n(iz)^{\alpha-in\tilde{c}}=(iz)^\alpha\sum_{n\in\mathbb{Z}}\omega_n e^{-in\tilde{c}\log(iz)} \tag{4.1}$$

is an analytic function in the lower half plane, where the series in (4.1) is absolutely convergent by Theorem 3.1 in [20], and ψ admits the representation

$$\psi(z)=\int_{-\infty}^0\left(e^{izx}-1-izx\right)\,d\phi(x). \tag{4.2}$$

Moreover, since $\omega_{-n}=\overline{\omega_n}$ for $n\in\mathbb{Z}$, the function

$$\psi(-ik)=k^\alpha\sum_{n\in\mathbb{Z}}\omega_n e^{-in\tilde{c}\log(k)}=:k^\alpha m(\log k) \tag{4.3}$$

for $k > 0$ is a real function such that m is $\log(c^{1/\alpha})$-periodic.

Lemma 4.1 *For any $s > 0$ there is a unique $z = z(s)$ in the lower half plane such that $s = \psi(z(s))$. Moreover, $z(s) = -i\,\xi(s)$ with $\xi(s) > 0$ lies on the negative imaginary axis.*

Proof From (4.2) it can be deduced that for z in the lower half plane $\psi(z) \in \mathbb{R}$ iff $z = -ik$ with $k > 0$. If we consider the real mapping $s(k) = \psi(-ik)$ for $k > 0$ then by (4.2)

$$s'(k) = \int_{-\infty}^{0} x\left(e^{kx} - 1\right) d\phi(x) > 0$$

and thus $k \mapsto s(k)$ is a continuously differentiable and strictly increasing function with $\lim_{k\downarrow 0} s(k) = 0$ and $\lim_{k\to\infty} s(k) = \infty$. Hence, for $s > 0$ there is a unique $\xi(s) > 0$ with $s = \psi(-i\,\xi(s))$.

Lemma 4.2 *The function ξ from Lemma 4.1 is continuously differentiable and for $s > 0$ we have $\xi(s) = s^{1/\alpha} g(\log s)$ for some $\log(c)$-periodic function g.*

Proof Since ξ is the inverse of the function $k \mapsto s(k) = \psi(-ik)$ appearing in the proof of Lemma 4.1, it is itself continuously differentiable and strictly increasing. By (4.3) we get

$$\begin{aligned}
\psi\left(-i\,c^{1/\alpha}\xi(s)\right) &= c\,\xi(s)^{\alpha} m\left(\log(c^{1/\alpha}) + \log \xi(s)\right) \\
&= c\,\xi(s)^{\alpha} m\left(\log \xi(s)\right) = c\,\psi(-i\,\xi(s)) \\
&= cs = \psi(-i\,\xi(cs))
\end{aligned}$$

and thus we have $c^{1/\alpha}\xi(s) = \xi(cs)$. Defining $g(x) = e^{-x/\alpha}\xi(e^x)$ we get

$$g(x + \log c) = e^{-x/\alpha} c^{-1/\alpha} \xi(c\,e^x) = e^{-x/\alpha}\xi(e^x) = g(x).$$

Proof of Theorem 3.1 Using equation (4.8.18) in [35], an inversion of the FT of $\bar{p}(k, s) = (s - \psi(k))^{-1}$ for fixed $s > 0$ gives

$$\widetilde{p}(x, s) = \frac{1}{2\pi} \lim_{T\to\infty} \int_{-T-i\xi_0}^{T-i\xi_0} \frac{e^{-ikx}}{s - \psi(k)}\, dk, \tag{4.4}$$

where we choose $\xi_0 \in (0, \xi(s))$. For large $T > 0$ consider the cut semicircle $C_T + L_T$ in the lower half plane as in the picture.

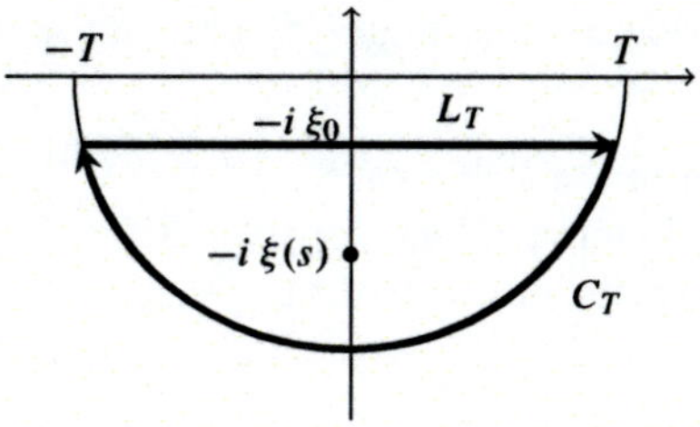

Letting $k = T\,e^{-i\varphi}$ we get

$$\left|\int_{C_T} \frac{e^{-ikx}}{s-\psi(k)}\,dk\right| \le \int_0^{\pi} \frac{T\exp(-Tx\sin\varphi)}{|s-\psi(T\,e^{-i\varphi})|}\,d\varphi \to 0$$

as $T \to \infty$ by dominated convergence, since we can easily derive $\operatorname{Re}\psi(T\,e^{-i\varphi}) \to \infty$ for $\varphi \in (0,\pi)$. By Lemma 4.1 and Cauchy's residue theorem we get from (4.4) with the function $s(k)$ from the proof of Lemma 4.1

$$\begin{aligned}\widetilde{p}(x,s) &= -i\ \operatorname{Res}(-i\,\xi(s)) = \frac{i\,e^{-x\xi(s)}}{\psi'(-i\,\xi(s))} = \frac{e^{-x\xi(s)}}{s'(\xi(s))}\\ &= \frac{e^{-x\xi(s)}}{\xi(s)^{\alpha-1}\,(\alpha\,m(\log\xi(s)) + m'(\log\xi(s)))}\\ &= \frac{1}{\alpha}\,\frac{\xi(s)e^{-x\xi(s)}}{\psi(-i\,\xi(s)) + \frac{1}{\alpha}\,\xi(s)^{\alpha}m'(\log\xi(s))} = \frac{1}{\alpha}\,\frac{\xi(s)e^{-x\xi(s)}}{s+f(s)},\end{aligned}$$

where $f(s) = \frac{1}{\alpha}\,\xi(s)^{\alpha}m'(\log\xi(s))$. Hence we have shown (3.4) and the denominator is strictly positive, since $s + f(s) = \alpha^{-1}\xi(s)\,s'(\xi(s)) > 0$. Note that due to the above approach f and g only depend on the parameters c, α and θ of the semistable distribution.

Lemma 4.3 *Let $f(s) = \frac{1}{\alpha}\,\xi(s)^{\alpha}m'(\log\xi(s))$ as above. Then we can write*

$$\frac{-f(s)}{s+f(s)}\,\xi(s) = s^{1/\alpha}\gamma(\log s)$$

for some $\log(c)$*-periodic and smooth function* γ.

Proof Write

$$\frac{-f(s)}{s+f(s)}\,\xi(s) = \frac{-g(\log s)^{\alpha}m'(\log\xi(s))}{\alpha + g(\log s)^{\alpha}m'(\log\xi(s))}\,s^{1/\alpha}g(\log s) = s^{1/\alpha}\gamma(\log s).$$

Since g is $\log(c)$-periodic, m is $\log(c^{1/\alpha})$-periodic and $\xi(cs) = c^{1/\alpha}\xi(s)$, the assertion follows easily.

Remark 4.4 Note that in the stable case we have $\psi(k) = (ik)^\alpha$ and thus $m \equiv 1$ in (4.3) and $g \equiv 1$ in Lemma 4.2 are constant. Thus $f \equiv 0$ in the above proof of Theorem 3.1 and (3.4) coincides with (2.5).

References

1. Baeumer, B., Meerschaert, M.M., Nane, E.: Space-time duality for fractional diffusion. J. Appl. Probab. **46**, 110–115 (2009)
2. Baeumer, B., Meerschaert, M.M., Nane, E.: Brownian subordinators and fractional Cauchy problems. Trans. Am. Math. Soc. **361**(7), 3915–3930 (2009)
3. Bergström, H.: On some expansions of stable distributions. Ark. Mat. **2**, 375–378 (1952)
4. Blumenthal, R.M., Getoor, R.K.: A dimension theorem for sample functions of stable processes. Ill. J. Math. **4**, 370–375 (1960)
5. Bonaccorsi, S., D'Ovidio, M., Mazzucchi, S.: Probabilistic representation formula for the solution of high-order heat-type equations. J. Evol. Equ. **19**(2), 523–558 (2019)
6. Butera, S., Di Paola, M.: A physically based connection between fractional calculus and fractal geometry. Ann. Phys. **350**, 146–158 (2014)
7. Ciesielski, Z., Taylor, S.J.: First passage times and sojourn times for Brownian motion in space and the exact Hausdorff measure of the sample path. Trans. Am. Math. Soc **103**, 434–450 (1962)
8. Csörgő, S., Dodunekova, R.: Limit theorems for the Petersburg game. In: Hahn, M.G. et al. (eds.) Sums, Trimmed Sums and Extremes. Progress in Probability, vol. 23, pp. 285–315. Birkhäuser, Boston (1991)
9. Das, S.: Functional Fractional Calculus. Springer, New York (2011)
10. Feller, W.: On a generalization of Marcel Riesz' potentials and the semi-groups generated by them. Meddelanden Lunds Universitetes Matematiska Seminarium Tome Supplémentaire, pp. 72–81. Scientific Research, Wuhan (1952)
11. Hendricks, W.J.: A dimension theorem for sample functions of processes with stable components. Ann. Probab. **1**, 849–853 (1973)
12. Hilfer, R.: Threefold introduction to fractional derivatives. In: Klages, R. et al. (eds.) Anomalous Transport: Foundations and Applications, pp. 17–74. Wiley, Weinheim (2008)
13. Hou, Y, Ying, J.: Sample path properties of a class of operator stable processes. Stoch. Anal. Appl. **25**, 317–335 (2007)
14. Kelly, J.F., Meerschaert, M.M.: Space-time duality for the fractional advection-dispersion equation. Water Resour. Res. **53**, 3464–3475 (2017)
15. Kelly, J.F., Meerschaert, M.M.: Space-time duality and high-order fractional diffusion. Phys. Rev. E **99**, 022122 (2019)
16. Kern, P., Wedrich, L.: The Hausdorff dimension of operator semistable Lévy processes. J. Theor. Probab. **27**, 383–403 (2014)
17. Kern, P., Wedrich, L.: Dimension results related to the St. Petersburg game. Probab. Math. Stat. **34**(1), 97–117 (2014)
18. Kern, P., Wedrich, L.: On exact Hausdorff measure functions of operator semistable Lévy processes. Stoch. Anal. Appl. **35**, 980–1006 (2017)
19. Kern, P., Meerschaert, M.M., Xiao, Y.: Asymptotic behavior of semistable Lévy exponents and applications to fractal path properties. J. Theor. Probab. **31**, 598–617 (2018)
20. Kern, P., Lage, S., Meerschaert, M.M.: Semi-fractional diffusion equations. Fract. Calc. Appl. Anal. **22**(2), 326–357 (2019)

21. Khoshnevisan, D.: Intersections of Brownian motions. Expos. Math. **21**, 97–114 (2003)
22. Kilbas, A.A., Srivastava, H.M., Trujillo, J.J.: Theory and Applications of Fractional Differential Equations. North-Holland Mathematical Studies, vol. 204, Elsevier, Amsterdam (2006)
23. Kiryakova, V.S.: A long standing conjecture failed? In: Transform Methods and Special Functions, Proceedings of 2nd International Workshop, Varna 1996, pp. 584–593. Institute of Statistical Mathematics, Bulgarian Academy of Sciences, Bulgaria (1998)
24. Kochubei, A.N.: General fractional calculus, evolution equations, and renewal processes. Integr. Equ. Oper. Theory **71**, 583–600 (2011)
25. Kochubei, A.N., Kondratiev, Y., da Silva, J.L.: From random times to fractional kinetics. Interdisciplinary Studies of Complex Systems **16**, 5–32 (2020)
26. Lévy, P.: Sur certains processus stochastiques homogènes. Compos. Math. **7**, 283–339 (1939)
27. Lukacs, E.: Stable distributions and their characteristic functions. Jahresber. Dtsch. Math.-Ver. **71**, 84–114 (1969)
28. Luks, T., Xiao, Y.: Multiple points of operator semistable Lévy processes. J. Theor. Probab. **33**, 153–179 (2020)
29. Martin-Löf, A.: A limit theorem which clarifies the "Petersburg paradox". J. Appl. Probab. **22**, 634–643 (1985)
30. Meerschaert, M.M., Scheffler, H.P.: Limit Distributions for Sums of Independent Random Vectors. Wiley, New York (2001)
31. Meerschaert, M.M., Scheffler, H.P.: Limit theorems for continuous time random walks with infinite mean waiting times. J. Appl. Probab. **41**(3), 623–638 (2004)
32. Meerschaert, M.M., Sikorskii, A.: Stochastic Models for Fractional Calculus. De Gruyter, Berlin (2012)
33. Meerschaert, M.M., Tadjeran, C.: Finite difference approximations for two-sided space-fractional partial differential equations. Appl. Numer. Math. **56**, 80–90 (2006)
34. Meerschaert, M.M., Xiao, Y.: Dimension results for sample paths of operator stable Lévy processes. Stoch. Process. Appl. **115**, 55–75 (2005)
35. Morse, P.M., Feshbach, H.: Methods of Theoretical Physics, Part I. McGraw-Hill, New York (1953)
36. Pruitt, W.E.: The Hausdorff dimension of the range of a process with stationary independent increments. J. Math. Mech. **19**, 371–378 (1969)
37. Pruitt, W.E., Taylor, S.J.: Sample path properties of processes with stable components. Z. Wahrsch. verw. Geb. **12**, 267–289 (1969)
38. Samko, S.G., Kilbas, A.A., Marichev, O.I.: Fractional Integrals and Derivatives. Gordon and Breach, London (1993)
39. Samorodnitsky, G., Taqqu, M.S.: Stable Non-Gaussian Random Processes. Chapman & Hall, New York (1994)
40. Sato, K.: Lévy Processes and Infinitely Divisible Distributions. Cambridge University Press, Cambridge (1999)
41. Sornette, D.: Discrete-scale invariance and complex dimensions. Phys. Rep. **297**, 239–270 (1998)
42. Sornette, D.: Why Stock Markets Crash: Critical Phenomena in Complex Financial Systems. Princeton University Press, Princeton (2017)
43. Tatom, F.B.: The relationship between fractional calculus and fractals. Fractals **3**(1), 217–229 (1995)
44. Taylor, S.J.: The Hausdorff α-dimensional measure of Brownian paths in n-space. Proc. Cambr. Philos. Soc. **49**, 31–39 (1953)
45. Taylor, S.J.: The exact Hausdorff measure of the sample path for planar Brownian motion. Proc. Cambr. Philos. Soc. **60**, 253–258 (1964)
46. Taylor, S.J.: Sample path properties of a transient stable process. J. Math. Mech. **16**, 1229–1246 (1967)
47. Wedrich, L.: Hausdorff dimension of the graph of an operator semistable Lévy process. J. Fractal Geom. **4**(1), 21–41 (2017)

48. Xiao, Y.: Random fractals and Markov processes. In: Lapidus, M.L., van Frankenhuijsen, M. (eds.) Fractal Geometry and Applications: A Jubilee of Benoit Mandelbrot, pp. 261–338. AMS, Providence (2004)
49. Zhou, W.X., Sornette, D., Pisarenko, V.: New evidence of discrete scale invariance in the energy dissipation of three-dimensional turbulence: correlation approach and direct spectral detection. Int. J. Modern Phys. C **14**(4), 459–470 (2003)
50. Zolotarev, V.M.: Expressions of the density of a stable distribution with exponent α greater than one by means of a frequency with exponent $1/\alpha$. Selected Translations in Mathematical Statistics and Probability, vol. 1, pp. 163–167. AMS, Providence (1961)
51. Zolotarev, V.M.: One-dimensional Stable Distributions. Translations of Mathematical Monographs, vol. 65. AMS, Providence (1986)

From Fractals in External DLA to Internal DLA on Fractals

Ecaterina Sava-Huss

Abstract We present an unified approach on the behavior of two random growth models (*external DLA* and *internal DLA*) on infinite graphs, the second one being an internal counterpart of the first one. Even though the two models look pretty similar, their behavior is completely different: while external DLA tends to build irregularities and fractal-like structures, internal DLA tends to fill up gaps and to produce regular clusters. We will also consider the aforementioned models on fractal graphs like Sierpinski gasket and carpet, and present some recent results and possible questions to investigate.

Keywords Random walks · Harmonic measure · Cluster models · Sierpinski gasket · Integer lattices · Trees · Hyperbolic plane · Fractal graphs

Mathematics Subject Classifications (2010) Primary: 60J10, 28A80; Secondary: 31A15, 05C81

1 Introduction

We consider two aggregation models initially introduced in physics in [51] and [43], and rigorously studied in mathematics over the last three decades, models for which we present a survey on the existing results and state several open problems. The models under consideration are *external diffusion limited aggregation* (shortly *external DLA*) and *internal diffusion limited aggregation* (shortly *internal DLA*). In the mathematical community, these two models started to gain interest only a couple of years after being introduced, with the first results on external DLA in [30, 32], and on internal DLA in [36]. Only recently, these models became interesting in the fractals community: few recent results concerning external DLA on the

E. Sava-Huss (✉)
Department of Mathematics, University of Innsbruck, Innsbruck, Austria
e-mail: Ecaterina.Sava-Huss@uibk.ac.at

U. Freiberg et al. (eds.), *Fractal Geometry and Stochastics VI*,
Progress in Probability 76, https://doi.org/10.1007/978-3-030-59649-1_12

m-dimensional pre-Sierpinski carpet as defined in [44], for $m \geq 3$ are available. For the internal DLA on the Sierpinski gasket graph, there are also some limit shape results, but other than these two examples, there is not much known about the two growth models on other fractal graphs where, according to simulations which we present towards to end of the paper, interesting behavior may be observed. With the current overview, we would like to draw the attention on the beauty of these models.

For the rest of the paper, G will be an infinite and locally finite graph, the reference state space, which will be replaced with concrete examples of graphs as needed. We denote by $o \in \mathsf{G}$ a fixed vertex, the *origin of the graph* G.

External DLA was initially introduced in physics by WITTEN AND SANDER [51] as an example to create ordering out of chaos due to a simple rule. Mathematically, this ordering is far away from being understood, and new methods and ideas are needed in order to move forward in this direction. External DLA is a model of random fractal growth which exhibits self-organized criticality and complex-pattern formation, and which produces scale-invariant objects whose Hausdorff dimension is independent of short-range details. Moreover external DLA has no upper critical dimension as shown in [51]; it is a model which builds a sequence of random growing sets $(\mathcal{E}_n)_{n\geq 0}$, starting with one particle $\mathcal{E}_0 = \{o\}$ at the origin of G. At each time step, a new particle starts a simple random walk from "infinity" (far away) and walks until it hits the outer boundary of the existing cluster, where it stops and settles. In this way, one builds a family $(\mathcal{E}_n)_{n\geq 0}$ of growing clusters; the set $\mathcal{E}_n$ consists of exactly $n + 1$ particles and it is called *external DLA cluster*. In spite of these very simple growth rules, only a few rigorous mathematical results about external DLA are available, results which will be surveyed below. A typical structure produced on a two-dimensional lattice is shown in Fig. 1. External DLA was found to well represent growth processes in nature such as growth of bacterial colonies, electrodeposition, or crystal growth.

Internal DLA is an attempt of a model which eliminates irregularities and fills gaps, as opposed to external DLA. It was proposed by MEAKIN AND DEUTCH [43] as a model of industrial chemical processes such as electropolishing, corrosion and etching. DIACONIS AND FULTON in [19] identified internal DLA as a special case of a "smash sum" operation on subsets of $\mathbb{Z}^2$. Internal DLA is a random growth model which builds a sequence of random growing clusters $(\mathcal{I}_n)_{n\geq 0}$ based on particles performing random walks, where all the particles start from the same fixed point o. Typically, one starts with $\mathcal{I}_0 = \{o\}$, and for each n, we let $\mathcal{I}_{n+1}$ be $\mathcal{I}_n$ plus the first point where a random walk started at o exits $\mathcal{I}_n$. There are several modifications of this model, where one can start the random walks uniformly at random in the already existing cluster, or one can start with an initial configuration of particles on the state space G. As in external DLA, understanding the shape of the limiting cluster $\mathcal{I}_n$, the *internal DLA cluster* with $n + 1$ particles, is the main question in this model. Also, of fundamental significance as mentioned in the initial paper [43], is to know how smooth a surface formed by internal DLA (processes) may be. These problems are well understood mathematically on many state spaces, and there are very precise results available. On the one hand, the limiting object formed from internal DLA does not show any fractal structure. On the other hand, when running

Fig. 1 External DLA cluster on $\mathbb{Z}^2$ with center initially occupied

internal DLA on a fractal graph, we have partial results that indicate the absence of fractal structure, though there remain many more fractal state spaces to be explored. The crucial difference between the above two models is that the dynamics of the external model roughens the cluster, whereas the dynamics of the internal model makes the cluster smoother.

Structure of the Paper After fixing the notation and the basic notions in Sect. 2, we focus on the external DLA model in Sect. 3, in which we survey the available results on the growth of arms in this model, number of holes, and variations of the standard model. The results will not be stated in the chronological order of publication, but according to the state space they evolve on. Finally, in Sect. 4 we survey the results for the limit shapes of the internal DLA cluster, and we include several questions through the whole paper.

2 Preliminaries

Graphs Let G be an infinite, locally finite graph (i.e. every vertex has finite degree denoted by $\deg(x)$). The neighborhood relation will be denoted by "$\sim$", and by $x \sim y$ we mean that (x, y) is an edge in G. Let $o \in \mathsf{G}$ be a fixed distinguished vertex, which will be called the *origin* or the *root*. For $x, y \in \mathsf{G}$, the distance $d(x, y)$ represents the minimal number of edges on the path connecting x with y. For a subgraph A of G, we denote by ∂A the outer boundary of A:

$$\partial A = \{y \in \mathsf{G} : y \notin A,\ \exists x \in A : x \sim y\}.$$

For $x \in \mathsf{G}$ and $n \geq 0$, we write $\mathcal{B}_n(x) = \{y \in \mathsf{G} : d(x, y) \leq n\}$ for the ball of radius n and center x in G. If the center of the ball is o, we write only $\mathcal{B}_n$.

Random Walks Let $(S_n)_{n\geq 0}$ be a random walk on G, and denote by $\mathbb{P}_x$ the probability measure of the random walk started at x. We do not fix yet the transition probabilities for the random walk, since those will change from case to case, and we will mention them as needed. For a subset $A \subset \mathsf{G}$, let $T(A)$ be the hitting time of A, defined as

$$T(A) = \min\{n \geq 0 : S_n \in A\}.$$

For a set $\{x\}$ consisting of a single vertex, we write $T(x)$ instead of $T(\{x\})$. The *heat kernel* of the random walk (S_n) is defined to be

$$p_n(x, y) = \mathbb{P}_x[S_n = y],$$

and the *Green function* $G(x, y)$ is defined as

$$G(x, y) = \sum_{n\geq 0} p_n(x, y),$$

which is well defined and finite precisely when the random walk is transient. For a subset $A \subset \mathsf{G}$, the *killed or the stopped Green function* $G_A(x, y)$ is defined as

$$G_A(x, y) = \sum_{n\geq 0} \mathbb{P}_x[S_n = y, T(A) > n].$$

The *hitting distribution* $H_A(x, y)$ is then

$$H_A(x, y) = \mathbb{P}_x[S_{T(A)} = y], \quad \text{for } y \in A$$

and $S_{T(A)}$ is the hitting position of A. If the random walk (S_n) starts at o, we write

$$h_A(y) = \mathbb{P}_o[S_{T(A)} = y], \quad \text{for } y \in A \tag{2.1}$$

for the probability of the random walk starting at o to first hit A in y, that is h_A is the harmonic measure (from o) of the set A, and $\sum_{y\in A} h_A(y) = 1$.

3 External DLA

We define here formally the external DLA model, by first explaining what it means to release a particle at infinity. Several variants of external DLA have been considered, but we refer here to the original, simplest model, which can be defined on any space where the notion of random walk or diffusion exists. If the Poisson boundary consists of one point and the random walk is recurrent (for instance the case of simple random walk on $\mathbb{Z}$ and $\mathbb{Z}^2$), external DLA can be defined so that the law of the location of a new particle is the harmonic measure of the existing aggregate with pole at infinity. If the random walk is transient (such as the case of simple random walk on $\mathbb{Z}^d$, with $d \geq 3$, or on regular trees $\mathbb{T}_d$ of degree $d \geq 2$, n-dimensional Sierpinski carpet graph, for $n \geq 3$), one can consider the harmonic measure with a pole far away from the aggregate, let the pole go to infinity and take limits (i.e., conditioning the random walk coming from infinity to hit the cluster). That is, in defining rigorously external DLA, we have to distinguish the cases when the random walk (S_n) on the infinite graph G is recurrent or transient; the Poisson boundary of the random walk also plays a role in this case. We recall that the *Poisson boundary* of a random walk is a measure space that describes the stochastically significant behavior of the walk at infinity. It provides an integral representation of the bounded harmonic functions of the random walk.

During the whole paper, when we speak about the n-dimensional Sierpinski carpet graph, we shall also use the notion *pre-Sierpinski carpet*, and we have in mind the construction introduced in [44].

We shall write $\mu_A(y)$ for the *harmonic measure from infinity*, that is, for the probability to start a random walk at infinity and to hit the finite subset $A \subset \mathsf{G}$ at the point y. Depending on whether the graph G is transient or recurrent, this measure can take different forms, and we cannot define it globally on any general graph here. This will be made precise in the concrete cases below.

Definition 3.1 Let G be an infinite graph, and (S_n) a discrete time random walk on it. *External DLA* on G is a Markov chain $(\mathcal{E}_n)_{n\geq 0}$ on finite subsets of G, which evolves in time in the following way. Start with a single vertex $o \in \mathsf{G}$, that is $\mathcal{E}_0 = \{o\}$. Given the state $\mathcal{E}_n$ of the chain at time n, let y_{n+1} be a random vertex in $\partial\mathcal{E}_n$ chosen according to the harmonic measure (from infinity) of $\partial\mathcal{E}_n$. That is,

$$\mathbb{P}[y_{n+1} = y|\mathcal{E}_n] = \mu_{\partial\mathcal{E}_n}(y), \quad \text{for } y \in \partial\mathcal{E}_n,$$

and we set $\mathcal{E}_{n+1} = \mathcal{E}_n \cup \{y_{n+1}\}$.

Definition 3.2 The *cluster at infinity* $\mathcal{E}_\infty$ for the external DLA process $(\mathcal{E}_n)$ on G is defined as

$$\mathcal{E}_\infty = \bigcup_{n=1}^{\infty} \mathcal{E}_n.$$

It is immediate that the external DLA cluster $\mathcal{E}_n$ at time n contains exactly $n+1$ vertices. This model is hard to study. The difficulty comes from the fact that the dynamics is neither monotone nor local (meaning that if big tentacles surround a vertex x, then x will never be added to the cluster). By non-monotonicity we mean that there is no coupling between the external DLA starting from a cluster C and another from a cluster $D \subset C$ such that, at each step, the inclusion of the clusters remains valid almost surely. Understanding the shape of $\mathcal{E}_n$ as $n \to \infty$ and the fractal nature of this object, are problems one would be typically interested in. While mathematically this is out of reach for the time being, there are other partial results concerning the growth of arms and the number of holes in external DLA.

3.1 Integer Lattices $\mathbb{Z}^d$

In this subsection the state space for the external DLA process is $\mathsf{G} = \mathbb{Z}^d$, $d \geq 1$. Even for $\mathbb{Z}^d$, there are no results that prove the fractal nature of the limiting object, or results that prove the zero density in the long run. The first rigorous results go back to Kesten [30, 32], who gives estimates on the growth of arms in external DLA. Since for $d = 1$, the behavior of standard external DLA is trivial, we consider $d \geq 2$, and let $(S_n)_{n \geq 0}$ be a simple random walk on $\mathbb{Z}^d$.

For $d = 2$, for any finite nonempty subset $A \subset \mathbb{Z}^2$, we have $T(A) < \infty$ with probability one, and we define the *harmonic measure (from infinity) of A*

$$\mu_A(y) = \lim_{|x| \to \infty} H_A(x, y), \tag{3.1}$$

where $|x|$ denotes the Euclidean norm of x. The limit $\lim_{|x| \to \infty}$ corresponds to "releasing the particle at infinity". In this case, (S_n) is recurrent, so that by [50, Theorem 14.1] the limit in (3.1) exists and $\sum_{y \in A} \mu_A(y) = 1$.

For $d \geq 3$, since the random walk (S_n) is transient, the limit $\lim_{|x| \to \infty} H_A(x, y)$ in (3.1) is identically zero (cf [50, Proposition 25.3]). So in order to obtain a nontrivial limit similar to the one in (3.1), we have to condition on $T(A)$ being finite. This conditioning gives the factor of the capacity of the set A in the denominator. In the case $d \geq 3$, we define the *harmonic measure (from infinity)* of a finite subset $A \subset \mathbb{Z}^d$ as

$$\mu_A(y) = \lim_{d(o,x) \to \infty} \frac{H_A(x, y)}{\sum_{z \in A} H_A(x, z)} = \lim_{d(o,x) \to \infty} \mathbb{P}_x[S_{T(A)} = y | T(A) < \infty], \quad \text{for } y \in A, \tag{3.2}$$

which is proportional to the so-called *equilibrium measure* associated to the set A. The limit in (3.2) exists again by [50, Proposition 26.2] for $d = 3$ (the same proof works also for $d > 3$) and satisfies $\sum_{y \in A} \mu_A(y) = 1$. Therefore, we have a valid definition for external DLA, and we let $r(\mathcal{E}_n)$ to be the radius of $\mathcal{E}_n$, defined as

$$r(\mathcal{E}_n) = \max\{|x| : x \in \mathcal{E}_n\}, \tag{3.3}$$

Theorem 3.3 ([31, Theorem] and [30, Corollary]) *There exist constants $C(d) < \infty$ such that with probability* 1

$$\limsup_{n\to\infty} n^{-2/3} r(\mathcal{E}_n) \le C(2), \quad \textit{if } d = 2$$

$$\limsup_{n\to\infty} n^{-2/d} r(\mathcal{E}_n) \le C(d), \quad \textit{if } d \ge 3.$$

The proof uses classical estimates for the harmonic measure (from infinity) as defined in (3.1) and (3.2) and for the hitting probabilities. Simulations actually indicate that for $d = 2$, $\mathbb{E}[r(\mathcal{E}_n)] \approx n^{10/17}$ but as far as the lower bound is concerned, nothing has been proven beyond $\sqrt{n}$ in the 35 years since the model has been introduced. It would be very interesting to prove even a logarithmic correction, i.e. to prove that $\mathbb{E}[r(\mathcal{E}_n)] \ge \sqrt{n}\log(n)$. On $\mathbb{Z}^d$, a lower bound on the number $N(n)$ of vertices in $\mathcal{B}_n$ which are occupied by the cluster E_∞ is known.

Theorem 3.4 ([32, Theorem 2]) *There exist constants $C(d) < \infty$ such that with probability* 1

$$N(n) \ge C(d) n^{d-1}, \quad \textit{for infinitely many } n.$$

Another non-trivial result on $\mathbb{Z}^2$ concerns the number of holes $\mathcal{H}_n$ in the external DLA cluster $\mathcal{E}_n$. A hole of $\mathcal{E}_n$ is a finite connected component of $\mathbb{Z}^2 \setminus \mathcal{E}_n$.

Theorem 3.5 ([21]) *For any finite connected subset e of $\mathbb{Z}^2$ we have*

$$\mathbb{P}[\mathcal{H}_n \textit{ converges to infinity as } n \to \infty | \mathcal{E}_0 = e] = 1.$$

Theorem 3.3 has been improved in [12], where upper bounds on the growth rate of arms in external DLA cluster are given on a big class of transient graphs with properties such as: transitive graphs of polynomial growth of degree ≥ 4; transitive graphs of exponential growth; $\mathbb{Z}^3$; non-amenable graphs; n-dimensional pre-Sierpinski gasket graphs ($n \ge 3$) as introduced in [44]. In particular, on $\mathbb{Z}^3$ the factor $n^{-2/3}$ from Theorem 3.3 has been improved to $n^{-1/2}/\log(n)$. On the class of transient graphs G considered in [12], the harmonic measure (from infinity) μ_A of a set A is defined as in (3.2).

A directed version of external DLA has been recently introduced on $\mathbb{Z}^2$ in [42]. In a series of three papers [1–3], a one-dimensional external DLA model based on random walks with long jumps (that depend on a parameter α) is proposed, which tries to capture the fractal nature of the standard DLA. Depending on the values of α, the random walk (S_n) with long jumps on $\mathbb{Z}$ may be recurrent or transient, and for the precise definition of harmonic measure from infinity we refer to those three papers. The main results of [2, 3] can be summarized into the following theorem.

Theorem 3.6 *Let (S_n) be a symmetric random walk on $\mathbb{Z}$ that satisfies $\mathbb{P}[|S_1 - S_0| = k] \sim ck^{-1-\alpha}$. Let $d(\mathcal{E}_n)$ be the diameter of the external DLA cluster $\mathcal{E}_n$. Then almost surely:*

(a) If $\alpha > 3$, then $n - 1 \le d(\mathcal{E}_n) \le Cn + o(n)$, where C depends only on α.
(b) If $2 < \alpha \le 3$, then $d(\mathcal{E}_n) = n^{\beta+o(1)}$, where $\beta = \frac{2}{\alpha-1}$.
(c) If $1 < \alpha < 2$, then $d(\mathcal{E}_n) = n^{2+o(1)}$.
(d) If $\frac{1}{3} < \alpha < 1$, then $n^{\beta+o(1)} \le d(\mathcal{E}_n) \le n^{\beta'+o(1)}$, where $\beta = \max(2, \alpha^{-1})$ and $\beta' = \frac{2}{\alpha(2-\alpha)}$.
(e) If $0 < \alpha < \frac{1}{3}$, then $d(\mathcal{E}_n) = n^{\beta+o(1)}$, where $\beta = \alpha^{-1}$.

The last one [1] from the series of three papers mentioned above deals with the cluster at infinity E_∞, and it is shown that for random walks (S_n) whose step size has finite third moment, E_∞ has a renewal structure and positive density. In contrast, for random walks whose step size has finite variance, the renewal structure no longer exists and E_∞ has zero density.

Theorem 3.7 ([1, Theorem 1]) *Assume that the step distribution ξ of the random walk (S_n) on $\mathbb{Z}$ satisfies $\mathbb{P}[\xi > n] \le Cn^{-\alpha}$ for any n and some $\alpha > 3$. There exists some $B > 0$ such that a.s. $\mathcal{E}_\infty$ has density B. Further, B is the limit density of $\mathcal{E}_n$:*

$$B = \lim_{m_1\to\infty, m_2\to\infty} \frac{|\mathcal{E}_\infty \cap [-m_1, m_2]|}{m_1 + m_2} = \lim_{n\to\infty} \frac{n}{d(\mathcal{E}_n)}.$$

Theorem 3.8 ([1, Theorem 2]) *Assume that there exist $2 < \alpha < 3$ and constants $c_1, c_2 > 0$ so that ξ satisfies $c_1 n^{-\alpha} \le \mathbb{P}[\xi > n] \le c_2 n^{-\alpha}$ for all n then a.s.*

$$|\mathcal{E}_\infty \cap [-n, n]| = n^{\frac{\alpha-1}{2}+o(1)}.$$

In particular, $\mathcal{E}_\infty$ has zero density in the sense that $\lim_{n\to\infty} \frac{|\mathcal{E}_\infty \cap [-n,n]|}{n} = 0$.

The results mentioned above are the only ones available for external DLA on $\mathbb{Z}^d$, and the limit shape and the density problem for $d \ge 2$ still resist a mathematical proof. There are many open problems and questions in this direction; see [12] for more details.

Conjecture 3.9 On $\mathbb{Z}^d$, the rate of growth of the radius of the external DLA cluster $\mathcal{E}_n$ started at $\mathcal{E}_0 = \{0\}$ is of order $n^{1/d}$:

$$\limsup_{n\to\infty} n^{-1/d} \mathbb{E}[r(\mathcal{E}_n)] = 0.$$

Question 3.10 What is the distribution of the number of ends of the cluster at infinity $\mathcal{E}_\infty$ on $\mathbb{Z}^d$?

Concerning recent progress on external DLA in a wedge of $\mathbb{Z}^d$, we refer to [47]. Furthermore, the reach of Kesten's idea is extended to non-transitive graphs in [46], where the (horizontally) translation invariant stationary harmonic measure on the upper half plane with absorbing boundary condition is defined and it is shown that the growth of such stationary harmonic measure in a connected subset intersecting x-axis is sub-linear with respect to the height; see also [45, 48] where the stationary harmonic measure as a natural growth measure for external DLA model in the upper planar lattice is investigated.

3.2 *Trees* $\mathbb{T}_d$

One reason that makes the lattice case $\mathbb{Z}^d$ hard to investigate is that there is no simple way to describe the harmonic measure (from infinity) for the boundary of an external DLA cluster on $\mathbb{Z}^d$. On other state spaces, such as trees, which have no loops, the model is more tractable and the harmonic measure (from infinity) can be understood. In [10], an adjusted version of external DLA on d-regular trees $\mathbb{T}_d$, where the fingering phenomenon occurs, was introduced. The dynamics of their model is as follows: the initial cluster $\mathcal{E}_0$ contains only the root. Vertices are then added one by one from among those neighboring the current subtree. The choice of which vertices to add is random, with vertices in generation n (i.e. distance n from the root) chosen with probabilities proportional to α^{-n} where $\alpha > 0$ is a fixed parameter. Then $\mathcal{E}_n$ is the subtree at step n and let $r(\mathcal{E}_n) = \max\{d(o,x) : x \in \mathcal{E}_n\}$ denote the maximum height of a vertex in $\mathcal{E}_n$, which is similar to the radius in (3.3). For this model, for a finite subtree $A \subset \mathbb{T}_d$ with boundary ∂A, its harmonic measure $\mu^{\alpha}_{\partial A}$ (from infinity) on ∂A, with parameter $\alpha > 0$ can be computed as

$$\mu^{\alpha}_{\partial A}(y) = \frac{\alpha^{-d(o,y)}}{\sum_{x \in \partial A} \alpha^{-d(o,x)}}, \qquad \text{for } y \in \partial A,$$

see Definition on page 4 in [10]. In the latter paper, the case $\alpha < 1$ is studied. The external DLA cluster $\mathcal{E}_n$ is the position at time n of the Markov chain defined in Definition 3.1, where $\mathcal{E}_{n+1}$ is obtained from $\mathcal{E}_n$ by adding a new vertex according to the harmonic measure defined in the previous equation. For $\alpha \geq 1$ it is easy to see that $\mathcal{E}_\infty$ is almost surely the entire tree. For $\alpha = 1$, one has the uniform measure $\mu^1_{\partial A}$ on $\partial \mathcal{E}_n$ (this corresponds to the Eden model). From the external DLA perspective, the case $\alpha < 1$ is the interesting one, where one obtains the so-called fingering phenomenon. For this external DLA model, [10] obtained a strong law and a central limit theorem for the height $r(\mathcal{E}_n)$ of the DLA cluster.

Theorem 3.11 *Let $\mathbb{T}_d$ be a d-regular tree and $0 < \alpha < 1$. There exist constants $r_0(\alpha, d) \in (0, 1)$ and $\sigma^2 = \sigma^2(\alpha, d) > 0$ such that*

(a) $\lim_{n\to\infty} \frac{r(\mathcal{E}_n)}{n} = r_0(\alpha, d)$ *a.s.*

(b) $\frac{r(\mathcal{E}_n)-nr_0(\alpha,d)}{\sqrt{n}} \xrightarrow{\mathcal{D}} N(0, \sigma^2)$*, as $n \to \infty$.*

The model considered here can be also interpreted as a model of first passage percolation on $\mathbb{T}_d$.

3.3 Hyperbolic Plane $\mathbb{H}^2$

In [22], external DLA on the hyperbolic plane $\mathbb{H}^2$ is considered, and it is shown that the cluster at infinity E_∞ almost surely admits a positive upper density. For completeness, we recall the definition of the *upper density* of a set, as used in [22]. In a metric measure space X whose diameter is infinite, we say that a locally finite set $A \subset X$ has an *upper density* greater or equal to c if there exist a point $p \in X$ and a sequence $R_1 < R_2 < \cdots$ such that $R_i \to \infty$ as $i \to \infty$, such that

$$\#\left(A \cap \mathcal{B}_{R_i}(p)\right) \geq c\mu(\mathcal{B}_{R_i}(p)), \quad \forall i \in \mathbb{N},$$

where $\mathcal{B}_r(p)$ is a metric ball centered at p with radius r and μ is the measure defined on X. On the hyperbolic plane, one can use this definition with the standard hyperbolic distance as a metric and the standard Riemannian volume of a set as a measure. In the hyperbolic setting the behavior of the aggregate is simpler to analyze than the Euclidean one; the rate of decay of the hyperbolic potential plays an important role in understanding the external DLA.

See Fig. 2 for a picture of external DLA model with 1000 particles, viewed on the Poincaré disc model. In his construction, particles are metric balls of radius 1, $\mathcal{E}_0 = \{p_0\}$, where p_0 is a fixed point in $\mathbb{H}^2$, and recursively $\mathcal{E}_{n+1} = \mathcal{E}_n \cup \{y_{n+1}\}$, where $\{y_{n+1}\}$ is added to the aggregate $\mathcal{E}_n$ according to a (harmonic) measure $\mu_{\partial\mathcal{E}_n}(y)$ with pole at infinity that has to be carefully constructed on $\mathbb{H}^2$, such that external DLA makes sense in this setting. For details on this construction, we refer to [22]; the main result of his paper reads as following.

Theorem 3.12 ([22, Theorem 1.1]) *The external DLA cluster at infinity E_∞ almost surely has an upper density greater than c, where $c > 0$ is an universal constant.*

We would like to point out the fact that the behavior of external DLA on the hyperbolic plane and on the regular tree $\mathbb{T}_d$ as considered in [10] is completely different, even though the hyperbolic plane has a tree-like structure.

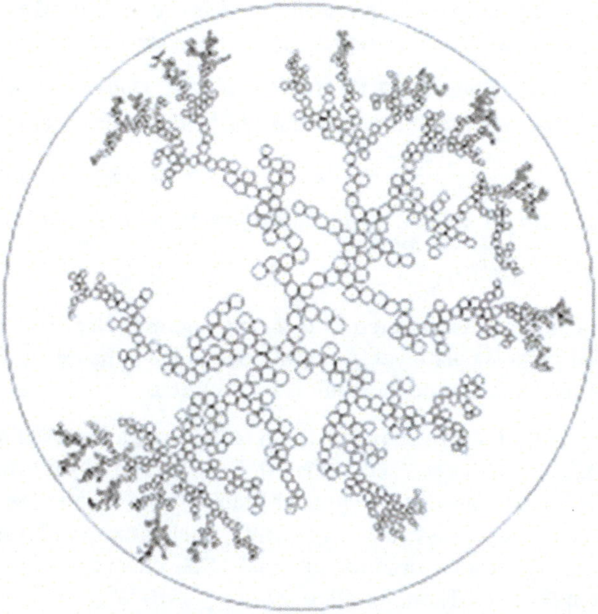

Fig. 2 External DLA with 1000 particles, viewed on the Poincaré disc model (by Ronen Eldan)

3.4 Cylinder Graphs

Other results on external DLA that are worth mentioning have been proven in [11] on cylinder graphs $\mathsf{G} \times \mathbb{N}$. Let us first fix the notation for the graphs we consider below. Let G be a finite, connected graph. The *cylinder graph* with base G, denoted by $\mathsf{G} \times \mathbb{N}$, is defined as: the vertex set of $\mathsf{G} \times \mathbb{N}$ is $V(\mathsf{G}) \times \mathbb{N} = \{(v, k) : v \in V(\mathsf{G}),\ k \in \mathbb{N}\}$, where $V(\mathsf{G})$ represents the vertex set of G. The edge set is defined by the following relations: for all $u, v \in V(\mathsf{G})$ and all $m, k \in \mathbb{N}$, $(u, m) \sim (v, k)$, that is between vertices (u, m) and (v, k) there is an edge in $\mathsf{G} \times \mathbb{N}$, if and only if $m = k$ and $u \sim v$ in G, or $|m - k| = 1$ and $u = v$. Equivalently, the cylinder with base G is obtained by just placing infinitely many copies of G one over the other, and connecting each vertex in a copy to its corresponding vertices in the adjacent copies.

On $\mathsf{G} \times \mathbb{N}$, particles perform simple random walks (S_n) from infinity. Since G is finite, such random walks are recurrent on $\mathsf{G} \times \mathbb{N}$, and the harmonic measure from infinity can be defined similar to the one on $\mathbb{Z}^2$, as in (3.1). That is, vertices are added to the existing cluster $\mathcal{E}_n$ according to the measure in (3.1). Denote by G_m the induced subgraph on the vertices of $\mathsf{G} \times \{m\}$, for all $m \in \mathbb{N}$, and call G_m the m-th level of the cylinder graph $\mathsf{G} \times \mathbb{N}$. One of the results proven in [11] is that external DLA on $\mathsf{G} \times \mathbb{N}$ grows arms if the base graph G mixes fast. Recall that the mixing time $t_{\text{mix}}(\mathsf{G})$ of the simple random walk on G is the time it takes for the random walk to come close in total-variation distance to the stationary distribution.

Theorem 3.13 ([11, Theorem 2.1]) *Let* $2 \le d \in \mathbb{N}$. *There exists* $n_0 = n_0(d)$, *such that the following holds for all* $n > n_0$: *let* G *be a* d*-regular graph of size* n, *and mixing time* $t_{mix}(\mathsf{G}) \le \dfrac{\log^2 n}{(\log\log n)^5}$. *Let* $(\mathcal{E}_t)$ *be the external DLA process on* $\mathsf{G} \times \mathbb{N}$ *with* $\mathcal{E}_0 = \mathsf{G}_0$, *and for* $m \in \mathbb{N}$, *let* T_m *be the first time the DLA cluster reaches* G_m. *Then, for all* m, $\mathbb{E}[T_m] < \dfrac{4mn}{\log\log n}$.

This phenomenon is often referred to as *the aggregate grows arms*, i.e. grows faster than order $|\mathsf{G}|$ particles per layer. As mentioned in [11], the result above is believed not to be optimal, and a stronger result is conjectured.

Conjecture 3.14 ([11, Conjecture 2.2]) Let $(\mathcal{G}^n)_{n\ge 0}$ be a family of d-regular graphs such that $\lim_{n\to\infty} |\mathcal{G}^n| = \infty$. There exists $0 < \gamma < 1$ and n_0 such that for all $n > n_0$ the following holds: consider the cylinder graph $\mathcal{G}^n \times \mathbb{N}$ with base $\mathcal{G}^n$ and let $(\mathcal{E}_t)$ be the external DLA process on $\mathcal{G}^n \times \mathbb{N}$ with $\mathcal{E}_0$ being the zero layer of the cylinder graph, and T_m be the first time the external DLA cluster reaches level m on the cylinder graph $\mathcal{G}^n \times \mathbb{N}$. Then, for all m, $\mathbb{E}[T_m] \le m|\mathcal{G}^n|^{\gamma}$.

Concerning the density of the limit cluster at infinity $\mathcal{E}_\infty$, for cylinder graphs $\mathsf{G} \times \mathbb{N}$ with base G, in the same paper there are two results. To state them, let us define the *empirical density of particles in the finite cylinder* $\mathsf{G} \times \{1, \dots, m\}$ as

$$D(m) = \frac{1}{mn} \sum_{i=1}^{m} |\mathcal{E}_\infty \cap \mathsf{G}_i|$$

and *the density at infinity* as $D = D_\infty = \lim_{m\to\infty} D(m)$. Using standard arguments from ergodic theory one can show that the above limit exists, and is constant almost surely. The next result relates the density at infinity to the average growth rate.

Theorem 3.15 ([11, Theorem 4.2]) *For the external DLA process on* $\mathsf{G} \times \mathbb{N}$, *where* G *is a* d*-regular graph of size* n, *we have*

$$D = \lim_{m\to\infty} \frac{1}{mn} \mathbb{E}[T_m].$$

In [11, Theorem 4.6] the previous result has been improved to $D \le \frac{2}{3}$ for the case when the base graph G is a vertex transitive graph. Finally, for a family of base graphs with small mixing time, the following holds.

Theorem 3.16 ([11, Theorem 4.8]) *Let* $(\mathcal{G}^n)_{n\geq 0}$ *be a family of d-regular graphs* $(d \geq 2)$ *such that* $\lim_{n\to\infty} |\mathcal{G}^n| = \infty$*, and for all n,*

$$t_{\mathsf{mix}}(\mathcal{G}^n) \leq \frac{\log^2 |\mathcal{G}^n|}{(\log\log |\mathcal{G}^n|)^5}.$$

Let $D(n)$ *be the density at infinity of the external DLA process on* $\mathcal{G}^n \times \mathbb{N}$*. Then* $\lim_{n\to\infty} D(n) = 0$.

We refer to the last section of [11] for several open questions and problems concerning external DLA on cylinder graphs. Many of the bounds from the previous three results can be improved, with some careful technicalities and assumptions on the base graph G.

3.5 Fractal Graphs

The appearance of fractal-like structures in DLA models (both internal and external) and their behavior on fractal graphs is the main theme of this paper, and we would like at this point to introduce two fractal graphs: the Sierpinski gasket graph and the Sierpinski carpet graph (called also pre-Sierpinski carpet).

Sierpinski gasket graph SG is a pre-fractal associated with the Sierpinski gasket, defined as follows. We consider in $\mathbb{R}^2$ the sets $V_0 = \{(0,0), (1,0), (1/2, \sqrt{3}/2)\}$ and

$$E_0 = \Big\{\big((0,0),(1,0)\big), \big((0,0),(1/2,\sqrt{3}/2)\big), \big((1,0),(1/2,\sqrt{3}/2)\big)\Big\}.$$

Now recursively define $(V_1, E_1), (V_2, E_2), \ldots$ by

$$V_{n+1} = V_n \cup \{(2^n, 0) + V_n\} \bigcup \Big\{\Big(2^{n-1}, 2^{n-1}\sqrt{3}\Big) + V_n\Big\}$$

and

$$E_{n+1} = E_n \cup \{(2^n, 0) + E_n\} \bigcup \Big\{\Big(2^{n-1}, 2^{n-1}\sqrt{3}\Big) + E_n\Big\},$$

where $(x, y) + S := \{(x, y) + s : s \in S\}$. Let $V_\infty = \cup_{n=0}^{\infty} V_n$, $E_\infty = \cup_{n=0}^{\infty} E_n$, $V = V_\infty \cup \{-V_\infty\}$ and $E = E_\infty \cup \{-E_\infty\}$. Then the doubly infinite Sierpinski gasket graph SG is the graph with vertex set V and edge set E. See Fig. 3 for a graphical representation of SG. Set the origin $o = (0, 0)$. External DLA on SG seems to be an approachable problem, due to the fact that SG is a post-critically finite fractal, and the existence of cut points simplifies the understanding of the harmonic measure from infinity, which can be defined again as in (3.1), since simple random walk on

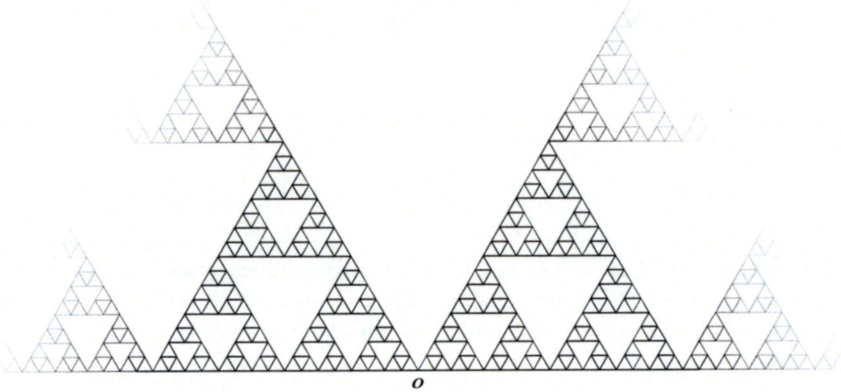

Fig. 3 The doubly-infinite Sierpinski gasket graph SG

SG is recurrent. We refer the reader to [8] and [33] for more details on analysis and diffusion on fractals.

Sierpinski carpet graph SC_m, called also m-dimensional pre-Sierpinski carpet, is an infinite graph derived from the *Sierpinski carpet*. SC_2 is constructed from the unit square in $\mathbb{R}^2$ by dividing it into 9 equal squares and deleting the one in the center. The same procedure is then repeated recursively to the remaining 8 squares. As mentioned in the introduction, we use the construction of the pre-Sierpinski carpet as in [44]. Recall that in this construction, the length scale factor is 3 and the mass scale factor is $3^m - 1$. For random walks on such graphs see [9] and the references therein. See Fig. 4 for a finite piece of Sierpinski carpet graph in dimension 2.

For $m \geq 3$, simple random walk on SC_m is transient, and the harmonic measure from infinity $\mu_A(y)$ for a finite subset $A \subset \mathsf{SC}_m$ is defined by using the capacity of A and the equilibrium measure of A, similar to (3.2). More details on the construction can be found in [12], where upper bounds for the arms $r(\mathcal{E}_n)$ of external DLA on a large class of transient graphs, including SC_m, $m \geq 3$, are proved. Their proofs are based on good control of heat-kernel estimates. The bounds for SC_m read as following.

Theorem 3.17 ([12, Theorem 5.5]) *Let* SC_m *be the m-dimensional Sierpinski carpet graph, and* $(\mathcal{E}_n)_{n\geq 0}$ *the external DLA process on* SC_m *started at* $\mathcal{E}_0 = \{o\}$ *(o is some fixed origin). Then almost surely,*

$$\limsup_{n\to\infty} n^{-\beta} r(\mathcal{E}_n) < \infty$$

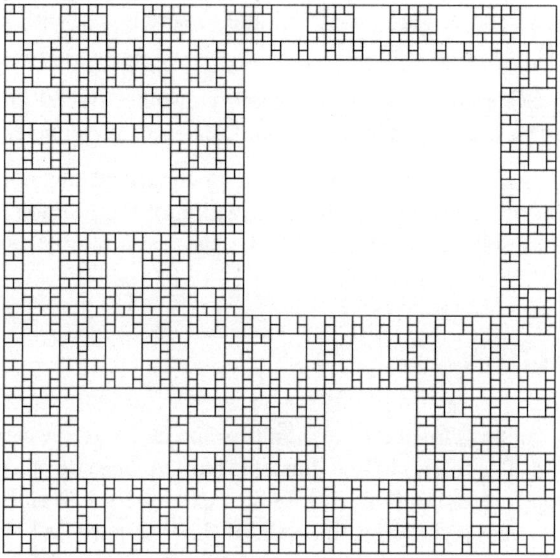

Fig. 4 Sierpinski carpet graph SC_2

where

$$\beta = \begin{cases} \frac{\log_2(13)-2}{3} = 0.5568, & \text{if } m = 3, \\ \frac{1}{2}, & \text{if } m = 4. \end{cases}$$

When $m \geq 5$, *we have almost surely,*

$$\limsup_{n\to\infty} (\log n)^{-1} n^{-\frac{2}{d(m)-2}} r(\mathcal{E}_n) < \infty,$$

where $d(m) = \dfrac{\log(3^m - 1)}{\log(3^m - 1) - \log(3^{m-1} - 1)}$.

We would like to conclude the section on external DLA with a couple of problems/questions.

Question 3.18 Can one find an upper bound for the growth of arms in external DLA on SG and on SC_2 (the random walk is strongly recurrent on these two graphs)? Can one extend the method Kesten used to upper bound the growth of arms in external DLA on $\mathbb{Z}^2$?

Question 3.19 Do we have zero density at infinity of the cluster $\mathcal{E}_\infty$ on the Sierpinski gasket graph SG?

Question 3.20 Does the external DLA cluster on the Sierpinski gasket graph and on the Sierpinski carpet graph have infinitely many holes, with probability one, as in the case of $\mathbb{Z}^2$ as proven in [21]?

Other than SG and SC_m there is a variety of other fractal graphs one can look at, and investigate the behavior of external DLA, which can be easier than $\mathbb{Z}^d$.

Question 3.21 Assuming that the Poisson boundary of the random walk on the graph G is non trivial, is there a characterization of the Poisson boundary in terms of the number of ends of the external DLA cluster at infinity $\mathcal{E}_\infty$ on G?

4 Internal DLA

Internal DLA can be defined on any infinite graph G; fix as above a vertex o of G and call it the origin. The internal DLA cluster is built up one site at a time, by letting the n-th particle perform a random walk until it exits the set of sites already occupied by the previous $n-1$ particles, the walk of the n-th particle being independent of the past. Similarly to external DLA, internal DLA is also a Markov chain on finite subsets of G.

Definition 4.1 Let G be an infinite graph, and (S_n) a simple random walk on G starting at o. *Internal DLA* on G is a Markov chain $(\mathcal{I}_n)_{n\geq 0}$ on finite connected subsets of G, which evolves in time in the following way. Start with a single vertex $o \in$ G and set $\mathcal{I}_0 = \{o\}$. Given the state $\mathcal{I}_n$ of the chain at time n, let y_{n+1} be a random vertex in $\partial\mathcal{I}_n$ chosen according to the harmonic measure (from o) of $\partial\mathcal{I}_n$, as defined in (2.1). That is, y_{n+1} is the first exist location from $\mathcal{I}_n$ of the simple random walk (S_n) starting from o, independent of the past:

$$\mathbb{P}[y_{n+1} = y|\mathcal{I}_n] = h_{\partial\mathcal{I}_n}(y), \quad \text{for } y \in \partial\mathcal{I}_n,$$

and we set $\mathcal{I}_{n+1} = \mathcal{I}_n \cup \{y_{n+1}\}$.

The set $\mathcal{I}_n$ is called the *internal DLA* cluster at time n, and it contains $n+1$ sites. As $n \to \infty$, we are interested in the asymptotic shape of internal DLA cluster $\mathcal{I}_n$, and the fluctuations of the cluster around the limiting shape. Due to the fact that the harmonic measure for "nice subsets" (for example balls) of G, when G is an Euclidean lattice, or a regular tree, is easier to understand than the harmonic measure from infinity as in the external DLA case, for the internal DLA model we have very precise estimates on many state spaces. Moreover, several variations of the classical internal DLA have been introduced.

4.1 Integer Lattices $\mathbb{Z}^d$

The first result concerning the internal DLA goes back to [36], where it is shown that the limit shape of internal DLA cluster is a ball, in the following sense. Let ω_d be the volume of the d-dimensional Euclidean ball of radius 1, and $\mathcal{B}_n$ be the d-dimensional "lattice ball" of radius n, that is, $\mathcal{B}_n = \{x \in \mathbb{Z}^d : |x| \leq n\}$, where $|x|$ denotes the Euclidean norm of x.

Theorem 4.2 ([36, Theorem 1]) *At time $\lfloor \omega_d n^d \rfloor$, internal DLA cluster occupies a set of sites close to a d-dimensional ball of radius n. More precisely, for any $\epsilon > 0$, with probability* 1

$$\mathcal{B}_{n(1-\epsilon)} \subset \mathcal{I}_{\lfloor \omega_d n^d \rfloor} \subset \mathcal{B}_{n(1+\epsilon)}, \quad \textit{for n large.}$$

In this first paper, a basic open question on fluctuations (deviation of $\mathcal{I}_n$ from the Euclidean ball) was asked: are the fluctuations of order $\sqrt{n}$, of order n^δ for some $\delta \in (0, \frac{1}{2})$, or even smaller? LAWLER [35] proved that for $d \geq 2$, the fluctuations are subdiffusive and they are of order at most $n^{1/3}$. While it was conjectured that the fluctuations are at most logarithmic in the radius, this resisted a mathematical proof for about 20 years. Two independent groups JERISON, LEVINE, AND SHEFFIELD [26–28] and ASSELAH AND GAUDILLIÈRE [4–6], and by different methods have shown that indeed, for $d = 2$ there are $\log(n)$ fluctuations, and for $d \geq 3$, there are $\sqrt{\log(n)}$ fluctuations in the radius. A summary of their results reads as following.

Theorem 4.3 *If $d = 2$, there is an absolute constant c, such that with probability* 1,

$$\mathcal{B}_{n-c\log n} \subset \mathcal{I}_{\lfloor \pi n^2 \rfloor} \subset \mathcal{B}_{n+c\log n}, \textit{ for all sufficiently large } n.$$

If $d \geq 3$, there is an absolute constant C, such that with probability 1,

$$\mathcal{B}_{n-C\sqrt{\log n}} \subset \mathcal{I}_{\lfloor \omega_d n^d \rfloor} \subset \mathcal{B}_{n+C\sqrt{\log n}}, \textit{ for all sufficiently large } n.$$

A generalization of the classical internal DLA on $\mathbb{Z}^d$ was treated in [39], where instead of running all particles from the origin, the authors run the process from an arbitrary starting configuration of particles (initial density of particles) on finer and finer lattices, all particles still performing simple random walks. They then show that, as the lattice spacing tends to zero, the internal DLA has a deterministic scaling limit which can be described as the solution to a certain PDE free boundary problem in $\mathbb{R}^d$. We do not state here the rigorous result, which requires more notation and definition, but refer to the lengthy and complex paper [39]. In order to study this general model, a new model called *divisible sandpile* was introduced in [38], which uses a continuous amount of mass instead of discrete particles.

The *divisible sandpile model* can be briefly described as following: start with an initial mass μ at the origin o. A vertex is called *full* if it has mass at least 1.

Any full site can topple by keeping mass 1 for itself and distributing the excess mass equally among its neighbors. At each time step, one chooses a full site and topples it. As time goes to infinity, provided each full site is eventually toppled, the mass approaches a limiting distribution in which each site has mass ≤ 1; this is proved in [38]. Individual topplings do not commute, but the divisible sandpile is *abelian* in the sense that any sequence of topplings produces the same limiting mass distribution; this is proved in [39, Lemma 3.1]. The set of sites with limit mass distribution equal to 1 is denoted by $\mathcal{S}_n$ and is called *the divisible sandpile cluster*. The asymptotic shape of the divisible sandpile cluster $\mathcal{S}_n$ is proven to be the same as the one of the internal DLA cluster on $\mathbb{Z}^d$ in [38], on regular trees in [37], on comb lattices in [24], and on Sierpinski gasket graphs in [25].

Random Walks with Drift on $\mathbb{Z}^d$ If one lets the particles which build up the internal DLA cluster $\mathcal{I}_n$ perform drifted random walk instead of simple random walk as in the classical model, one can again ask about the shape of the limit cluster on any state space. On $\mathbb{Z}^d$, this was open for several years, and the cluster was believed to be represented by the level sets of the Green function for the drifted random walk. This fact has been disproved, and with the help of the divisible sandpile model, in [41] it was proven that the internal DLA cluster is a true heat ball, because it gives rise to a mean-value property for caloric functions. The author introduced there the *unfair divisible sandpile*, where the mass is not distributed equally to the neighbors, but according to the one-step transition probabilities of the drifted random walk; the limit shape for the unfair divisible sandpile on $\mathbb{Z}^d$ was also described there. The main result for the limit shape for drifted internal DLA can be found in [41, Theorem 1.1], and for the limit shape of the unfair divisible sandpile cluster in [41, Theorem 3.3].

Uniform Starting Points To my knowledge, the most recent result for internal DLA on $\mathbb{Z}^d$, concerns the limit shape for the cluster when the particles do not all start from the same vertex o. Instead the starting position is chosen uniformly at random in the existing cluster. Formally, one can define the internal DLA as in Definition 4.1, starting with $\mathcal{I}_0 = \{0\}$, and given the process at time n, let y_{n+1} be the first exit location from $\mathcal{I}_n$ of the simple random walk $S_n^{X_n}$ starting at X_n, where X_n is a point chosen uniformly on $\mathcal{I}_n$, independent of the past. Set $\mathcal{I}_{n+1} = \mathcal{I}_n \cup \{y_{n+1}\}$. It turns out, as shown in [13], that this additional source of randomness arising from the choice of the initial position of the random walk, does not change the limit shape of the process, as the result below shows. Let $b_n := |\mathcal{B}_n|$.

Theorem 4.4 ([13, Theorem 1.1]) *Let $d \geq 2$. There exist constants c_1, c_2, C_1 and C_2 depending only on the dimension d such that, almost surely, the internal DLA cluster $\mathcal{I}_n$ on $\mathbb{Z}^d$ with uniform starting points satisfies*

$$\mathcal{B}_{n(1-C_1 n^{-c_1})} \subseteq \mathcal{I}_{b_n} \subseteq \mathcal{B}_{n(1+C_2 n^{-c_2})}, \quad \text{for } n \text{ large enough.}$$

Question 4.5 What can we say about the fluctuations of the internal DLA cluster on $\mathbb{Z}^d$ with uniform starting points around the limit shape? Are they bigger (smaller) that the fluctuations for internal DLA when all particles start their random walk from the same vertex o?

Supercritical Percolation Cluster on $\mathbb{Z}^d$ In [49], the underlying state space for the internal DLA model is the supercritical bond percolation cluster on $\mathbb{Z}^d$, with the origin conditioned to be in the infinite cluster. It is shown in [49, Theorem 1.1] that an inner bound for the internal DLA cluster is a ball in the graph metric. The picture for the outer bound was completed in [20, Theorem 1.1], where the authors show that also in this case the limit shape is a ball. The results in their paper hold in a more general setting: given the existence of a "good" inner bound for internal DLA, one can also prove a matching outer bound by using their methods. An interesting problem in the context of internal DLA model on a random graph is to understand the fluctuations.

4.2 *Comb Lattices C_2*

The 2-dimensional comb lattice C_2 is the spanning tree of $\mathbb{Z}^2$ obtained by removing all horizontal edges except the ones on the x-axis. While C_2 is a simple graph, see Fig. 5 (left), it has some remarkable properties in what concerns the behavior of random walks: no form of the so-called Einstein relation for exponents associated with random walks hold on C_2, see [14]. PERES AND KRISHNAPUR [34] showed that on C_2 two independent simple random walks meet only finitely often. The comb C_2 is an example where the limit shape of internal DLA is not a ball in the graph metric or in another standard metric. Indeed, the diameter of the internal DLA cluster with n particles grows like $n^{2/3}$ in the y-direction, and like $n^{1/3}$ in the x-direction. See Fig. 5 (right) for a picture of the internal DLA cluster with 100, 500, and 1000 particles, respectively.

Let

$$\mathcal{D}_n = \left\{ (x, y) \in C_2 : \frac{|x|}{k} + \left(\frac{|y|}{l}\right)^{1/2} \leq n^{1/3} \right\} \tag{4.1}$$

where the constants k and l are given by

$$k = \left(\frac{3}{2}\right)^{2/3}, \quad l = \frac{1}{2}\left(\frac{3}{2}\right)^{1/3}.$$

The inner bound for the limit shape of internal DLA cluster on C_2 was proven in [24, Theorem 4.2], while the outer bound together with the fluctuations was proven in [7].

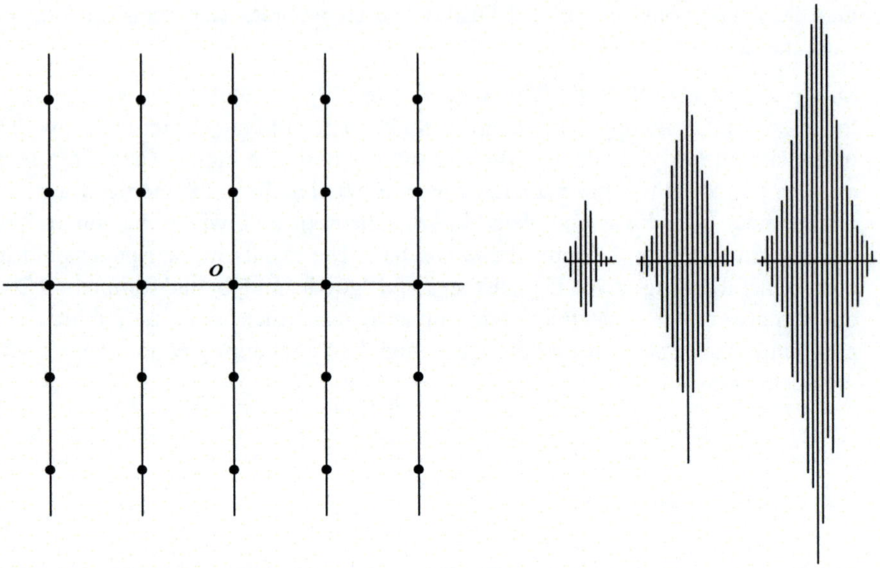

Fig. 5 The comb C_2 (left) and internal DLA clusters on C_2 (right)

Theorem 4.6 ([7, Theorem 1.2]) *There is a positive constant a such that with probability* 1*, and n large enough*

$$\mathcal{D}_{n-a\sqrt{\log n}} \subset \mathcal{I}_n \subset \mathcal{D}_{n+a\sqrt{\log n}}.$$

Remark that this result does not mean that the fluctuations are sub-logarithmic, but rather Gaussian; see [7, Theorem 1.2] and the comments afterwards. In [24, Theorem 3.5] we also prove that the limit shape for the divisible sandpile cluster on C_2 is given by the set $\mathcal{D}_n$.

4.3 Trees $\mathbb{T}_d$

Internal DLA on discrete groups with exponential growth has been studied in [16]. The homogeneous tree $\mathbb{T}_d$ is a particular case (as a Cayley graph of a free group) of these state spaces, for which the authors have proven that the limit shape of internal DLA cluster is a ball in the graph metric, and they give lower bounds for the inner and outer error. The more general result is the following.

Theorem 4.7 ([16, Theorem 3.1]) *Let* G *be a finitely generated group of exponential growth, and consider the internal DLA model* $(\mathcal{I}_n)$ *on* G*, built up with symmetric* *random walks with finitely supported increments, starting at the identity o of* G. *Then, for any constants $C_O > 2$ and $C_I > 3/K$,*

$$\mathbb{P}\big[\exists n_0 \ s.t. \ \forall n > n_0 : \quad \mathcal{B}_{n-C_I \ln n} \subset \mathcal{I}_{|\mathcal{B}_n|} \subset \mathcal{B}_{n+C_O\sqrt{n}}\big] = 1,$$

where K is a constant that ensures that the ball $\mathcal{B}_n$ contains the boundary $\partial\mathcal{B}_{n-1}$, and $\mathcal{B}_n$ is the ball of radius n centered at the identity in the word metric on G.

An extension of this result to non-amenable graphs for a wide class of Markov chains was considered in [23]. On discrete groups with polynomial growth, internal DLA has been considered in [15].

4.4 Cylinder Graphs

Like in Sect. 3.4, we consider here cylinder graphs $\mathsf{G} \times \mathbb{Z}$, and we let G to be the cycle graph $\mathbb{Z}_N$ on N vertices. Internal DLA on cylinder graphs $\mathbb{Z}_N \times \mathbb{Z}$ was investigated in [29], for the following initial setting. For $k \in \mathbb{Z}$, the set $\mathbb{Z}_N \times \{k\}$ is called the k-th level of the cylinder, and $R_k = \{(x, y) \in \mathbb{Z}_N \times \mathbb{N} : \ y \leq k\}$ the *rectangle of height* k. Let $\mathcal{I}_0 = R_0$, and given the cluster $\mathcal{I}_n$ at time n, let y_{n+1} be the first exit location from $\mathcal{I}_n$ of a random walk that starts uniformly at random on level zero of the cylinder, independent on the past, that is, the starting location is chosen with equal probability among the N sites $(x, 0)$, for $x \in \mathbb{Z}_N$. We then set $\mathcal{I}_{n+1} = \mathcal{I}_n \cup \{y_{n+1}\}$. It has been proven in [29, Theorem 2] that the limit shape of internal DLA clusters on $\mathbb{Z}_N \times \mathbb{Z}$ is logarithmically close to rectangles, result that we do not state in complete form here, but instead we state a more recent result due to LEVINE AND SILVESTRI [40, Theorem 1.1] which generalizes the previous one [29] (here the fluctuations are described in terms of the Gaussian Free Field exactly). Remark that in the cylinder setting, there are two parameters, the size N of the cycle base graph, and the time n.

Theorem 4.8 ([40, Theorem 1.1]) *Let $(\mathcal{I}_n)_{n\geq 0}$ be the internal DLA process on $\mathbb{Z}_N \times \mathbb{Z}$ starting from $\mathcal{I}_0 = R_0$. For any $\gamma > 0$, $m \in \mathbb{N}$ there exist a constant $C = C(\gamma, m)$ such that*

$$\mathbb{P}[R_{\frac{n}{N}-C\log N} \subseteq \mathcal{I}_n \subseteq R_{\frac{n}{N}+C\log N},\ n \leq N^m] \geq 1 - N^{-\gamma},\ \textit{for N large enough.}$$

For other results concerning the fluctuations and the behavior of internal DLA clusters on $\mathbb{Z}_N \times \mathbb{Z}$, we refer to [40].

4.5 Fractal Graphs

We would like to conclude the section about internal DLA with the behavior of the model on Sierpinski gasket graphs SG. Recall the definition of the Sierpinski gasket graph SG and of the Sierpinski carpet graph SC_2, as given in Sect. 3.5. Due to the symmetry of SG, it is clear that the limit shape of the internal DLA cluster on SG is a ball in the graph metric, a result proved in [18].

Theorem 4.9 ([18, Theorem 1.1]) *On* SG, *the internal DLA cluster of* $|\mathcal{B}_n|$ *particles occupies a set of sites close to a ball of radius n. That is, for all* $\epsilon > 0$, *we have*

$$\mathcal{B}_{n(1-\epsilon)} \subset \mathcal{I}_{|\mathcal{B}_n|} \subset \mathcal{B}_{n(1+\epsilon)}, \text{ for all } n \text{ sufficiently large}$$

with probability 1.

A limit shape for the divisible sandpile on SG was described in [25]. Concerning the fluctuations for internal DLA, it is conjectured that they are sub-logarithmic.

Conjecture 4.10 ([17, Conjecture 4.1]) There exists $C > 0$ such that

$$\mathcal{B}_{n-C\sqrt{\log n}} \subset \mathcal{I}_{|\mathcal{B}_n|} \subset \mathcal{B}_{n+C\sqrt{\log n}}.$$

Many other questions concerning internal DLA on fractal graphs can be found in the final section of [17].

Question 4.11 Is the limit shape for the internal DLA model with uniform starting points on SG, again a ball in the graph metric? What about the fluctuations in this case?

A reason why SG is easier to work with is because (1) it is a finitely ramified fractal graph, and (2) we have a precise characterization of the divisible sandpile model on SG, thanks to the finite ramification and the symmetries it possesses. In contrast, SC_2 is infinitely ramified, and characterizing the harmonic measure thereon is a challenging open question in the study of analysis on fractals. So at the moment it is very difficult to analyze growth models on SC_2. See Fig. 6 for the behavior of internal DLA on SC_2.

Question 4.12 Does the internal DLA cluster on the 2-dimensional Sierpinski carpet graph SC_2 have a (unique) scaling limit? What can one say about the boundary of the limit shape, which according to simulations appears to be of fractal nature?

Question 4.13 What is the limit shape of internal DLA on fractal graphs, other that SG (which is understood) and SC_2 (which seems hard to investigate)?

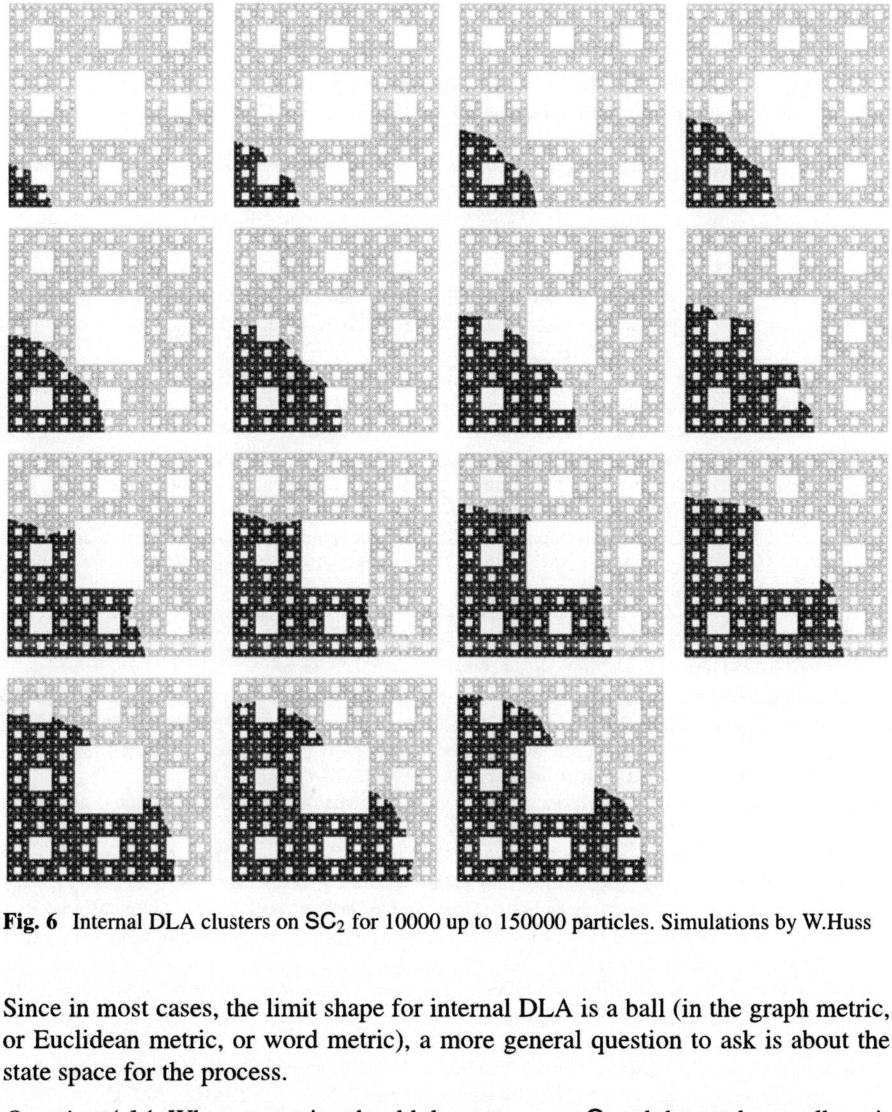

Fig. 6 Internal DLA clusters on SC_2 for 10000 up to 150000 particles. Simulations by W.Huss

Since in most cases, the limit shape for internal DLA is a ball (in the graph metric, or Euclidean metric, or word metric), a more general question to ask is about the state space for the process.

Question 4.14 What properties should the state space G and the random walk on it have, in order for the internal DLA cluster on G to have a ball as limit shape?

We would like to conclude this survey with the remark that fractals provide a class of state spaces with intriguing properties, both for the behavior of the external and internal DLA model, respectively. This behavior is definitely not fully understood on such graphs, and we hope to attract more people from the fractal community into the beauty of these topics.

Acknowledgments I am very grateful to the anonymous referee for a very careful reading of the manuscript and for several useful comments that improved the paper substantially.

References

1. Amir, G.: One-dimensional long-range diffusion-limited aggregation III—The limit aggregate. Ann. Inst. Henri Poincaré Probab. Stat. **53**(4), 1513–1527 (2017). https://doi.org/10.1214/15-AIHP731
2. Amir, G., Angel, O., Benjamini, I., Kozma, G.: One-dimensional long-range diffusion-limited aggregation I. Ann. Probab. **44**(5), 3546–3579 (2016). https://doi.org/10.1214/15-AOP1058
3. Amir, G., Angel, O., Kozma, G.: One-dimensional long-range diffusion limited aggregation II: The transient case. Ann. Appl. Probab. **27**(3), 1886–1922 (2017). https://doi.org/10.1214/16-AAP1248
4. Asselah, A., Gaudillière, A.: From logarithmic to subdiffusive polynomial fluctuations for internal DLA and related growth models. Ann. Probab. **41**(3A), 1115–1159 (2013). http://dx.doi.org/10.1214/12-AOP762
5. Asselah, A., Gaudillière, A.: Sublogarithmic fluctuations for internal DLA. Ann. Probab. **41**(3A), 1160–1179 (2013). http://dx.doi.org/10.1214/11-AOP735
6. Asselah, A., Gaudillière, A.: Lower bounds on fluctuations for internal DLA. Probab. Theory Related Fields **158**(1–2), 39–53 (2014). http://dx.doi.org/10.1007/s00440-012-0476-6
7. Asselah, A., Rahmani, H.: Fluctuations for internal DLA on the comb. Ann. Inst. Henri Poincaré Probab. Stat. **52**(1), 58–83 (2016). https://doi.org/10.1214/14-AIHP629
8. Barlow, M.T.: Diffusions on fractals. In: Lectures on Probability Theory and Statistics (Saint-Flour, 1995), Lecture Notes in Math., vol. 1690, pp. 1–121. Springer, Berlin (1998). https://doi.org/10.1007/BFb0092537
9. Barlow, M.T., Bass, R.F.: Random walks on graphical Sierpinski carpets. In: Random Walks and Discrete Potential Theory (Cortona, 1997), Sympos. Math., vol. XXXIX, pp. 26–55. Cambridge Univ. Press, Cambridge (1999)
10. Barlow, M.T., Pemantle, R., Perkins, E.A.: Diffusion-limited aggregation on a tree. Probab. Theory Related Fields **107**(1), 1–60 (1997). https://doi.org/10.1007/s004400050076
11. Benjamini, I., Yadin, A.: Diffusion limited aggregation on a cylinder. Comm. Math. Phys. **279**(1), 187–223 (2008). https://doi.org/10.1007/s00220-008-0424-6
12. Benjamini, I., Yadin, A.: Upper bounds on the growth rate of diffusion limited aggregation (2017). Preprint. arXiv:1705.06095
13. Benjamini, I., Duminil-Copin, H., Kozma, G., Lucas, C.: Internal diffusion-limited aggregation with uniform starting points. Ann. Inst. Henri Poincaré Probab. Stat. **56**(1), 391–404 (2020)
14. Bertacchi, D.: Asymptotic behaviour of the simple random walk on the 2-dimensional comb. Electron. J. Probab. **11**(45), 1184–1203 (2006). http://www.math.washington.edu/~ejpecp/EjpVol11/paper45.abs.html
15. Blachère, S.: Internal diffusion limited aggregation on discrete groups of polynomial growth. In: Random Walks and Geometry, pp. 377–391. Walter de Gruyter GmbH & Co. KG, Berlin (2004)
16. Blachère, S., Brofferio, S.: Internal diffusion limited aggregation on discrete groups having exponential growth. Probab. Theory Related Fields **137**(3–4), 323–343 (2007). http://dx.doi.org/10.1007/s00440-006-0009-2
17. Chen, J.P., Kudler-Flam, J.: Laplacian growth and sandpiles on the Sierpinski gasket: limit shape universality and exact solutions. Ann. Inst. Henri Poincaré Comb. Phys. Interact. **7**, 585–664 (2020)
18. Chen, J.P., Huss, W., Sava-Huss, E., Teplyaev, A.: Internal DLA on Sierpinski gasket graphs. Analysis and Geometry on Graphs and Manifolds. In: M. Keller et al. (Eds.) London Mathematical Society Lecture Note, pp. 126–155. Cambridge University Press (2020)
19. Diaconis, P., Fulton, W.: A growth model, a game, an algebra, Lagrange inversion, and characteristic classes. Rend. Sem. Mat. Univ. Politec. Torino **49**(1), 95–119 (1993) (1991). Commutative algebra and algebraic geometry, II (Italian) (Turin, 1990)

20. Duminil-Copin, H., Lucas, C., Yadin, A., Yehudayoff, A.: Containing internal diffusion limited aggregation. Electron. Commun. Probab. **18**(50), 8 (2013). https://doi.org/10.1214/ECP.v18-2862
21. Eberz-Wagner, D.M.: Discrete growth models. Ph.D. thesis, University of Washington (1999)
22. Eldan, R.: Diffusion-limited aggregation on the hyperbolic plane. Ann. Probab. **43**(4), 2084–2118 (2015). https://doi.org/10.1214/14-AOP928
23. Huss, W.: Internal Diffusion-Limited Aggregation on non-amenable graphs. Electron. Commun. Probab. **13**, 272–279 (2008). http://www.math.washington.edu/~ejpecp/EcpVol13/paper27.abs.html
24. Huss, W., Sava, E.: Internal aggregation models on comb lattices. Electron. J. Probab. **17**(30), 21 (2012). http://dx.doi.org/10.1214/EJP.v17-1940
25. Huss, W., Sava-Huss, E.: Divisible sandpile on Sierpinski gasket graphs. Fractals **27**(3), 1950,032, 14 (2019). https://doi.org/10.1142/S0218348X19500324
26. Jerison, D., Levine, L., Sheffield, S.: Logarithmic fluctuations for internal DLA. J. Am. Math. Soc. **25**(1), 271–301 (2012). http://dx.doi.org/10.1090/S0894-0347-2011-00716-9
27. Jerison, D., Levine, L., Sheffield, S.: Internal DLA in higher dimensions. Electron. J. Probab. **18**(98), 1–14 (2013)
28. Jerison, D., Levine, L., Sheffield, S.: Internal DLA and the Gaussian free field. Duke Math. J. **163**(2), 267–308 (2014). http://dx.doi.org/10.1215/00127094-2430259
29. Jerison, D., Levine, L., Sheffield, S.: Internal DLA for cylinders. In: Advances in Analysis: The Legacy of Elias M. Stein, Princeton Math. Ser., vol. 50, pp. 189–214. Princeton Univ. Press, Princeton, NJ (2014)
30. Kesten, H.: Hitting probabilities of random walks on $\mathbf{Z}^d$. Stochastic Process. Appl. **25**(2), 165–184 (1987). https://doi.org/10.1016/0304-4149(87)90196-7
31. Kesten, H.: How long are the arms in DLA? J. Phys. A **20**(1), L29–L33 (1987). https://doi.org/10.1088/0305-4470/20/1/007
32. Kesten, H.: Upper bounds for the growth rate of DLA. Phys. A **168**(1), 529–535 (1990). https://doi.org/10.1016/0378-4371(90)90405-H
33. Kigami, J.: Analysis on fractals. Cambridge Tracts in Mathematics, vol. 143. Cambridge University Press, Cambridge (2001). https://doi.org/10.1017/CBO9780511470943
34. Krishnapur, M., Peres, Y.: Recurrent graphs where two independent random walks collide finitely often. Electron. Commun. Probab. **9**, 72–81 (2004)
35. Lawler, G.: Subdiffusive fluctuations for internal diffusion limited aggregation. Ann. Probab. **23**, 71–86 (1995)
36. Lawler, G.F., Bramson, M., Griffeath, D.: Internal diffusion limited aggregation. Ann. Probab. **20**(4), 2117–2140 (1992)
37. Levine, L.: The sandpile group of a tree. Eur. J. Combin. **30**(4), 1026–1035 (2009). https://doi.org/10.1016/j.ejc.2008.02.014
38. Levine, L., Peres, Y.: Strong spherical asymptotics for rotor-router aggregation and the divisible sandpile. Potential Anal. **30**(1), 1–27 (2009). http://dx.doi.org/10.1007/s11118-008-9104-6
39. Levine, L., Peres, Y.: Scaling limits for internal aggregation models with multiple sources. J. d'Analyse Math. **1**, 151–219 (2010)
40. Levine, L., Silvestri, V.: How long does it take for internal dla to forget its initial profile? Probab. Theory Related Fields **174**, 1219–1271 (2019)
41. Lucas, C.: The limiting shape for drifted internal diffusion limited aggregation is a true heat ball. Probab. Theory Related Fields **159**(1–2), 197–235 (2014). https://doi.org/10.1007/s00440-013-0505-0
42. Martineau, S.: Directed diffusion-limited aggregation. ALEA Lat. Am. J. Probab. Math. Stat. **14**(1), 249–270 (2017)
43. Meakin, P., Deutch, J.: The formation of surfaces by diffusion limited annihilation. J. Chem. Phys. **85**, 2320 (1986)
44. Osada, H.: Isoperimetric constants and estimates of heat kernels of pre Sierpiński carpets. Probab. Theory Related Fields **86**(4), 469–490 (1990). https://doi.org/10.1007/BF01198170

45. Procaccia, E., Zhang, Y.: On sets of zero stationary harmonic measure. Stochastic Process. Appl. **131**, 236–252 (2021)
46. Procaccia, E.B., Zhang, Y.: Stationary harmonic measure and DLA in the upper half plane. J. Stat. Phys. **176**(4), 946–980 (2019)
47. Procaccia, E., Rosenthal, R., Zhang, Y.: Stabilization of DLA in a wedge. Electron. J. Probab. **25**(42), 22 (2020)
48. Procaccia, E., Ye, J., Zhang, Y.: Stationary harmonic measure as the scaling limit of truncated harmonic measure (2019). arXiv:1811.04793
49. Shellef, E.: IDLA on the supercritical percolation cluster. Electron. J. Probab. **15**(24), 723–740 (2010). http://dx.doi.org/10.1214/EJP.v15-775
50. Spitzer, F.: Principles of Random Walk, 2nd edn. Graduate Texts in Mathematics, vol. 34. Springer, New York, Heidelberg (1976)
51. Witten, T.A., Sander, L.M.: Diffusion-limited aggregation. Phys. Rev. B **27**, 5686–5697 (1983). https://link.aps.org/doi/10.1103/PhysRevB.27.5686

Contents of the Previous Volumes of the 'Fractal Geometry and Stochastics' Series

Fractal Geometry and Stochastics I

Birkhäuser, Progress in Probab. 37, 1995

CONTENTS

Preface ... vii

Introduction ... ix

1. Fractal sets and measures ... 1

Kenneth J. Falconer
Probabilistic methods in fractal geometry ... 3

Benoit B. Mandelbrot
Measures of fractal lacunarity: Minkowski content and alternatives ... 15

Pertti Mattila
Tangent measures, densities, and singular integrals ... 43

2. Iterated function systems ... 53

Ka-Sing Lau
Self-similarity, L^p-spectrum and multifractal formalism ... 55

R. Daniel Mauldin
Infinite iterated function system: theory and applications ... 91

3. Random fractals ... 111

Lincoln Chayes
Aspects of the fractal percolation process ... 113

Yuri Kifer
Fractals via random iterated function systems
and random geometric constructions ... 145

U. Freiberg et al. (eds.), *Fractal Geometry and Stochastics VI*,
Progress in Probability 76, https://doi.org/10.1007/978-3-030-59649-1

4. Fractals and dynamical systems 165

Manfred Denker
Remarks on weak limit laws for fractal sets 167

Mariusz Urbański and Alexander Volberg
A rigidity theorem in complex dynamics 179

5. Harmonic analysis on fractals 189

Palle E. T. Jørgensen and Steen Pedersen
Harmonic analysis and fractal limit-measures induced
by representations of a certain C^*- algebra 191

Jun Kigami
Laplacians on self-similar sets and their spectral distributions 221

List of participants 239

Fractal Geometry and Stochastics II

Birkhäuser, Progress in Probab. 46, 2000

CONTENTS

Preface ... vii

Introduction ... ix

1. Fractal Sets and Measures ... 1

L. Olsen
Multifractal Geometry ... 3

Yuval Peres, Wilhelm Schlag, and Boris Solomyak
Sixty Years of Bernoulli Convolutions ... 39

2. Iterated Function Systems ... 67

Manuel Moran
Problems on Self-similar Geometry ... 69

Yuval Peres and Boris Solomyak
Problems on Self-similar Sets and Self-affine Sets: An Update ... 95

3. Stochastic Processes, Random Fractals ... 107

John E. Hutchinson and Ludger Rüschendorf
Selfsimilar Fractals and Selfsimilar Random Fractals ... 109

Jean-Pierre Kahane
Random Coverings and Multiplicative Processes ... 125

Jacques Peyriére
Recent Results on Mandelbrot Multiplicative Cascades ... 147

Vladas Pipiras and Murad S. Taqqu
The Weierstrass-Mandelbrot Process Provides a Series
Approximation to the Harmonizable Fractional Stable Motion ... 161

4. Fractals and Dynamical systems ... 181

Gerhard Keller
An Ergodic Theoretic Approach to Mean Field Coupled Maps ... 183

Yakiv B. Pesin
Entropy and Dimension Families Associated with
Equilibrium Measures for Hyperbolic Dynamical Systems ... 209

5. Harmonic Analysis on Fractals ... 225

Masatoshi Fukushima
On Limit Theorems for Brownian Motions
on Unbounded Fractal Sets ... 227

B.M. Hambly

Heat Kernels and Spectral Asymptotics
for some Random Sierpinski Gaskets 239

Umberto Mosco
Lagrangian Metrics and Fractal Dynamics 269

List of Participants 285

Fractal Geometry and Stochastics III

Birkhäuser, Progress in Probab. 57, 2004

CONTENTS

Preface ... vii

Introduction ... ix

1. Fractal Sets and Measures

Andrzej Lasota, Józef Myjak and Tomasz Szarek
Markov Operators and Semifractals ... 3

Jacques Lévy Véhel and Claude Tricot
On Various Multifractal Spectra ... 23

Zhi-Ying Wen
One-Dimensional Moran Sets and the Spectrum of Schrödinger Operators ... 43

2. Fractals and Dynamical Systems

Albert M. Fisher
Small-scale Structure via Flows ... 59

Károly Simon
Hausdorff Dimension of Hyperbolic Attractors in $\mathbb{R}^3$ ... 79

Bernd O. Stratmann
The Exponent of Convergence of Kleinian Groups;
on a Theorem of Bishop and Jones ... 93

Amiran Ambroladze and Jörg Schmeling
Lyapunov Exponents Are not Rigid
with Respect to Arithmetic Subsequences ... 109

3. Stochastic Processes and Random Fractals

Ai Hua Fan
Some Topics in the Theory of Multiplicative Chaos ... 119

Peter Mörters
Intersection Exponents and the Multifractal Spectrum
for Measures on Brownian Paths ... 135

Davar Khoshnevisan and Yimin Xiao
Additive Lévy Processes: Capacity and Hausdorff Dimension ... 151

4. Fractal Analysis in Euclidean Space

Hans Triebel
The Fractal Laplacian and Multifractal Quantities ... 173

Jenny Harrison

Geometric Representations of Currents and Distributions 193

Maria Agostina Vivaldi
Variational Principles and Transmission Conditions for Fractal Layers 205

5. Harmonic Analysis on Fractals

Takashi Kumagai
Function Spaces and Stochastic Processes on Fractals 221

Alf Jonsson
A Dirichlet Form on the Sierpinski Gasket,
Related Funktion Spaces, and Traces 235

Alexander Teplyaev
Spectral Zeta Function of Symmetric Fractals 245

Fractal Geometry and Stochastics IV

Birkhäuser, Progress in Probab. 61, 2009

CONTENTS

Preface vii

Introduction ix

Part 1: Analysis on Fractals

A. Grigor'yan, J. Hu and K.-S. Lau
Heat Kernels on Metric Spaces with Doubling Measure 3

V.A. Kaimanovich
Self-similarity and Random Walks 45

Part 2: Conformal Dynamics and Schramm-Loewner Evolution

G.F. Lawler
Multifractal Analysis of the Reverse Flow for the Schramm-Loewner Evolution 73

Part 3: Random Fractal Processes

D. Khoshnevisan
From Fractals and Probability to Lévy Processes and Stochastic PDEs 111

Part 4: Emergence of Fractals in Complex Systems

J.E. Steif
A Survey of Dynamical Percolation 145

J. Blath
Measure-valued Processes, Self-similarity and Flickering Random Measures 175

G. Miermont
Random Maps and Their Scaling Limits 197

Part 5: Iterated Function Schemes and Transformations of Fractals

M.F. Barnsley
Transformations Between Fractals 227

M. Furukado, S. Ito and H. Rao
Geometric Realizations of Hyperbolic Unimodular Substitutions 251

M. Dekking
Random Cantor Sets and Their Projections 269

Appendix 285

Fractal Geometry and Stochastics V

Birkhäuser, Progress in Probab. 70, 2015

CONTENTS

Part I Geometric Measure Theory

Sixty Years of Fractal Projections 3
Kenneth Falconer, Jonathan Fraser, and Xiong Jin

Scenery Flow, Conical Densities, and Rectifiability 27
Antti Käenmäki

The Shape of Anisotropic Fractals: Scaling of Minkowski Functionals 27
Philipp Schönhöfer and Klaus Mecke

Projections of Self-Similar and Related Fractals: A Survey of Recent Developments 53
Pablo Shmerkin

Part II Self-Similar Fractals and Recurrent Structures

Dimension of the Graphs of the Weierstrass-Type Functions 77
Krzysztof Barański

Tiling $\mathbb{Z}^2$ by a Set of Four Elements 93
De-Jun Feng and Yang Wang

Some Recent Developments in Quantization of Fractal Measures 105
Marc Kesseböhmer and Sanguo Zhu

Apollonian Circle Packings 121
Mark Pollicott

Entropy of Lyapunov-Optimizing Measures of Some Matrix Cocycles ... 143
Michał Rams

Part III Analysis and Algebra on Fractals

Poincaré Functional Equations, Harmonic Measures on Julia Sets, and Fractal Zeta Functions 157
Peter J. Grabner

From Self-Similar Groups to Self-Similar Sets and Spectra 175
Rostislav Grigorchuk, Volodymyr Nekrashevych, and Zoran Šunić

Finite Energy Coordinates and Vector Analysis on Fractals 209
Michael Hinz and Alexander Teplyaev

Fractal Zeta Functions and Complex Dimensions: A General Higher-Dimensional Theory 229
Michel L. Lapidus, Goran Radunović, and Darko Žubrinić

Part IV Multifractal Theory

Inverse Problems in Multifractal Analysis 261
Julien Barral

Multifractal Analysis Based on p-Exponents and Lacunarity Exponents 279
Patrice Abry, Stéphane Jaffard, Roberto Leonarduzzi, Clothilde Melot, and Herwig Wendt

Part V Random Constructions

Dimensions of Random Covering Sets 317
Esa Järvenpää and Maarit Järvenpää

Expected Lifetime and Capacity 327
András Telcs and Marianna E.-Nagy

Springer INdAM Series

Volume 50

This series will publish textbooks, multi-authors books, thesis and monographs in English language resulting from workshops, conferences, courses, schools, seminars, doctoral thesis, and research activities carried out at INDAM - Istituto Nazionale di Alta Matematica, http://www.altamatematica.it/en. The books in the series will discuss recent results and analyze new trends in mathematics and its applications.
THE SERIES IS INDEXED IN SCOPUS